T0270663

Father Nature

Father Nature

The Science of Paternal Potential

James K. Rilling

The MIT Press
Cambridge, Massachusetts
London, England

The MIT Press would like to thank the anonymous peer reviewers who provided comments on drafts of this book. The generous work of academic experts is essential for establishing the authority and quality of our publications. We acknowledge with gratitude the contributions of these otherwise uncredited readers.

This book was set in Stone Serif and Stone Sans by Jen Jackowitz. Printed and bound in the United States of America.

Library of Congress Cataloging-in-Publication Data

Names: Rilling, James K., author.
Title: Father nature : the science of paternal potential / James K. Rilling.
Description: Cambridge, Massachusetts : The MIT Press, [2024] | Includes
    bibliographical references and index.
Identifiers: LCCN 2023054433 (print) | LCCN 2023054434 (ebook) | ISBN
    9780262048934 (hardcover) | ISBN 9780262378222 (epub) | ISBN
    9780262378239 (pdf)
Subjects: LCSH: Fathers—Psychology. | Men—Psychology. | Male caregivers.
Classification: LCC HQ756 .R55 2024 (print) | LCC HQ756 (ebook) | DDC
    306.874/2—dc23/eng/20231205
LC record available at https://lccn.loc.gov/2023054433
LC ebook record available at https://lccn.loc.gov/2023054434

10 9 8 7 6 5 4 3 2 1

To the memory of Robert John Rilling

Contents

Preface

In college, I became fascinated by the study of human origins and human evolution. The fascination was strong enough to dissuade me from applying to medical school and instead pursue a PhD in anthropology. I had the good fortune of attending graduate school at a time when new noninvasive brain imaging methods had just become available, providing a golden opportunity to address big anthropological questions with the methods of modern neuroscience. After subsequent postdoctoral training in neuroimaging, I began investigating the neural basis of human behavioral specializations such as cooperation and language. Sometime later, it occurred to me that paternal caregiving was also a human specialization and a neglected and worthy topic of investigation. Spurred on by a generous grant from the John Templeton Foundation, I redirected my research focus to investigating the biology of fatherhood. In the midst of that Templeton grant, my wife gave birth to our first child, a delightfully plump boy named Toby. Raising him while researching and learning about fatherhood presented me with a remarkable opportunity for my home life to inform my research and vice versa. My daughter Mia was born five years later and provided a whole new set of lessons that helped to shape my knowledge of fatherhood beyond the mere academic. They are now twelve and six years old, and I feel I have learned enough about fatherhood over those twelve years, both at home and at work, to have some useful knowledge to pass on to others. This book is my effort to do so.

In the academic realm, I have many mentors to thank for having shaped me and my research and scholarship, including Ted Golos, Karen Strier, Carol Worthman, Melvin Konner, Thomas Insel, Clinton Kilts, Jonathan Cohen, Todd Preuss, and Larry Young. I am also grateful to the many friends and colleagues who have devoted the time to read and offer valuable feedback on different chapters of the book, including Andrew Engell, Sher-ryl Goodman, Peter Gray, Michael Gurven, Barry Hewlett, Sarah Hrdy, Glenn Hutchinson, Melvin Konner, Michael Lamb, Minwoo Lee, Peter Lit-tle, Elizabeth Lonsdorf, Daniel Paquette, Jodi Pawluski, and Larry Young. Finally, I thank all of the former students and staff in my lab, especially Jennifer Mascaro, who has been an important collaborator on the research that I describe in this book, and Arianna Ophir, who produced several of the beautiful animal drawings in chapter 2.

I thank my editor at MIT Press, Matthew Browne, for his thoughtful and prudent guidance throughout this process.

I have been blessed to be raised by two of the world's best parents, Rich-ard and Carolyn Rilling. Not only have they been tremendous parental role models, they have been relentlessly supportive in my academic pursuits, including this book.

I am especially grateful to my wife, Barbara, for encouraging and enabling me to dedicate the time and mental focus needed to complete a project of this magnitude. I am also grateful for her close reading and helpful critiques of several chapters. I could not have written the book without her love and support.

I would also be remiss if I did not thank my two children, Toby and Mia, for the lessons they have taught me about fatherhood and for allowing me to include anecdotes about them throughout the book.

Finally, I dedicate the book to the memory of my dear brother Bob, who passed away unexpectedly as I was writing the book. He was a wonderful husband and father, and I like to think he would have enjoyed the book. I miss him every day.

Introduction

Xiao Qiang was born in southwest China's Sichuan province with twisted arms and legs and a hunched back that prevented him from walking. At age three, his parents separated, and his father, Yu Xukang, decided to raise him alone. The only school that would accept his son and accommodate his handicap was nearly five miles away and inaccessible by public transportation. But Xukang was determined for his son to have an education. He would wake up at 5:00 a.m. to prepare Qiang's lunch and then carry him on his back to school. He would then walk back home to go to work to earn the money needed to provide for the two of them, only to later retrieve his son from school and carry him home again. In all, he traveled almost twenty miles every day to provide Qiang with an education. His efforts seemed to be paying off: Xukang said, "I am proud of the fact that he is already top of his class and I know he will achieve great things. My dream is that he will go to college."[1]

Now contrast this devoted human father with fathers of the primate species that is most closely related to us, the chimpanzee. Adult males aggressively compete for dominance and for the opportunity to mate with as many females as possible. They make war with neighboring communities of chimps and rarely care for their young.[2] In fact, they occasionally kill infants sired by other males.[3] Chimpanzees are not our ancestors, but prominent scholars believe they are a reasonable model for the ancestral ape species that gave rise to both humans and chimpanzees approximately 6 million years ago.[4] Paternal caregiving is also lacking in our other closest primate relatives, the bonobo, the gorilla, and the orangutan.[5] All the evidence suggests that human evolution involved a striking transformation of our male ancestors to endow them with paternal capabilities that we see in Xukang and so many other modern human fathers.

This book tells the story of how that happened and why fathers have become important for our children's well-being.

Chimpanzee males are not an anomaly. Paternal caregiving is rare among mammals.[6] In most species, evolution has produced males who choose to pursue additional mates rather than stay committed to just one and help raise her offspring. This is because in the cold, hard calculus of natural selection, promiscuous males are usually better at sending copies of their genes into the next generation; they have greater reproductive success. However, in about 5% of mammals, males instead seem to maximize their reproductive success by parenting.[7] Male parenting, or paternal caregiving, has evolved multiple times independently in mammals. There are two major types of care: direct and indirect care. Direct care encompasses hands-on caregiving activities such as holding, carrying, cleaning, and playing. Indirect care offers care that benefits the infant but does not involve direct contact with them; it primarily includes provisioning and protection. Human males specialize in indirect caregiving, especially provisioning, and they adopt this role consistently across human societies.[8] In contrast to indirect care, direct care is highly variable across and within human societies.[9] In many societies, fathers do little to no direct caregiving. Yet direct paternal caregiving is extensive in others, and fathers' potential to be highly involved, sensitive, and nurturing caregivers is without question.[10] In fact, highly committed fatherhood is demonstrated most convincingly by the growing phenomenon of stay-at-home fathers, gay fathers, and single fathers, illuminating men's capacity to adopt even the primary caregiving role.[11]

Chapter 1 of this book focuses on the impact that fathers have on children's development. In many traditional human societies, they help to keep children alive.[12] That is, children are more likely to survive when their father is present compared to when he is not. We might think this could never apply to a modern, developed, affluent society like the United States, but one recent study showed that children born in the state of Georgia in 1990 were more likely to survive if the mother listed a father's name on the birth certificate compared to if she did not.[13] Beyond basic survival, a considerable body of evidence shows that children with positively engaged fathers have better social, psychological, behavioral, and academic outcomes.[14] Fathers are particularly involved in cultivating the development of emotion regulation skills in their children, with myriad downstream benefits and in preparing children for the world outside the home.[15] This is not to say that other caregivers cannot do these things as well, only that fathers tend to excel at them.

Chapter 2 focuses on the evolution of paternal caregiving. We can't go back in time to determine exactly how and why committed fatherhood evolved in human males, but some compelling theories have been offered. While this uncertainty can be unsettling for those of us interested in human origins, to me it has always imbued the study of human evolution with an intriguing sense of mystery. One theory asserts that climate change led to a drying of the African savannas and that the presence of large mammals on the savanna caused men to begin specializing in hunting game and women to begin specializing in gathering savanna plant foods. Women provided mainly carbohydrates, while men provided mainly protein. Children needed both of these complementary resources to survive, so men had to provision their offspring if those offspring were to survive.[16] Another theory is that men's provisioning hinged on knowing who their offspring were, and this depended on female choice. According to this theory, men began provisioning their offspring once women offered men paternity confidence by selectively mating with them. That is, male provisioning depended on the evolution of pair bonding.[17] It also seems clear from studies of contemporary hunter-gatherer societies, whose lifestyle is believed to be similar to that of our ancestors throughout most of human evolution, that fathers were only one of a roster of alloparents who were potentially available to help mothers with direct caregiving.[18] Collectively, this alloparental support allowed mothers to shorten their interbirth intervals while simultaneously improving their offspring's survival prospects. The result was increased reproductive success for mothers and fathers and, by extension, increased inclusive fitness of grandmothers and other related alloparents.[19]

However it happened, the evolution of paternal caregiving required that natural selection adapt male physiology to this new role. This is the focus of chapters 3 to 5. In women, profound hormonal changes during pregnancy help to ensure they will be motivated to provide care of their offspring. We now know that men experience subtler hormonal changes that also prepare them for caregiving.[20] One such physiological adaptation is a decrease in testosterone levels to support direct caregiving. Across species, testosterone consistently biases males to direct energy toward mating effort and away from parenting effort. Human males experience a decline in testosterone when they become involved in direct caregiving,[21] but a rise in the peptide hormone oxytocin with fatherhood apparently biases males in the opposite direction.[22] These hormones act on an ancient neural system in the brain that originally evolved to support maternal caregiving but that natural selection also utilized to support paternal caregiving in the minority of mammals where it evolved.[23] Research in other paternally investing mammals shows that males experience significant changes in their brain as they transition to fatherhood, and these changes prepare them for involvement in caregiving.[24] Similar research in humans is just beginning, but preliminary results hint at analogous neurological changes in men as they become fathers.[25]

Paternal caregiving is highly beneficial to children and, by extension, to society at large, yet it is highly variable both across and within human societies. Chapter 6 considers how and why paternal caregiving varies across and within human societies, as well as how we might endeavor to increase positive paternal engagement where it is lacking.

Finally, chapter 7 distills the main messages that I would like expectant fathers to take away from this book and closes with a consideration of what fathers *get from* rather than *give to* their children.

1 Paternal Influence

One father is more than a hundred schoolmasters.
—George Herbert

In addition to being genocidal dictators, Joseph Stalin, Adolf Hitler and Mao Tse-tung all shared at least one thing in common: they hated their fathers. All three men had abusive and strongly authoritarian fathers. Mao Tse-tung's father beat him when he did not work hard enough. When Mao later took merciless revenge on his political enemies, he told the men who were torturing them that he would like to have seen his father treated similarly.[1] Joseph Stalin bitterly resented his father, a violent alcoholic who beat him severely.[2] Hitler's father ruled the family "with tyrannical severity and injustice," and he viewed him as the enemy.[3] Remarkably, all three seemed to have loved their mothers, and at least Hitler and Mao saw themselves in alliance with their mother against their father.[4]

It would be foolish to conclude that these fathers were responsible for the devastation and suffering their sons unleashed on the world. Most people who were abused by their fathers as children do not become tormenters, and some even thrive. There are surely a host of genetic, environmental, historical, and sociopolitical factors that collectively conspired to produce these horrific developmental outcomes. Yet we can still ask how things might have been different if each of these men had been raised instead by a warm, nurturing father who provided strong moral guidance and set appropriate limits without harsh discipline.

Other famous world leaders had more commendable fathers. Teddy Roosevelt revered and idolized his father. As Roosevelt tells it, his father nurtured him through early bouts of severe asthma, prepared him for the world, and cultivated his morality. Roosevelt's own words say it best:

> I was fortunate enough in having a father whom I have always been able to regard as an ideal man. . . . I was a sickly and timid boy. He not only took great and untiring care of me . . . but he also most wisely refused to coddle me, and made me feel that I must force myself to hold my own with other boys and prepare to do the rough work of the world. . . . and alike from my love and respect, and in a certain sense, my fear of him, I would have hated and dreaded beyond measure to have him know that I had been guilty of a lie, or of cruelty, or of bullying, or of uncleanness or cowardice. Gradually I grew to have the feeling on my account, and not merely on his.[5]

Of course, there are also a great many ordinary, everyday men who are wonderful fathers. Some are even heroes. Qiang, for example, the father described in the Introduction, carried his disabled son on his back for five miles to school and then back again each day.

What if Hitler had been raised by Theodore Roosevelt Sr.? What if Mao Tse-tung had been raised by Yu Qiang. Is it possible we could have averted the loss of millions of innocent lives?

I've described some bad men with bad fathers and suggested that there may be a link. However, these observations are really just anecdotes, not proof, and I am indulging in speculation—which may be interesting and intriguing but it's not legitimate science. What do formal research studies based on large samples conclude about the role of fathers in shaping who their children become?

Do Fathers Help Keep Children Alive?

The most obvious way to assess the impact of fathers on their children is to look at what happens when fathers are absent. We can begin by asking a fundamental question: "Do fathers help keep children alive?" This may sound silly to many of us living in modern, developed nations where rates of child mortality are very low. However, it is a relevant question among people without access to modern medicine, where child mortality rates are much higher. A number of such "natural mortality" populations are found around the world today. Because this type of population characterized almost all of human evolution, studying today's natural mortality populations may shed light on the role of fathers throughout our evolutionary past.

Children in natural mortality populations live at the edge of survival. People in these societies obtain food by hunting, gathering, foraging, herding livestock, planting crops, and fishing, for example. Obtaining enough food is a persistent challenge, and children struggle to get enough calories. Children need energy to grow, of course, but also to fuel their immune systems so that they can fight off pathogens and remain healthy. When food is limited and energy is insufficient, malnourishment, growth faltering, illness, and even death can result.

One natural mortality population are the Ache hunter-gatherers from Paraguay. Prior to 1970, when the Ache lived in the forest rather than on reservations, fathers did keep their children alive. It was not uncommon for Ache fathers to die in adulthood or to leave their child's village after a divorce. Children of such absent fathers were three times more likely to die compared with other children. The Ache people were heavily dependent on meat from hunting for their sustenance. Children were reliant on their fathers for food, as men hunted and were enmeshed in food-sharing networks that helped ensure their families were provided for. Without a father, a child's access to these networks was jeopardized. Ache children without fathers were often seen as burdens, and infanticide of such children was not uncommon, so having a father present protected Ache children from this fate.[6]

Despite this evidence that Ache fathers kept their children alive, data from other natural mortality populations do not always find effects of fathers on child survival. In fact, one study concluded that the presence of a father improved child survival in less than half of the populations considered. This is in stark contrast to mothers, whose death was associated with increased infant mortality in all twenty-eight populations in the study. Interestingly, grandmothers improved survival in more than half of the populations, eclipsing the beneficial effects of fathers on infant survival.[7]

So why don't fathers have a more consistent impact on child survival across societies? One possibility is that other relatives can compensate for father absence, and other relatives are often available in natural mortality populations. For example, if a substantial share of calories in a population comes from gathered foods, which is typically true in foraging societies, grandmothers may respond to father absence by foraging more to provide for their grandchildren. In societies like the Ache, where most of the calories come from meat hunted by men, father absence has a bigger impact on children. Still, even in societies that are heavily dependent on male hunting for sustenance, male relatives, or even unrelated men, may step up to compensate for a father's absence and improve a child's survival prospects.

But why then do grandmothers overall appear more essential to child survival than fathers? One possibility is that grandmothers are more consistently involved with essential direct, hands-on caregiving activities than fathers are and that this care helps keep children alive. For example, a grandmother might tend to an infant while the mother is sick or to a sick toddler while the mother tends to her new infant. Grandmothers might also provide critical expertise and guidance to new mothers as they learn how to effectively care for their infant. We often talk of parental instincts, but people need to learn how to parent, and the older generation might facilitate this learning. Another possibility is that grandmothers provide another set of attentive, experienced eyes to watch out for danger and prevent infant accidents.

Given the above evidence for limited effects of fathers on child survival in natural mortality populations, coupled with the lower rates of infant mortality in the United States, it may surprise you to learn that one study uncovered evidence to suggest that fathers helped keep infants alive in the United States as recently as 1990.[8] In Georgia, mothers have the option of listing the father's name on the birth certificate but are not required to do so. Therefore, the absence of a father's name on the birth certificate may indicate the lack of a supportive father. It turns out that Georgia infants without a father's name listed on their birth certificate were more than twice as likely to die in the first year of life as compared to infants with a father's name listed, and this effect persisted after controlling for a range of potential confounding variables.

What Georgia fathers are doing to help keep their babies alive was not addressed in the study, but we can speculate. We know that partnered mothers are more likely to breast-feed than single mothers and that breast-feeding has myriad health benefits for babies. So perhaps the paternal survival benefit is explained by increased breastfeeding by mothers when fathers are present. Infants with no father named were more likely to have inadequate prenatal care and to be low birthweight, so another possibility is that fathers sometimes support mothers' efforts to receive adequate care during their pregnancy and this leads to a healthier baby. Another possibility is that fathers, similar to grandmothers, may become critical if mothers are ill. Perhaps they step up to assume hands-on caregiving responsibilities or assist mothers with other tasks so they can allocate limited energy to caregiving. Or perhaps, as also suggested above for grandmothers, they help monitor the infant to prevent accidents. When our son was an infant, he put everything—absolutely everything—he could find in his mouth, and I was sure he would choke to death if we were not constantly monitoring him. With both me and his mother watching, we were less likely to miss something.

*   *   *

As parents, we are of course not only concerned that our children survive, but that they are healthy. In some natural mortality populations, fathers appear to help children avoid growth faltering and illness. Among the Bondongo fisher-horticulturalists from the Republic of Congo, fathers who are judged by their peers as being more involved in caring for their children have plumper children, a good thing for those living on the edge of survival.[9] Similarly, among the Mayangna and Miskito horticulturalists of Nicaragua, fathers who provide more direct and indirect care have heavier children.[10] Finally, in poor, rural Mexican communities, father absence due to labor migration is associated with increased risk of illness, diarrhea in particular. Parents can prevent diarrheal disease in children by boiling water and ensuring that children wash their hands regularly, and fathers may normally be involved with these responsibilities when present.[11]

Fathers and Psychosocial Development

Developmental psychologists study how parents affect their children's psychological and social development, among other things. For decades, this research was understandably primarily focused on maternal influences, but in the 1970s, Michael Lamb (now a professor emeritus at Cambridge University) and other researchers began examining how fathers affect their children's development.[12] Beyond the limited body of research on fathers' influences over children's physical health and survival, there is now abundant research showing that fathers influence children's social, psychological, and behavioral development. Note that most of this research has been done in high-income settings such as the United States and Europe, and we should not uncritically generalize these conclusions to people living in other parts of the world or other circumstances. It is important that more research on fatherhood be conducted among non-Western, developing, and low- and middle-income countries.

Once again, the most obvious place to start when considering the influence of fathers on psychosocial development is to compare the outcomes of children raised with and without a father. These statistics are by now well known to many people both within and outside academia. In fact, President Barack Obama summarized them in his Father's Day speech to the Apostolic Church of God in Chicago in 2008: "We know the statistics—that children who grow up without a father are five times more likely to live in poverty and commit crime; nine times more likely to drop out of schools and 20 times more likely to end up in prison. They are more likely to have behavioral problems, or run away from home or become teenage parents themselves."[13] Obama did not mention that father absence is also associated with increased risk of infant mortality, substance abuse, and obesity in children. Furthermore, children raised without fathers in the home are likely to earn less money as adults and more likely to experience a divorce.[14]

Before we take a closer look at the evidence, there are a couple of important caveats. The first is that father absence doesn't affect all children the same way. Human beings come into the world with genes that bias them to be either more or less influenced by their parents.[15] In his insightful book, *The Orchid and the Dandelion*, Thomas Boyce describes "orchids" as children who are highly susceptible to their environment, including their parental environment.[16] They struggle when confronted with stress and adversity but thrive in positive environments such as a caring and nurturing family. "Dandelions" are hardy children who tend to be resilient to stress and adversity, but capitalize less on nurturing environments than orchids do. It is imperative that we grasp this concept. It means that some children who grow up with absent, or even abusive, fathers will not be crippled by this situation because they are genetically predisposed to be less influenced by their environment. It also means that some children with superb parents will not thrive.

I came to appreciate this concept during a research study I conducted in 2016 and 2017 in which I interviewed 120 fathers from the Atlanta area about the rewards and challenges of fatherhood. One of my standard battery of interview questions was whether the father's own father was present in his life when he was a child. If he said "no," I would ask if he felt that

negatively impacted him in any way. I was struck by the starkly contrast-
ing answers I heard. Some men spoke of how devastating it was that their
father did not spend time with them and how much they longed to see
more of them. Others were angry with their father and felt betrayed by his
absence. But still others, although a minority, convincingly told me that
their father's absence did not bother them; they had plenty of other relatives who spent time with them and cared about them, and they harbored
no resentment toward their father. Father absence doesn't impact all the
children the same way.

Nevertheless, on average, fathers do make a difference in multiple
domains of children's psychosocial development. But before we examine
the evidence, my second caveat is that throughout this chapter, I present
evidence that fathers contribute to positive child development in several
ways. In some cases, I argue that fathers tend to specialize in certain aspects
of raising children—things like helping children regulate their emotions or
preparing them to succeed in the world outside the home. This is not to
imply that mothers and other caregivers cannot, or do not, also do these
things, only that fathers tend to gravitate toward these roles in societies like
ours and there is evidence that this benefits their children. Moreover, the tendency of fathers to take on these roles is not necessarily about their gender.
For example, it could be attributable to their more typical role as a secondary rather than a primary caregiver. Perhaps secondary caregivers attempt to
differentiate from primary caregivers and complement rather than duplicate
what they are providing. I say this in part because multiple studies show
that children raised by lesbian couples are as psychologically and socially
well adjusted as children raised by heterosexual couples.[17] This suggests that
fathers may be substitutable—that others are capable of providing children
with most of the helpful things that fathers do. Many people have the reasonable intuition that fathers are uniquely helpful to boys in teaching them
how to become men. Since they have experienced it themselves, they are
likely to have special knowledge in this domain. As former NBA superstar
Dwyane Wade put it, "All children need their fathers, but boys especially
need fathers to teach them how to be men. I remember wanting that so
badly before I went to live with my dad. I wanted someone to teach me
how to tie a tie and walk the walk, things only a man could teach a boy."[18]
Nevertheless, for many of the outcomes that we think are most important—
behavioral problems, peer relationships, social competence, self-esteem,

academic achievement—the evidence indicates that children raised by two mothers are indistinguishable from those raised by a mother and father.[19] Further support for the idea that paternal contributions are often substitutable is provided by the many small-scale, nonindustrial societies where fathers normatively have minimal involvement in hands-on caregiving, yet children reliably grow to be well-adjusted adults of their culture.[20] In these societies, mothers often receive help from relatives other than the father— grandmothers, aunts, uncles, older children—in raising their children. In summary, good fathers can be very valuable to children, but so can others, and children raised without a father are still capable of thriving.

*   *   *

It's important that we stop and think critically about the statistics President Obama mentioned in his speech before we leap to conclusions. He described a number of associations, but those associations are not necessarily causal. This is a critical point and best illustrated with some examples. Imagine a father living in poverty with limited employment opportunities who turns to criminal theft to improve his economic situation. He gets arrested and goes to prison, and his son is therefore raised without a father in the home. His son eventually finds himself in a similar setting as the father and also turns to crime. In this case, there is an association between father absence and criminality, but it is a spurious one that is not caused by the father's absence. Instead, it is caused by father and son responding similarly to the same set of grim circumstances. Or imagine an alcoholic father whose addiction leads to marital conflict, divorce, and separation of the parents. Perhaps his daughter inherits a genetic predisposition to alcoholism and also becomes substance dependent. Once again, father absence is associated with substance abuse, but it is not the cause. As a final example, imagine a child exposed to an acrimonious divorce who develops an anxiety disorder in response to this trauma. Again, there will be an association between father absence and poor child mental health, but it was not caused by the father's absence. It was caused by the conflict surrounding the divorce. These are all just hypothetical examples meant to illustrate the problem with assuming causation from mere associations; they are not necessarily the true explanations behind the associations.

Researchers have attempted to address the question of causation with a variety of strategies. The easiest, but weakest, approach is to control for potentially confounding variables. So if you suspect that the relationship between father absence and drug dependence can be explained by their mutual relationship with low socioeconomic status (SES), you can control for SES in your statistical models. In other words, you determine how much of drug dependence can be attributed to low SES, and after accounting for that effect, ask if there is still an association between father absence and drug dependence. If an association remains, then you have better (but not very good) evidence for a causal relationship between father absence and drug dependence. Another strategy is to conduct longitudinal studies in which children are studied both before and after their father leaves to see if they experience changes in well-being. This allows you to better isolate the effect of father absence. A third approach is to ask if children exposed to more years of father absence suffer more severe consequences than children exposed to fewer years of absence. That is, does a larger dose of father absence lead to more severe consequences for children? A final strategy is to look specifically at children whose fathers have died, rather than children whose fathers have left due to separation or divorce. The rationale here is that death is a more random phenomenon than divorce in terms of the kinds of people who are likely to experience it, although this seems questionable. While none of these methods can definitively establish causation, they can all collectively contribute evidence in support of it. Overall, these types of studies show that the best evidence for a causal effect of father absence are for (1) behavioral problems (e.g., aggression, hostility) and delinquency, especially when father absence occurred early in development and especially in boys; (2) substance use, including smoking and drug and alcohol use; (3) poor adult mental health; (4) not completing high school; (5) low employment; and (6) early childbearing.[21]

There is one more method of determining if there is a causal relationship between father absence and children's development, and this is the gold standard. The most scientifically rigorous approach to establish causality is to conduct an experiment in which the researcher randomly assigns individuals to different conditions. In other words, the experimenter would randomly assign children to either be raised with or without a father. We don't want to do this experiment in people, but it has been done in some animal species where fathers are involved in caregiving. In mandarin voles, a socially monogamous rodent that lives in China, experimental removal of the father makes pups more anxious and less interactive with other voles.[22]

It also reduces the tendency of adult offspring to form a social bond with their opposite-sex partner and to care for their offspring.[23]

These are weighty consequences. However, voles are not people. Mandarin voles are classified as biparental, which means paternal care is obligate, required for offspring to survive. Humans are more accurately described as cooperative breeders, in which paternal care is facultative and other caregivers are often available to help in raising children (see chapter 2). Nonetheless, we may infer from these studies that there are likely to be negative consequences of paternal absence when there is no one else available to help the mother, a circumstance that is not unheard of in high-income nations.

Another important caveat to these findings is that even when the associations are causal, that doesn't mean that every child raised without a father will suffer these negative outcomes; it is not inevitable. Many will not; some will triumph in spite of this disadvantage. Some will be dandelions. Oprah Winfrey, for example, was born to a single teenage mother living in poverty in rural Mississippi; she was sexually abused during her childhood and early teenage years yet became a multibillionaire talk show host, actress, and philanthropist. She was rated by *Life* magazine as the most influential woman of her generation, and in 2013, she received the Presidential Medal of Freedom, the nation's highest civilian honor.[24] The data only suggest that negative outcomes are more likely among children who are raised without a father.

A final important qualification is that being raised without a father officially in the home is not necessarily the same thing as father absence. This is a crucial point. As we will see later in the chapter, many nonresident fathers are significantly involved with their children, and their children can benefit from such involvement.

Nevertheless, on balance, children seem to do better when they have a positively engaged father in the home. What is it exactly that fathers provide their children that benefits them?

Paternal Provisioning

Throughout most of human evolution, people produced and procured their own food through hunting and gathering, and later sometimes through planting crops and herding animals. Some modern human populations still do so, and we call these *subsistence-level societies* or *small-scale societies*. In such societies, fathers' most consistent contribution to children is the food that they provide.

What does paternal provisioning look like in such societies? The Aka people live in the dense tropical rainforest of the Central African Republic. Traditionally, groups of Aka men used spears to hunt large mammals, including elephants, antelopes, and hogs. Lone men also hunted monkeys with cross-bows and poison arrows. Traditional Inuit men living in the Arctic used harpoons, lances, and bows and arrows to hunt land mammals like caribou, as well as marine mammals like seals and whales that were hunted by boat. The Ainu people from Japan relied heavily on fishing. Men would fish for salmon using spears, nets, clubs and fishing traps. They also hunted deer and bear with bows and poison-tipped arrows.[25]

Among the Hadza people, a group of nomadic hunter-gatherers from Tanzania, men traditionally hunt mammals and birds with bow and arrow. The meat that they bring back to their camp goes not only to their nuclear family. Hadza cultural norms dictate that it is shared evenly across households. Men also collect wild honey, a delicacy among hunter-gatherers, and honey can be more easily targeted to their own household without sharing. Women forage for food throughout the day and typically bring more calories back to camp than men do (57% for women versus 43% for men). However, men step up their provisioning when they have an infant, apparently to offset decreases in the mother's foraging returns as she carries, tends to, and nurses a young infant. Simultaneously, she is burning more calories due to lactation. In this situation, men focus more of their effort on collecting honey that can be more easily directed toward their partner and child. The result is that men are providing 69% of calories when they have an infant under one year of age.[26]

In contrast to these subsistence level, small-scale societies, the most obvious and consistent contribution that fathers from modern, high-income societies make is the money they earn for their family. One way to isolate the specific impact of paternal economic contributions to child well-being is to examine the effect of child support payments made by fathers who do not live with their children, also known as nonresidential fathers. When these fathers pay child support, their children tend to do better academically and tend to have fewer behavioral problems, such as aggression and delinquency.[27] These payments can provide children with adequate housing in safe neighborhoods, as well as commodities such as books, computers, and private lessons that promote children's academic success. When child support payments are lacking, single mothers may face economic distress that interferes with their ability to deliver the sensitive and responsive care that is best for their children's development. Imagine an economically strained single mother with a demanding toddler, struggling to balance her budget, look for a job, find affordable day care, manage the day-to-day instrumental caregiving, take care of her own health, AND sensitively respond to her child—quite a tall order. This helps us to understand why fathers' economic contributions matter for child development.

Paternal Sensitivity

In addition to economic provisioning, many fathers are directly involved in hands-on child care. What specifically do fathers provide in terms of child care, and how does that benefit children? Parental sensitivity is the ability to accurately interpret infant signals and communications and to respond to them promptly and appropriately. For example, when your infant smiles and coos at you, you notice this and promptly reciprocate. Or your infant is fussy, and you quickly and reliably respond by trying to comfort her. In both mothers and fathers, sensitive parenting is associated with the development of secure social attachment in children. Your parent is typically the first person with whom you form a social relationship as an infant. According to attachment theory, children generalize from this initial, formative relationship to subsequent relationships. Infants with sensitive and responsive parents come to trust that their parent will reliably respond to their needs and social signals. They find this first relationship comforting, reciprocal, and rewarding, and they approach other relationships with the same expectations. This sets them up for success in their future relationships. Children who are securely attached tend to exhibit increased sociability, fewer behavioral problems, and better mental health.[28] As adolescents, they also tend to have later onset and less risky sexual behavior.[29] So one of the more important things a parent, mother or father, can do for their child's psychosocial development is to respond to their cues and signals with sensitivity. There is also evidence that fathers can promote secure attachment in their children through sensitive and challenging play behavior.[30] Nevertheless, keep in mind that the association between parental sensitivity and attachment security, as well as the association between attachment security and developmental outcomes, is not a strong association and leaves much room for other developmental influences on children. That is, some children have negative outcomes despite having sensitive parents due to these other influences.

Depressed parents show reduced parental sensitivity, and their children are at increased risk for both anxiety disorders and behavioral problems. In one study, the children of severely depressed mothers had four times the rate of psychopathology as children of nondepressed mothers. However, among these children, fathers who were high in sensitivity were able cut their child's risk of psychopathology in half. Thus, fathers seemed to be compensating for deficits in maternal sensitivity, and their children benefited from it.[31]

The reciprocal interactions between sensitive parents and their infants undoubtedly also foster the development of empathy, which is fundamental to so much of human kindness and cooperation (figure 1.1). The parent demonstrates awareness of and concern for the infant's mental states, and the infant thereby learns to do the same.

As children get older, optimal psychological and social development depends not only on parental sensitivity but also on parents' imposing limits on children's behavior. This combination of parental sensitivity and parental control is known as *authoritative* parenting. In contrast, *permissive* parenting involves sensitivity but reluctance to impose rules, standards,

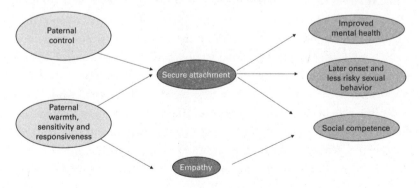

Figure 1.1
Effect of paternal warmth and sensitivity on child developmental outcomes. Paternal behavior behaviors are on the left, child developmental outcomes are on the right, and mediators are in the middle.

and discipline. A third parenting style, *authoritarian*, involves setting limits but without warmth and sensitivity. Finally, parents who both lack sensitivity and fail to set limits are categorized as *neglectful*. As described in the introduction to this chapter, Hitler, Stalin, and Mao Tse-tung had authoritarian fathers. In contrast to authoritarian parenting, authoritative parents avoid the use of harsh or arbitrary punishments to shape their children's behavior. Rather than inspiring fear in their children, these parents aim to promote cooperative behavior by reasoning with children about their behavior. Children of authoritative parents tend to fare the best, reporting less depression and anxiety and exhibiting fewer behavioral problems such as delinquency and drug use.[32] Many of us parents have suffered through a child becoming upset with us after we impose a limitation or restriction on them, but the research suggests that doing so in the appropriate circumstance benefits their long-term psychosocial development, as long as the limit setting is accompanied by warmth.

Paternal Activation

We have discussed how both mothers and fathers can help cultivate social competence in their children through authoritative parenting, but is there anything special that fathers provide to their children? Daniel Paquette from the University of Montreal has presented a compelling theory of the special role that fathers play in child development. According to Paquette, while mothers often provide children with safety, security, and comfort, fathers are often involved in "activating" their children through encouraging and supervising their exploration of the world outside the home and family. Activating involves parental stimulation of risk taking and control during children's exploration of their environment. This is part of a larger responsibility of preparing children to succeed in the external world. Critically, the role of the father depends on the temperament and behavioral tendencies of the child. For children who are shy, withdrawn, and reluctant to take risks, fathers can help cultivate and encourage exploration. For children who are impulsive and hypercompetitive, fathers can help children learn to control their aggressive tendencies.[33]

As a young child, my son had a natural aversion to novel experiences that pushed him outside a fairly narrow comfort zone. Once when he was about four years old, we took him to a children's museum in downtown

Atlanta. I remember standing with him amid myriad enticing activities, with the other children excitedly swarming about, seizing every opportunity to try every activity before another kid got to it. Meanwhile, Toby stood clinging to my leg. I felt bad that he wasn't enjoying the museum like the other kids and thought to myself, *You gotta get in there, buddy, or you are going to miss out on opportunities.* I knew he would need to learn to compete if he wanted to succeed, at least in the society that we live in. So it was clear to me from an early age that my job was to activate Toby—to help him learn to become more comfortable with novelty, exploration, and competition.

When Toby was four, I signed him up for a gymnastics class. The first day, he was terrified. He wouldn't let go of me. He kept wailing, telling me how he didn't want to do it and that he just wanted to go home. He was making a scene. All the other kids seemed okay, and I was embarrassed. But I knew, down to my very core, that this was the moment to dig in and begin teaching him the importance of not going home every time he felt uncomfortable. I somehow managed to calmly say to him, "Toby, I know you are scared and you don't want to do this, but we are not going home." He kept wailing and protesting, and I kept saying softly but resolutely, "We're not going home." I know he could sense my determination because eventually he stopped asking to go home, and I could see him begin contemplating how he was going to manage the courage to try this. Eventually the teacher let me take him by the hand over to the group and hold his hand while he walked on the balance beam. The second time we did it, he kind of liked it. After the third time, he was smiling, and then he was on his own . . . and my stress hormones began normalizing. We repeated episodes like this several times in the future whenever he began a new activity, including a first week of school I will never forget. We also had him in a sport every season once he turned four. Sports are a great way to expose kids to the concept of competition if they haven't encountered it yet elsewhere. Now twelve years old, he has grown immeasurably more comfortable with both novelty and competition and I attribute that partially to my efforts in activating him.

Of course, fathers can also activate their daughters. Along with two other dads, I am currently coaching my daughter's basketball team. I love coaching them because they are great kids, and they listen a little bit better than the boys did when I coached them at that age. A couple of weeks ago, we were defeated badly and the girls looked demoralized afterwards. Of course, sports at this age (seven to eight years) should be mostly about having fun and learning lessons in the process, but its hard to have fun when you get walloped. We lost in part because the other team was more aggressive—they played tougher defense, rebounded better and hustled more after loose balls, whereas our kids were reluctant and tentative. After the game, we told our team that they would need to do all those things better if they wanted to win, and we worked hard at it the next week in practice. There was a palpable difference in their play the next game, and we won a nail-biter, with a final score of 10–8. What an amazing lesson. If you work harder, you are more likely to succeed. And, perhaps more generally, success sometimes depends on competing.

Paquette argues that underactivated children are at risk for anxiety, but that fathers can help such children by activating them (figure 1.2). This is highly intuitive to me based on personal experience, but more important, it is supported by data insofar as fathers who playfully challenge their children to push their limits seem to buffer those children from developing social anxiety.[34] Note, however, that the quality of this paternal involvement is crucial, since overcontrolling paternal behavior is associated with *increased* child anxiety.[35] Activating children involves stimulating, surprising, destabilizing, and encouraging the child to take risks, but all in a warm, supportive, often playful manner that does not unduly restrict autonomy.[36]

Importantly, mothers can also activate children. It's just that fathers tend do more of it in societies where it has been studied.[37] The model I

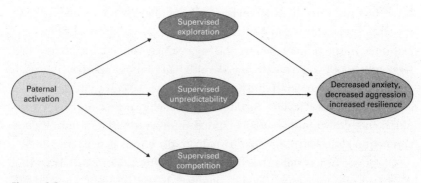

Figure 1.2
Effect of paternal activation on child developmental outcomes. Paternal behavior is on the left, child developmental outcomes are on the right, and mediators are in the middle.

have outlined is obviously a heteronormative one. Research on parental activation in gay and lesbian couples is so far limited, but one reasonable hypothesis is that just as in heterosexual couples, one member of the pair gravitates toward providing safety, security, and comfort while the other complements them by becoming more involved in exploration of the world outside the home and family.

Fathers and Emotion Regulation

Fathers can also play a crucial role in helping children with the opposite problem: being overactivated. In these children, fathers are important for helping them to control their aggressive impulses and related behavioral problems.[38] Several high-quality, longitudinal studies that control for socioeconomic status have shown that paternal involvement, and especially positive paternal engagement, protects against the development of aggressive behavior and delinquency in children.[39] In one study, paternal cohabitation at age ten predicted lower levels of teacher-reported aggression at age twelve among boys.[40] In another study of families living in poverty, adolescents who reported having more supportive fathers whom they felt close to were less likely to fall into delinquency in young adulthood.[41] Another study among low-income, minority youth found that for adolescents with high rates of delinquency at baseline, more frequent contact with their father protected against criminality one to two years later.[42] These were all US studies, but this effect is not uniquely American. In a large British study, paternal engagement with boys at age seven, as reported by the mother, protected against trouble with the police at age sixteen.[43]

We can't say for sure that these relationships are causal (that positive paternal engagement causes fewer behavioral problems). But if they are, how might fathers accomplish this? How are they able to decrease problematic aggression in their children? There are at least three potential pathways.

One is by physical play, especially rough-and-tumble play. Play is very important for psychosocial development in mammals. We know this because mammals that are deprived of play exhibit abnormal social behavior. For example, rhesus monkeys that are raised by their mothers but deprived of play with peers are hyperaggressive and are later rejected by their peers.[44] In many modern, developed nations, rough-and-tumble play is something that fathers tend to specialize in. Fathers spend considerably more of their time engaged in it than mothers do, and infants generally respond to play initiated by fathers with more positive emotion than when their mothers initiate play.[45]

Consider what a child learns from rough-and tumble play with his father. First, he learns that his father, much larger and much stronger than he is, is capable of hurting him, yet he does not. So the child learns that he can trust his father, and this helps build a secure attachment with myriad downstream benefits. Fathers cannot be too rough with their child, so the child also sees the father modeling the control of his own aggressive impulses during the play, something that the child is likely to emulate. The child also learns that if he can't control his own aggression, his father may stop playing with him and may even become upset. This teaches the child to regulate his own aggressive impulses. He also feels how powerful his father is and may fear the consequences of upsetting him. Finally, he wants to keep his father engaged and enjoying the play so that it can continue, so he monitors his father's emotions throughout so he can try to sustain this mutually enjoyable interaction. In short, it is difficult to imagine a more effective means of cultivating emotion regulation and empathy than by embedding the learning in a highly arousing and rewarding hypersocial interaction like play. We now know that such interactions shape the developing brain. For example, when rats play, they release a neurochemical known as brain-derived neurotrophic factor (BDNF), a type of brain growth factor, into brain areas that are known to be involved in impulse control.[46] As the neuroscientist Jaak Panksepp, one of the first scientists to take play seriously, put it, "Play constructs the social brain."[47]

Rough-and-tumble play needs to be done the right way. Children benefit when fathers assume dominance in these interactions (e.g., *I show that I am stronger than you without hurting you*), but not overly so (*I let you tackle me too every once in a while*). Fathers who are good at rough-and-tumble play allow their child to experience a little bit of fear and excitement—but not too much. The child feels that fear, but then learns the result was not catastrophic—indeed, it was fun—and that helps them grow comfortable with the anxiety that may accompany novelty and exploration. In this way, fathers are preparing their children for the unexpected so they will be able to react confidently to a wide range of circumstances they experience in life.[48] Skilled fathers are sensitive to their child's emotional reactions during play and will modulate their play as a function of the child's level of arousal.

I have an older brother who used to periodically hold his toddler son upside
down by his feet while playing. We were all amused by it because his son,
Jack, seemed so content in that position, but that certainly wouldn't be the
preferred position of many other kids. My brother knew Jack's threshold
for novelty and was good at taking him up to that point but not beyond.

Across human societies, boys are more physically aggressive than girls on
average.[49] So if one function of rough-and-tumble play is to help children
learn to manage their aggressive impulses, it makes sense that fathers do
more of it with sons than with daughters.[50] My own son craves rough-and-
tumble play. He needs it. Sometimes he will look me straight in the eye and
say flatly, "Wrestle me." At other times, he adopts a more subtle approach
that involves gently annoying me until I am forced to playfully retaliate. He
might poke me in the back with a toy sword while I am concentrating on
something, for example. Or if that doesn't get my attention, he might poke
me in the neck instead. He also likes to lob playful insults at me like, "You
have no muscles," or the more devastating, "You're bad at math." Eventually he usually succeeds in drawing me into a wrestling match. It feels to
me that he likes the reassurance he gets that this stronger, more powerful
person who takes care of him will not hurt him. So I am normally Toby's
primary playmate, at least in the realm of rough-and-tumble play, and I
am sure many other fathers have a similar relationship with their children.
However, there is one notable context in which I lose my status as primary
playmate, and that is when his male cousins come to visit. In fact, I have
had the experience of feeling superfluous at those times. He just doesn't
need me as much in these situations. I bring this up because the primary
playmates of children living in more traditional societies, such as foraging
societies, are not fathers but rather other children. In their everyday lives,
children in these societies are surrounded by other children of different
ages.[51] Through their play, younger children learn many of the same lessons from older children that children in nuclear families learn from their
fathers. Thus, throughout human evolution, it was probably not the case
that father–child rough-and-tumble play was essential for normative social
development. In most cases, children could get this from others.

Physical play is one pathway by which fathers might help children regulate their emotions and control their aggression, but there are others. Some
children refrain from misbehaving because they are afraid of their father.
One father I interviewed told me an interesting story that illustrates the

point. He admitted to me that he misbehaved in school as a child and that he liked to see what he could get away with in class. He mentioned that if his teachers had only threatened to call his father, that would have struck fear in his heart and made him behave. But as it was, there was nothing to keep his mischief in check—no threat that he found worrisome enough to deter his antics. Another father I interviewed told me that all his father had to do was look at him in a certain way, and the fear this engendered would similarly make him fall in line.

It is possible that children are innately afraid of large, powerful adults when they lower their voice or express disapproval. Or perhaps some children learn to fear their father. Another father I interviewed told me that his father spanked him only once, and "that was enough." Spanking is of course highly controversial and officially discouraged by the American Academy of Pediatrics due to evidence that it is associated with many negative child outcomes.[52] Nonetheless, spanking is a common form of discipline around the world, and most children are spanked at some point during childhood.[53] Why is it so common? One of the fathers I interviewed had this to say about his own father's use of corporal punishment: "Because he beat me a lot, I remember key moments of my life where my friends were doin' this and I didn't, because my initial thought was, 'My dad'll beat my ass.'"[54]

On a more positive note, fathers also can serve as a model to their children, perhaps especially their sons, as to how men should behave. When a boy sees his father skillfully controlling his own aggressive impulses, he may aspire to do the same. In his book, *Father Hunger*, child psychiatrist James Herzog recounts his experience with a group of boys who upon losing their fathers to divorce, began having nightmares with violent themes. Herzog concluded, "Without a father to help him integrate and modulate it, a boy's aggression typically appeared as a foreign force in his dreams and fantasies." Absent a father, boys may struggle to manage their aggressive impulses.[55]

Another theory that attempts to explain why paternal involvement decreases child aggression problems and why father absence precipitates them is known as *protest masculinity*. According to this idea, if a boy's father is unavailable during childhood, he will come to identify with his mother and form what is known as a cross-sex identification. Later, when he tries to identify with males, he will feel considerable conflict with his primary feminine identification and will resort to violent and aggressive behavior to reject it.[56] In a sense, he overcompensates for his early feminine identification with hypermasculine behavior. Cross-culturally, rates of violence are higher in societies where fathers have less physical proximity to their sons, and some have attributed this to protest masculinity.[57] Others scholars have instead argued that boys have a natural proclivity toward aggression and that involved fathers help children to control those impulses,[58] as discussed above. Uninvolved fathers would permit these aggressive impulses greater expression. Low father–son proximity may even be a cultural mechanism by which some societies produce warriors—that is, through fathers maintaining physical and emotional distance from their sons and thereby allowing those aggressive impulses greater expression. It may also partially explain gang violence, which some fatherless boys turn to for male companionship. Still, the direction of causality linking warfare and father absence has always been questionable to me. What if fathers who are warriors distance themselves from sons out of compassion, so that their sons will not be emotionally crippled should the father be killed in battle?

The ability of fathers to cultivate emotion regulation skills in their children has important consequences (figure 1.3). For example, children who can better control their emotions and their attention perform better academically.[59] They also seem to have better social skills. Children need to be able to control their anger and aggression in order to make friends, and

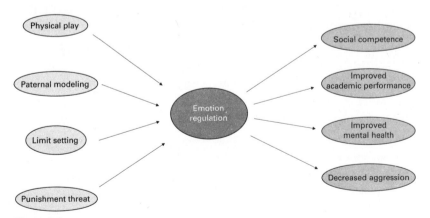

Figure 1.3
Paternal influence over children's emotion regulation abilities and downstream developmental outcomes. Paternal behaviors are on the left, child developmental outcomes are on the right, and mediators are in the middle.

father–child play can teach children how to do that. In one study, researchers observed fathers playing with their three- and four-year-old children at home and then asked their preschool teachers to rate their popularity with their peers. Researchers also observed the children during interactions with their preschool peers. Children with more physically playful fathers and with fathers who made them laugh during play were rated as more popular by their teachers. Furthermore, boys with engaged and playful fathers had more harmonious, relaxed, and dominant interactions with their peers. Here, *dominant* does not mean domineering. It just means taking initiative and making suggestions during peer interactions. Girls with physically playful fathers had more dominant but also more abrasive interactions with peers. As noted, the quality of the father–child play interaction seems crucial, since fathers who issued more commands to their children during play had children who were rated as less popular by their teacher, and boys of these fathers had more abrasive peer interactions. Curiously, maternal directiveness was instead positively associated with peer popularity.[60]

Paternal Influences over Sexual Decision Making

Emotion regulation and impulse control may also be involved in sexual decision making. In both the United States and New Zealand, girls raised without fathers were more likely to have sex before age sixteen and to become pregnant before age eighteen compared with father-present girls, and this association held after controlling for several potential confounding variables such as race and socioeconomic status. Strengthening the argument for a causal effect of father absence, girls whose fathers left when they were younger and therefore had more years of exposure to father absence were more at risk for these outcomes than girls whose fathers left when they were older.[61] Even when fathers are *physically* present, girls with more *psychologically* distant fathers are more likely to engage in early sexual activity compared with those who feel closer to their fathers.[62] One explanation offered for these findings is that father absence serves as a developmental cue to girls that male parental investment is unreliable or unimportant and that this reduces their reticence in forming sexual relationships. In other words, there is no point in waiting to find a man who will be a good husband and father if those men don't appear to exist or if they are so rare that it's futile to try to find one. From an evolutionary perspective, the better strategy in this situation is to find a man with good genes, start reproducing right away, and hope for the best.

There are also other possible explanations for the association between father absence and early sexual activity in girls. For example, involved fathers may supervise or constrain their adolescent daughter's social and sexual opportunities, and this constraint is absent when fathers are not physically or psychologically present.

Although most of this research has focused on girls, there is also evidence that boys raised without fathers have an earlier sexual debut.[63] Similar to girls, boys raised in these circumstances may be less inclined to aspire to a single, stable monogamous partnership and may therefore enter into an initial sexual relationship more readily.

Preparing Children for the World

I noted previously that fathers often help prepare children for the world outside the home by supporting children's exploration and by gently challenging them to move beyond their comfort zone. Fathers prepare children for the world in other ways too. For example, they seem to help get children ready to converse with people outside the family, though perhaps not intentionally. Fathers tend to be more challenging conversational partners than mothers, particularly insofar as they make more clarification requests and ask more what, where, and why questions that encourage children to provide longer responses.[64] This helps build children's vocabulary and fosters verbal reasoning abilities.[65] Interestingly, fathers', but not mothers', vocabulary use with two-year-olds predicted children's language development at age three.[66]

Many fathers teach their children skills or provide knowledge that will help them succeed as adults. One British study showed that increased paternal involvement was associated with increased child IQ at age eleven, as well as upward social mobility at age forty-two.[67] In many traditional societies, fathers are charged with teaching sons subsistence skills.[68] For example, among the !Kung San hunter-gatherers, adolescent boys accompany their fathers on hunts in order to learn these skills.[69] Some fathers try to prepare their children for the larger world by being tough on them. They think, "Not everyone is going to be as nice to you as your mother is, and I need to get you ready for that." During the 1998 World Cup, the rising

British soccer star David Beckham was given a red card by the official and ejected from the game for retaliating against an opponent who had fouled him. That meant England had to play with one fewer player for the rest of the match, and they ultimately lost to their archrival Argentina. The English, like the fans in many countries around the world, take their soccer (football) seriously, and Beckham suffered serious abuse from fans and the media for a long time afterwards. In the 2023 Netflix series *Beckham*, he was asked how he made it through that period; how he managed to continue playing in front of thousands of fans who were brutally harassing him.[70] He responded, "I think I could handle it because of the way my dad had been to me." His father had pushed him relentlessly in soccer when he was a child. I don't know whether this negatively affected their relationship, but it seems to have cultivated a resilience in Beckham that helped him succeed in the face of enormous stress.

More generally, we will see in chapter 6 that there are many societies where this type of authoritarian parenting prevails despite evidence that the kinder and gentler authoritative parenting is often better for children's psychological and behavioral development. It may be that fathers in these places are preparing their children for a society where they will be expected to follow rules without question and do as they are told, or else there can be severe consequences. My own *authoritative* style, in which my children incessantly negotiate with me about absolutely everything, would probably not be very beneficial to my children in some of these places. One might even argue that authoritarian fathers in these societies are sacrificing the potential for a warm and intimate relationship with their child in order to help them succeed in the future. That is, their harshness can (sometimes) come from a place of love.

Fathers, and mothers, also provide children with social, educational, and job opportunities. Consider the father who helps pay for his child to attend a prestigious private college, or even any college at all, and all the future career opportunities this opens. Other opportunities emerge from fathers' efforts at social networking and cultivating positive relationships with other adults. This can be as simple as increasing the likelihood that your five-year-old gets invited to a birthday party, which isn't trivial for a five-year old or as consequential as a job opportunity. This isn't just a Western phenomenon. Among the Martu aborigines from Australia, adolescent boys typically progress through a ritual initiation into adulthood that involves expensive ceremonies typically paid for by the father. The elder males of the group collectively decide which boys are ready to be initiated. Boys who have a father present are initiated more than a year earlier, on average, probably because fathers are able to successfully advocate for their sons to be chosen. An earlier initiation benefits the young man insofar as he is able to marry and begin reproducing earlier.[71]

Both fathers and mothers also teach their children social norms that they will need to abide by in order to become competent members of their society (figure 1.4). If you are a parent of a young child, pay attention sometime to just how much of what you say to your child is aimed at teaching social norms—for example, "Don't take things from others without asking," "Share," "Say please when you ask for something," "Say thank you when someone does something for you," "Listen to and respect your teacher," "Be kind to elderly people," "Sit in your chair when eating dinner," "Don't hurt other people's feelings," "Don't discriminate against people based on their appearance," "Don't chew with your mouth open," "Don't skip other people in line," "You have to wear clothes in public," "Shake hands when you meet someone" (unless you are in the midst of a global pandemic), "Apologize if you hurt someone," "Don't interrupt someone when they are talking," and, finally, "Don't look at your iPad when you are talking to people, especially your parents!" Children need to learn all of these social norms and many more, and parents are the main agents of socialization in early childhood. Look back at the Teddy Roosevelt quote at the beginning

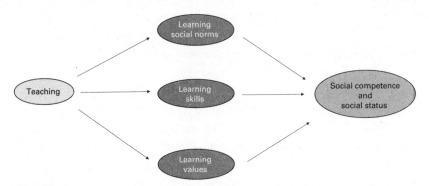

Figure 1.4
Effect of paternal teaching on child developmental outcomes. Paternal behavior is on the left, child developmental outcomes are on the right, and mediators are in the middle.

of the chapter where TR vividly explains the process by which the social norms his father taught him became internalized or embodied within him. This is the often invisible work that parents do on behalf of society.

The Family System

In recent years, developmental psychologists have begun emphasizing that fathers do not operate in isolation but are instead part of a family system that children are embedded in and that children benefit when that whole system is working well, whereas they suffer when the system is strained or broken. Maternal and paternal influences are not independent of one another. This means that part of being a good father is providing mothers with emotional and instrumental support so that mothers can be more available and attentive in their interactions with infants and children. High marital quality is associated with more sensitive maternal and paternal behavior, as well as better developmental outcomes for children. Conversely, marital conflict has consistently harmful effects on children's socioemotional development.[72] So fathers can also help their children by trying to ensure that the whole family is functioning smoothly.

Is That All?

Figures 1.1 to 1.4 capture much of what fathers contribute to child development, but my own personal reflection suggests that this model, and the research that supports it, fails to include other key aspects of paternal contributions. One of these is what I would call managing your child's psychology. My son is half me, genetically speaking, plus he has been shaped by my parenting and the environment we share. So his psychology is something I believe I understand quite well. I often feel that one of my most important roles is what I say to him when he fails. Last year, he lost a tennis match 8–0. At about 6–0, I started thinking about what I could say to him on the way home that would not leave him discouraged and feeling that he didn't want to play again next season. I knew how my own ten-year-old self would have reacted—sullen, discouraged, pouting, with plummeting self-confidence. I also knew what would have made me feel better, so I made a mental note of everything he had done well in that match. After the match, I immediately told him how well he played, which was true (the other kid was just better). Then I went through my list with him. I also said he was only a point away from winning several of the games (also true). And then I told him how much more he learned from playing against someone better than him like this. Finally, I told him how proud I was that he didn't give up. Then we got some ice cream. He seemed pretty upbeat after all that, and he won his next three matches. He's still playing and enjoying tennis—and he is learning how to overcome adversity. When parents understand their children's psychology well, they can say the right thing at the right time and help put them in a frame of mind that increases their chance of success.

Fathers can also teach their sons how to be good fathers through the example they set. When my son sees me interacting sensitively and responsively with his four-year-old sister, he is internalizing how to act as a father. On rare occasion, when he is not arguing with her and when it really matters, I have even caught him treating her in the same way. Boys raised without fathers lack this template. In his book *Becoming Dad*, Leonard Pitts Jr. describes how the absence of his father had lasting consequences for his own paternal self-concept: "I have two stepchildren and three biological children, all from the same wife. I fancy myself a decent father to them. But that fact is, there's never a day that passes when I'm not wondering whether I am doing it all wrong."[73]

Cross-Cultural Paternal Impact

In this chapter, I have presented evidence that fathers are important for their children's psychological and social development. However, nearly all of these studies have been conducted in high-income nations in which families tend to live in an evolutionarily atypical family context. More traditionally, children were surrounded by a "village," including a wide range of relatives who are available to assist with caregiving. In many of these societies, fathers are minimally involved in direct caregiving, yet children typically grow up to be competent and well-adjusted adults.[74] Thus, it is quite possible that other caregivers are able to substitute for limited paternal involvement and that for this reason, the importance of fathers in child development varies cross-culturally as a function of family context. Fathers may be particularly important in isolated nuclear families that live apart from their extended families.

In addition to fathers, grandmothers can be invaluable caregivers. In fact, child survival is more strongly associated with the presence of grandmothers than with the presence of fathers among natural mortality populations. Among the Hadza hunter-gatherers from Tanzania, grandmothers provision their grandchildren with difficult-to-extract tubers, and this improves grandchild growth.[75] In rural Pakistan, grandmothers are involved in daily child care such as feeding, bathing, and play, and high grandmaternal involvement at age one is associated with improved cognitive, fine motor, and socioemotional development at age two.[76] In sub-Saharan Africa, children with a coresident grandmother are 40% more likely to attend school. By caring for young children or performing household tasks, a grandmother might free the mother to work outside the home to pay for schooling. Or she might enable an older child to attend school rather than stay home and perform household tasks.[77] In another study of low birth weight, preterm infants born to adolescent mothers in the United States, grandmaternal coresidence at one year of age predicted higher IQ at age three.[78]

In situations where mothers are surrounded by grandmothers and other relatives who can help with child care, we might expect fathers to be less important for child development. Nevertheless, recent evidence suggests that their importance is not strictly limited to high-income countries where isolated nuclear families are common. One large study of thirty-eight low- and middle-income countries spanning South America, Africa, Eastern Europe, and Asia asked primary caregivers to report on paternal involvement with their three- to four-year-olds over the past three days. Overall levels of paternal involvement were quite low, with around 50% of fathers not having engaged in any stimulation activities with their children over the past three days. However, fathers who more often read books, told stories, sang songs, and played with their children had children with better socioemotional and literacy and numeracy outcomes. This positive association with paternal involvement was stronger when the mother was not engaged with the child, again suggesting that fathers can compensate for lack of maternal stimulation.[79]

Nontraditional and Underrepresented Fathers

When we think of fathers, many of us may imagine a biological, coresident, heterosexual, mentally healthy, secondary-caregiving father. However, many fathers do not fit this mold, and studies are beginning to examine the impact of these nontraditional types of fathers, as well as fathers from underrepresented groups. In the discussion that follows, please be aware that my own experience as a traditional father imposes limitations on my ability to fully and accurately represent the perspectives of nontraditional fathers. More generally, despite my best efforts, my own background and perspective may at times unconsciously bias my interpretation of findings throughout the book.

Nonresidential Fathers

Many people feel that fathers have a moral responsibility to live with and raise their children and that those who do not are shirking their responsibility. No doubt this is sometimes the case. However, in situations where spouses become hostile to one another, parental separation is often the best thing a father, or mother, can do for their child. Children of separated parents often live with their mother, resulting in the father being nonresidential. Doubtless, many nonresidential fathers care deeply about their children and are strongly committed to having a positive impact on their lives. One way they can do so is by continuing to contribute economically to the household where their child is living. Studies show that when nonresidential fathers make child support payments, children have better academic, psychological, and behavioral outcomes. When nonresident fathers contribute financially, mothers may experience less stress, and the quality of their parenting may improve.[80] Direct care from nonresident fathers also has an impact, but the data suggest that the quality of the father–child interaction is more important than the frequency of visits. In particular, children do better when fathers adopt an authoritative parental role, as opposed to when they act more like an adult companion. This means not just taking children to the movies and other fun events, but continuing to help with homework, talk about problems, and discipline for misbehavior.[81]

Gay Fathers

Imagine a man coming to the realization that he will not be able to have children and become a father because he is attracted to men and that the only way out of this dilemma is to conceal his preference and pair with a woman. Until recently, that was the situation many gay men faced. Some gay men have children within heterosexual marriages, often before coming out as gay. But more recently, gay couples have begun adopting children or having biological children with egg donors and surrogate mothers. Gay men face many obstacles in having children, including stigma and discrimination by adoption agencies and others, the cost of assisted reproductive technologies such as surrogacy and in vitro fertilization, and the cost of legal fees to ensure assignment of custody. For this reason, gay men who do become fathers are likely to be highly motivated and committed to parenting. Gay men don't have a child by accident. A study of adopted children between four and eight years old found that gay fathers showed more warmth, greater amounts of interaction, and lower levels of disciplinary aggression than heterosexual fathers did. Adopted children raised by gay fathers also had fewer behavioral problems compared with those raised by heterosexual couples.[82] Other studies have similarly shown that children of gay fathers tend to have better outcomes than children of heterosexual parents on average.[83] This is not to say that it is better for children to be raised by gay than heterosexual couples, but that children benefit from having highly motivated and committed parents. Children also benefit when caregivers provide warmth, sensitivity, and appropriate discipline and control (i.e., authoritative parenting), no matter who is providing that care. Men and women are both capable, and having two authoritative parents is better than having one.

Despite these findings, there has been only limited research on children raised by gay fathers and how their developmental outcomes compare with other children. One can reasonably ask whether fathers are able to fully and adequately prepare daughters for their gender-specific experiences. Girls might especially benefit from maternal guidance during puberty and childbirth, since their fathers will not have experienced these important life events. *What does it mean when I have my first menstrual period and what do I need to do? What about these other anatomical and psychological changes I am experiencing? Is this normal? What will labor be like? How do I nurse and care for my baby?* Probably many gay fathers anticipate these challenges and find female role models who can compensate for their own inexperience in these domains.

Single Fathers

It is difficult to generalize about single fathers, since the circumstances by which they become single parents are highly variable. For example, well-known celebrities like Brad Pitt and Lenny Kravitz reportedly became single fathers by divorce, Liam Neeson on the tragic death of his wife, and Ricky Martin and Cristiano Ronaldo by surrogacy. The experience of a man who seeks out single fatherhood by surrogacy is likely to be quite different from that of a man who has single parenthood thrust on him due to widowhood or divorce.

What is clear is that single fatherhood is becoming more common around the world, and particularly in the United States, where the proportion of single-parent households headed by fathers has steadily increased from 10% in 1980 to 24% in 2020.[84] These statistics suggest that many fathers have both the motivation and the ability to serve as primary caregivers. But how do these children fare? As discussed at length earlier in this chapter, children raised by single parents in general are at higher risk for negative developmental outcomes compared with children raised in two-parent households. However, researchers have asked how single fathers differ from single mothers and how children raised by single fathers compare with those raised by single mothers.

Single fathers tend to have higher incomes than single mothers on average and are therefore less likely to be living in poverty, which is expected to benefit children. However, single fathers also tend to supervise and monitor their children less closely. In terms of academic performance and mental health, children of single fathers do about as well as children of single mothers. However, children of single fathers are more likely to exhibit antisocial and violent behavior and are more likely to have problems with substance abuse. This need not imply that single fathers are worse parents, since fathers may be more likely to take custody of children when the mother has a serious mental illness or substance use disorder. If a child has suffered from severe maternal neglect or abuse and consequently comes to live with their father, the child's problems should not be attributed to the fact that they now live with their father.[85]

Single fathers may face social stigma and generally have worse mental health than married fathers.[86] One study of South Korean fathers showed that single fathers had poorer quality of life, more depressive symptoms, and more stress than did married fathers, even after controlling for socioeconomic status.[87] However, it is difficult to know if this reflects the challenges of single parenting, the lingering consequences of divorce or death of the spouse, or both.

Black Fathers

Voices both within and outside the African American community have called on Black fathers to become more involved in raising their children.[88] This is based on statistics showing that more Black children are raised in single-mother households compared with children of other races, combined with evidence that children raised in single-mother households are at increased risk for negative developmental outcomes. However, other voices have appropriately cautioned against simplistic interpretations of these data. First, there is concern that the data will be interpreted to suggest that Black men are not committed fathers, and this contributes to damaging stereotypes. Imagine someone assuming that you will not be a good father simply because of the way you look. In fact, among fathers who do live with their children, a Centers for Disease Control study found that Black fathers reported being more involved in instrumental caregiving activities like bathing, diapering, and dressing their children, as well as helping with homework, compared with White and Hispanic fathers.[89] Other research shows that nonresidential Black fathers are at least as involved as their White counterparts.[90] Although nonresidential fathers tend to be less involved than residential fathers, some nonresidential fathers are quite involved.[91] I interviewed a number of Black fathers who were not living with their children full time but who had arrangements with the mother to regularly spend time with them. A father might pick the kids up from school every day and have them every Thursday night and every other weekend. What struck me is that they were the primary caregiver during these times—all the diapers, all the meals, all the baths—and I thought about how infrequently I, as a residential father of two children, adopted that primary caregiving role myself. Nonresidential or noncustodial does not mean uninvolved.

The other concern related to these statistics is with the simplistic assumption that all of the problems of the Black community would vanish if more fathers would simply live with their children. However, the Black community is burdened by many structural disadvantages that also contribute to worse developmental outcomes for children, including inequality in educational opportunities, housing segregation, mass incarceration, and chronic unemployment. So while the data show that having fathers in the home helps, these structural inequalities must also be addressed for Black children to have the same chance of success as their White peers.

Depressed Fathers

Most of us are aware that postpartum depression is a common challenge for new mothers and has a negative impact on both mothers and their infants. What most people don't realize is that men also face an increased risk of depression surrounding the birth of their child. The baseline rate of depression among men living in developed nations is around 5%. In contrast, the rate of perinatal depression (depression during pregnancy or the first postnatal year) in men is around 10%, so the transition to fatherhood approximately doubles the risk. For comparison, the rate of perinatal depression in women is around 25%. Parents seem to be at particularly high risk three to six months postpartum, when the rate increases to about 25% for fathers and to about 40% for mothers. Interestingly, rates of paternal depression vary considerably across developed nations. Rates in the United States are particularly high, where 14% of fathers are afflicted across the perinatal period.[92] The United States is one of the only high-income countries without state-supported paternity leave, so it is natural to wonder if this could be a contributing factor (see chapter 6).

Similar to maternal depression, depression in fathers is associated with a host of negative outcomes in children. Children with depressed fathers tend to do worse socially and academically, and they tend to have more behavioral and psychological problems.[93] There are at least three possible explanations for this association. The first is that children of depressed fathers inherit a tendency to have poor mental health from their fathers and that this leads to psychological and behavioral problems, which in turn jeopardize social and academic performance. Twin studies have established that depression is significantly heritable (passed from parent to child genetically), so this is a plausible explanation for the association. These studies show that both genetics and the environment have an impact on the likelihood of depression and that they have comparable influences.[94] A second possible explanation is that depression has a negative impact on how fathers interact with their children and that this in turn has a negative impact on the child's development. Indeed, depressed fathers tend to show less paternal warmth and sensitivity, as well as increased hostility and disengagement with their children. Finally, paternal depression is associated with lower marital quality, and this could have a negative impact on maternal caregiving, with downstream consequences for the child.[95]

Why do men become more vulnerable to depression as they transition to fatherhood? Stress is a well-known precipitant of depression. For many men, especially for those who are unemployed or living in poverty, the birth of a child causes an increase in economic stress. This may be why both low income and unemployment are risk factors for paternal depression.[96] In my interviews with fathers from the Atlanta area, I asked them if they felt pressure to provide economically for their children. Here are some of their responses.

"On a scale of 0 to 100%, like, I mean, like, 100%."

"A ton," "it's, it's enormous."

"Immense pressure. Immense pressure."

"All the pressure. Just all the f*cking pressure, if I can curse. All of it. All of it, all of it."

"Yes, yes and yes."

"Yeah, um, tremendous pressure."

"Oh yea. A ton."

Another source of stress is not knowing how to take care of an infant. My brother told me that when he and his wife drove home from the hospital with their first child, they placed him on the floor and then looked at each other and said, "What do we do now?" First-time fathers and fathers who judge themselves as low in parental efficacy are at increased risk of depression, probably because they feel unprepared and overwhelmed by the task in front of them.[97] No one magically knows how to be a parent. We all have to learn. In traditional human societies, children and adolescents are surrounded by younger children, and they often help with their care. This is good preparation for parenthood. For contrast, as the last-born child in my family, I had no exposure to younger siblings and never interacted with an infant for more than a few seconds until I had my own at age forty. I had a lot to learn, and it wasn't easy, and this pales in comparison to what first-time *primary* caregivers face. Our society often segregates children from others, and this limits opportunities for adolescents and young adults to learn about infants and small children so they can be more prepared as new parents.

Yet another source of stress is work–family conflict, also a risk factor for paternal depression.[98] As American mothers have become more involved in working outside the home, fathers have become more involved in direct caregiving.[99] However, this has not been without some strain, as fathers also report more difficulty with work–family balance than in the past.[100] As one father I interviewed put it, "A challenge has been to put the family first yet make sure that work is there at a high enough level that I'm going to succeed, and sometimes that means you gotta put family second, and that's been a hard concept for me to deal with."

Fathers with less social support are also more likely to experience depression.[101] As discussed at length in chapter 3, humans are naturally cooperative breeders. This means that mothers typically receive help from a variety of others in raising their offspring. Raising children without help is probably something that most humans are psychologically ill equipped for. Everyone needs a break from child care. It is stressful for couples, not to mention single parents, to raise children all on their own without any assistance. This support may be particularly important during emergencies. My in-laws live near us, and their availability to pick up a sick child from school and tend to them in a pinch has helped us preserve stability in our careers, as well as our sanity.

Fathers are also more likely to become depressed when they have a bad marriage, and many fathers report decreases in relationship quality after the birth of a child. [102] Forty-three percent of the fathers I interviewed indicated that having a child had a negative impact on their relationship with their partner. A common theme in these interviews was that the attention and affection fathers had previously received from their partner shifted to the child after the birth. One fifty-three-year-old immigrant father explained that after having children, "She is spending more time on child. Yeah, sometime he the center. I feel it, personally. I feel less care, care about me." A fifty-nine-year-old accountant and father of two described this attentional shift similarly: "So that first year, I mean, she really doted on me, but then my daughter was born . . . the focus for her became the kids . . . in a lot of ways, I am looking forward to him going off to college." Another father described an asymmetry between he and his wife: "In my heart I still think . . . that's [your spouse] the most . . . close person to you. But for my wife . . . she didn't think the same . . . kids are more important than me and it's the difference I think."

Fathers also experience hormonal changes that may contribute to depression, and these are discussed at length in chapters 3 and 4. Among involved fathers, testosterone levels decrease across the transition to fatherhood, and low levels of testosterone are known to be a risk factor for depression in men.[103] Testosterone replacement therapy can also alleviate depressive symptoms in men with low testosterone levels.[104] So the natural decline in testosterone with fatherhood may render men more vulnerable to depression. It may also decrease libido, which, combined with the typically reduced libido of his postpartum partner, means less sex in the relationship.

Fathers who feel their infant has a difficult temperament are also more likely to be depressed.[105] We might surmise that depressed fathers are simply more irritable and react more negatively to infant crying and fussing. However, it is also possible that some infants come into the world with objectively difficult temperaments that pose special challenges to their parents' mental health. I had one such infant. There is great variation in the amount of crying that infants do. Especially during the first three months of life, many infants cry inconsolably and for many hours every day, especially in the evening when mom or dad gets home from work.[106] This is understandably frustrating to well-intentioned parents. You are trying your very hardest to do whatever it is your infant needs to be calmed, but to no avail—and the crying persists. And you begin to feel helpless and to wonder if you are a bad parent. I remember my infant son crying—actually more like screaming—throughout the night. I remember what it did to my sleep, my mood, my work. I had nights where I thought, *This is hell on earth.* It doesn't surprise me one bit that difficult infants are more likely to have depressed parents.

One very strong predictor of paternal postnatal depression is maternal depression.[107] This could be because maternal depression negatively impacts relationship quality, which in turn makes paternal depression more likely. Or it could be that maternal depression severely disrupts the stability of the family system, and fathers find that instability stressful. In other words, dads know that children need a warm and attentive mother and when that is missing, they realize the threat to the whole family; mother, father and children. While maternal depression may pose a risk for paternal depression, other fathers may respond by attempting to compensate for deficits in maternal sensitivity. As mentioned previously, sensitive fathers can ameliorate the negative impact that maternal depression has on their children.

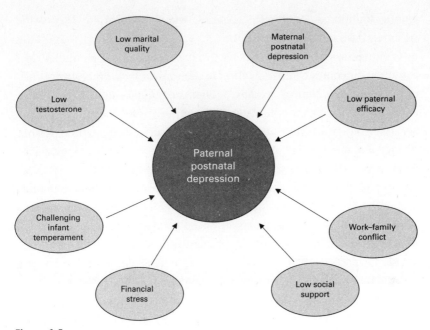

Figure 1.5
Predictors of paternal depression.

Finally, one of the strongest predictors of paternal perinatal depression is having had a mental illness diagnosis previously in life, prior to the pregnancy (figure 1.5). From this perspective, paternal depression can often be viewed as recurrence of a mental illness that is triggered by the stress of becoming a first-time parent.[108]

Stepfathers

Many stepfathers are confused as to exactly what their role should be. Should they try to be an authoritative parent to their stepchild? A friend? Or just a nonthreatening companion to their mother?

Most stepchildren are raised by stepfathers versus stepmothers. This is because children more commonly live with their mother after divorce or separation. While the number of stepfamilies in the United States has grown in recent years due to higher rates of divorce and nonmarital births, stepfamilies are not unique to developed societies. They are also common among traditional, small-scale societies, where divorce also occurs and parental death is more common.[109]

Children raised in stepfamilies tend to have worse social, emotional, and academic outcomes compared with children raised by both biological parents for several potential reasons.[110] First, stepfamilies tend to be poorer, and as we have seen, economic stability is important for child development. Second, children who live in stepfamilies often suffer through their parents' divorce and any conflict associated with it. Parental separation, especially when acrimonious, is known to have a negative impact on child social, emotional and academic outcomes. The formation of the new stepfamily is also often stressful for children as they figure out the new rules, expectations, and routines they need to follow. Finally, there is unfortunately abundant evidence that, on average, stepfathers do not treat their stepchildren as well as they treat their own biological children. First, there is the startling statistic that children are between forty and one hundred times more likely to be abused or murdered if they live in a stepparent household than if they live with both biological parents.[111] Of course, most stepfathers are neither homicidal nor abusive, but on average, fathers invest less in their stepchildren as compared to their biological children. Fathers have been shown to spend less time and money on stepchildren and are also less likely to provide parental supervision, engage children in interaction or provide emotional support.[112] Children feel this. One study demonstrated higher levels of the stress hormone cortisol in stepchildren versus biological children.[113]

Why is it that men tend to treat their biological children better than their stepchildren? As discussed in the next chapter, one theory is that this tendency is an evolved predisposition that was selected for because it enhanced male reproductive success. Men who invested in other men's offspring at the expense of investing in their own would have been selected against. On the other hand, men who selectively invested in their own offspring would have been favored by evolution. However, there is an important caveat here: men may sometimes succeed in attracting a mate by caring for her existing offspring who are not his. That is, investment in stepchildren may actually be a form of mating effort. This may help to explain why some stepfathers invest in stepchildren at all, even if it is less than what they would typically invest in biological children.[114]

Regardless of theoretical predictions and statistics about the average stepfather, we should not lose sight of the fact that some stepfathers genuinely love their stepchildren, often just as much as they love their biological children. In fact, one study found that 52% of stepfathers disagreed with the statement, "It is harder to love stepchildren than it is to love your own children."[115]

Conclusion

So after wading through all of these studies, what is the bottom line? What can I recommend to a new father about how best to raise a child?

First, be present—not just physically but also mentally. That means being attentive and responsive to your child from their very first weeks. Return their smile. Reciprocate their coo. Let them know that you care about the contents of their mind. Be warm and nurturing, but set necessary limits, and be firm when needed. Encourage and supervise their exploration of the world beyond the home. Play with them. Physical play, especially rough-and-tumble play, is an excellent way to cultivate their social skills, from empathy to emotion regulation. Model for them how to control their aggression. Gently push them outside their comfort zone. Lovingly tease them. Destabilize them a bit. Help them face unpredictability and learn how to navigate it. Let them compete and fail, and help them cope with that failure. All of this will build their resilience. Teach them social norms and values. You are one of their main socialization agents, especially before peers take over once school starts. Build your own social networks to provide them with opportunities. Remember that your family is a system, and do your best to keep that system running smoothly. This includes tending to your own and your partner's mental health, for example, by arranging for others to give you a break from child care.

But after all of that, remember that you don't fully control their destiny. They come to the world with genetic predispositions and are shaped by other influences that you can't always control. If you try your best and they struggle nonetheless, know that it was not your fault but that you gave them the best shot you possibly could. If you are lucky, your efforts will be rewarded with the greatest gift imaginable, their love.

Highlights

1. Although humans are cooperative breeders, many modern couples live as isolated nuclear families in which fathers take on added importance as the mother's primary helper.

2. In both modern and more traditional societies, fathers are normally expected to be involved in provisioning their children.

3. Children tend to do better when they have fathers (and mothers) who are warm and responsive but also exercise control and set appropriate limits.

4. Fathers often specialize in supervising children's exploration of and preparation for the world outside the home and family.

5. Heterosexual families are not the only configuration in which children can thrive.

2 Transformations

Most mammalian fathers do not take care of their young. Instead, they leave caregiving to the mother and go off seeking other mates. But there are exceptions. About 5% of mammalian species exhibit male parental care, and humans are one of them.[1]

How did we come to be one of those exceptions? In this chapter, I tell a story about how paternal caregiving evolved in the human species—how men like Yu Qiang and other highly invested fathers came to be. It was far from inevitable. In fact, it was a very improbable outcome that arose only because of a unique set of ecological circumstances that confronted our hominin ancestors. Since none of us were able to directly witness our evolutionary past, the story I tell should not be assumed as absolute truth but instead as one scholar's best estimate of what happened based on the available scientific evidence.

*　　*　　*

I am an anthropologist, and this professional identity influences the way that I think and the way that I understand and write about fatherhood. Occasionally, parents of my kids' friends ask me what I do for a living. I always cringe a bit because I know how the conversation will (or, more likely, will not) unfold. When I tell them I am an anthropologist, I get a lot of blank stares, and it's usually a pretty good conversation stopper. Most people don't know what anthropology is. So let's start there.

There are many definitions of *anthropology*, most of them valid in their own way, but one of my favorites is simply "the study of human nature." Anthropologists are committed to a broad, holistic approach to understanding human beings. We believe that humanity can be fully understood only through cross-cultural and cross-species comparisons and through appreciation of our evolutionary origins. As an anthropologist, I believe that a complete understanding of fatherhood requires consideration of how it varies across species, how humans fit into the general zoological pattern, and how and why it evolved in our species. That is the focus of this chapter. In chapter 6, we consider how and why fatherhood varies across human cultures.

Evolutionary Theory

How do we explain why traits—paternal caregiving, for example—evolve in some species but not others? A number of processes influence the traits that evolve in different species, but one of the most influential is natural selection, the mechanism that Charles Darwin described in 1859. Darwin was brilliant in deducing the mechanism from his meticulous observations of nature at the time, but the logic behind natural selection is quite simple in retrospect. Individual members of a species vary in their appearance or their behavior. If those traits are heritable (i.e., transmitted to their offspring), the traits that most improve an individual's ability to survive or reproduce will increase in frequency over time. Evolution by natural selection is a logical necessity when these criteria are satisfied. Beyond that, all that is needed is a source of variation. Variation comes from random genetic mutations and from the reshuffling of genes during the production of sperm and egg cells, known as *sexual recombination*. The end result of natural selection is that species tend to become better adapted, or fit, to their environment over time—until the environment changes again.

As a simple example, imagine you have a population of brown mice that find themselves living on snow-covered terrain. A fortuitous mutation occurs that produces a mouse with a slightly lighter coat color. Hawks don't see that mouse as well against the background, and so this mouse survives better than the others and produces offspring that inherit the lighter coat color. These light brown individuals begin to spread in the population, and eventually, by chance, another mutation occurs that results in an even lighter-colored coat. The process continues until eventually the population of mice all have white coats. Mutations are not directional; they are random mistakes that occur when DNA is copied. Therefore, some mutations will be in the direction of darker coat color, but these will be selected against (eliminated from the population) since those mice will be more visible to predators.

This example shows how a species' *appearance* can change over time. Critically, *behavior* can also evolve if the same three principles apply: (1) the behavior varies among individuals in the population, (2) it is heritable, and (3) it contributes to differential survival or reproductive success. Reproductive success can be thought of as the number of offspring one produces that survive to reproduce themselves. To extend our example to behavioral traits, we can imagine that mice that stop feeding periodically to check the sky for predators will be selected for over those that continually feed without doing so.

Paternal caregiving is a behavioral trait that evolved in some evolutionary lineages but not others, and we would like to understand why. By and large, we expect it to evolve when it increases male survival or reproductive success. Evolutionary biologists often use the term *fitness* to encompass both survival and reproductive success. Your biological fitness is basically your ability to pass on your genes to the next generation. So we can say that we expect paternal caregiving to evolve when it increases a male's fitness. The example with mice focused on a trait that affected survival. However, the primary impact of paternal caregiving is on a male's reproductive success. Reproductive success is a bit more complicated and a bit more interesting than survival. It depends on both success in finding and keeping a mate and success in producing and raising offspring to reproductive maturity. This means that we generally expect evolution to select for traits that improve an organism's mating success or parenting ability, or both, in addition to traits that improve an organism's survival insofar as that will allow them to reproduce again.

But here is the rub: organisms cannot maximize survival, mating success, and parenting success all at the same time. In other words, life is a zero-sum game. This idea is the foundation of a branch of evolutionary theory known as life history theory, which posits that organisms have a finite amount of energy that they can allocate among three competing categories: growth, maintenance, and reproduction. Before reaching adulthood, organisms put most of their energy into growth. Maintenance is the business of keeping yourself alive. It includes energy invested in the immune system to protect against pathogens and disease, as well as energy needed to avoid predators and energy invested in slowing the aging process and preventing senescence. Reproduction involves both mating effort (effort devoted to finding and keeping mates) and parenting effort. If we hold investment in growth and maintenance constant, then organisms face a trade-off between mating and parenting effort. Many of us can intuitively grasp this trade-off in our everyday lives. Perhaps you have found yourself exhausted at the end of a long day of tending to your children, knowing that you really should exercise, prepare a healthy meal, and make love with your spouse, but you really can't muster the energy for any of that so you instead eat some ice cream and watch TV until you fall asleep. Energy is finite.

In male animals, mating effort typically involves competing with other males for access to female mates or for social status that provides access to female mates. In other (nonhuman) animals, male competition often involves overt aggression. Bigger, stronger, and more aggressive males win fights and achieve higher social status and greater access to female mates. However, it doesn't always work this way. In some species, males instead compete to attract females. The perfect example are male birds of paradise, from New Guinea and Australia, which attract female mates with dramatic displays of their strikingly beautiful plumage. The male with the best display attracts and therefore mates with the most females. Thankfully, competition among human males for mates and status often takes nonaggressive forms as well. Human males can achieve status, and attract mates, in a variety of ways, including through possession of special skills or abilities, knowledge, intellect, resources, or even kindness, generosity, and cooperation.

I should be clear about what I mean by parenting effort or the related term, *parental investment*. In his classic 1972 article, the evolutionary biologist Robert Trivers defined parental investment in a manner that is consistent with the idea of life history trade-offs. Trivers's definition was "any investment by the parent in an individual that increases the offspring's chance of surviving (and hence reproductive success) at the cost of the parent's ability to invest in other offspring."[2] This cost includes forgone mating effort as well as forgone maintenance since both are required in order to make future offspring. For example, it requires a lot of energy for female mammals to nurse their young, and the suckling stimulus normally inhibits estrogen, the hormone that supports both female sexual motivation and ovulation. So female mammals do not typically mate while nursing their young or have additional offspring until the current one is weaned. As another illustration of the costs of parental care, consider the case of male birds that incubate eggs and feed and protect their chicks. While they are in the nest parenting, they cannot be off courting other females.

Trivers's definition of parental investment is a good starting point, but many of you may be reflecting on your own parenting and thinking that most of what you do for your kids is not about simply keeping them alive, but more about cultivating skills and values so that they will be competent and successful adults. We discussed these and other important influences that human parents have on their children in the previous chapter, but Trivers's definition suffices for other animals.

Parental effort can take many forms. One useful distinction is between direct and indirect forms of caregiving. Direct care involves direct physical contact with the offspring and includes behaviors such as holding, carrying, huddling, incubating, feeding, grooming, cleaning, playing, teaching, and supervising or babysitting. Indirect care, which is often as or more beneficial to offspring than direct care, involves behaviors performed at a distance from offspring. Indirect caregiving includes preparation of nests or shelters for young, physical protection from predators or enemies, and provisioning of both offspring and the mother. In modern humans, indirect care includes earning wages outside the home that are then used for food, housing, health care, and so on.

Organisms allocate their energy among these competing life demands, and different organisms balance their energy portfolio in different ways. These different ways of investing energy are known as *life history strategies*. Life history strategies are a zero-sum game, meaning that if one invests more energy into one category, there will be less available for another category. So the expectation is that organisms allocate their energy among growth, maintenance, and reproduction, including both mating effort and parenting effort, in such a way as allows them to maximize their reproductive success (figure 2.1). *In sum, they are expected to invest heavily in parenting when that is the best strategy for maximizing their reproductive success.* Those circumstances do not always apply. They are most likely to occur when parents are able to improve the survival or health of their young and when their mating opportunities are limited. However, when parents have a limited ability to reduce their offspring's mortality risk and when mating opportunities abound, we do not expect parental behavior to evolve. In fact, as we will see, parental behavior is uncommon or rare in many major animal taxa.

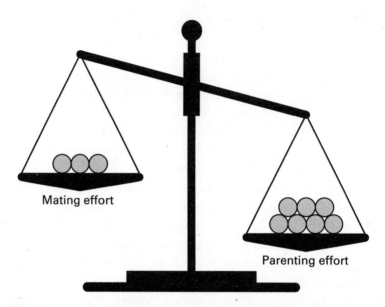

Figure 2.1
Organisms face a trade-off between mating and parenting effort.

Life history theory is powerful for explaining variation in paternal care-giving across nonhuman species. However, humans are more complicated than other animals for a variety of reasons that will be discussed in chapter 6. Consequently, it will not take us all the way toward understanding why human males often care for their children, but it is relevant and a valuable starting point.

*   *   *

So how did the capacity for paternal caregiving ultimately evolve in our species?

When I teach college students about human evolution, I start by trying to orient them to the timescale of human evolution and how that relates to the longer time horizon of planet Earth. We can better understand our origins if we can conceptualize how much time elapsed on Earth before our arrival relative to how long we have existed. The best approach I have found for conveying this idea is to imagine compressing the history of Earth into a single 24-hour day (figure 2.2). The Earth formed around 4.5 billion years ago (BYA), so we set that as the beginning of our 24-hour day, 00:00. The

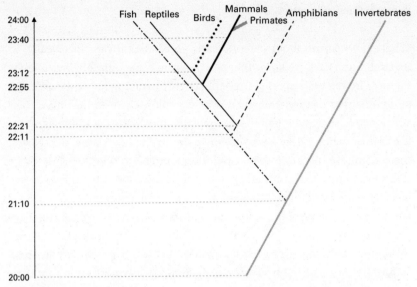

Figure 2.2
Simplified phylogenetic tree for animals, based on 24-hour day analogy. 20:00 is equivalent to 750 million years ago (MYA).

present moment in which you are reading this sentence is the very end of the day, 24:00. On this timescale, 1 second is equivalent to roughly 52,000 years. Your lifetime is about 2 milliseconds.

When did various important evolutionary events happen in our 24-hour day? It took a long time for life to get started on Earth, but we have evidence that it was present by 05:20 (3.5 BYA). The earliest life forms were simple prokaryotic cells. Prokaryotes are basically packages of DNA surrounded by a plasma membrane. Eukaryotes are more complex cells that consist of various organelles with specialized functions, such as nuclei and mitochondria. We have to wait a very long time for the evolution of eukaryotes. In fact, eukaryotes do not appear on Earth until the early afternoon, at about 13:20 (2 BYA). So more than half of our day has passed, and all we have are single-celled organisms. It was not until after dinner, at roughly 19:12 (0.9 BYA or 900 million years ago; MYA), that multicellular life emerges. The first animals appear, in the form of ocean-dwelling invertebrates, at around 20:00 (750 MYA). Invertebrates are animals lacking a backbone and include species such as jellyfish, worms, clams, octopus, and insects.

Parental Care in Invertebrates

What do we know about parental care, and paternal care specifically, in invertebrates? First, we know that parental care is uncommon. Presumably, for most invertebrate species, investing energy in caring for offspring is not an optimal life history strategy; it does not allow them to maximize their reproductive success. A better strategy is to simply mate and move on, in the case of males, or lay eggs and move on, in the case of females. Nevertheless, among the more than 1 million invertebrate species living today, several species do care for their eggs or their young, and this tends to evolve when environments are particularly harsh or involve severe predation risk.[3] Under these circumstances, parental care may be needed to give the eggs even a fighting chance of survival. In most cases, invertebrate parental care is provided by females rather than males. This seems potentially interesting and relevant since human females often provide more child care than human males do. Perhaps we can learn some general principles about the circumstances under which maternal versus paternal care evolves by studying invertebrates and other non-mammalian taxa.

Biologists have offered two explanations for the preponderance of female caregiving among the minority of invertebrate species that care for their young. Both hinge on the fact that, similar to mammals, fertilization of female egg cells in invertebrates occurs internally, inside the female's body. This presents males with a fundamental problem in terms of their reproductive strategy. If a female mates with multiple males, the male may or may not be the father of the fertilized eggs that she eventually deposits.

Imagine two types of males within the same population. Type A males care for eggs laid by a female mate, even when they did not fertilize those eggs. Type B males refrain from caring for the eggs laid by their female mate and use the energy saved to instead seek additional mating opportunities. Type B males that find additional mates are likely to have greater reproductive success than type A males because type A males are often caring for eggs that they did not fertilize. They are using valuable energy to enhance the reproductive success of other males instead of their own. In fact, in this scenario, they are likely to be caring for the eggs of type B males. The result seems obvious: successful type B males will be selected for over type A males. Eventually the population will consist mainly of type B males. Critically, females do not have the same problem. The eggs that they deposit always belong to them. This may be why males are less likely than females to provide care when fertilization is internal.

Nevertheless, one can imagine circumstances in which type A males will not be selected against despite their lack of paternity certainty. One such scenario is when offspring have little or no hope of survival without paternal care. In that case, if males don't parent, they have no reproductive success. Another scenario is one in which males have no better option than to stay and parent. For example, if a male will not be able to find another mate (i.e., they can't implement the type B strategy), better to stay and help the offspring that *might* be theirs than to futilely pursue other mates or do nothing at all. This helps explain why male uniparental care, while rare, has in fact evolved in several instances in insects.

This simple modeling exercise also reveals a second explanation for the preponderance of maternal relative to paternal care in invertebrates. In many cases, invertebrate offspring can be raised effectively by a single parent or the contribution of a second parent has diminishing returns to offspring well-being. In this situation, it may often be in the evolutionary interest of each parent for their partner to provide care rather than themselves, so that they can instead go off and seek additional mates. In this way, they produce the maximum number of surviving offspring. Since fertilization takes place inside the female's body before she lays her eggs, males are able to leave her and her fertilized eggs before she can desert him. When later faced with the decision of abandoning the fertilized eggs she has laid and letting them die or caring for them so they can live, the behavior that will evolve by natural selection is obvious. She will provide care.

So paternal care is rare among invertebrates due to low paternity certainty and because males are able to desert their offspring earlier and more easily than females can, and these are both a consequence of internal fertilization.

Remember that parental care is generally uncommon among invertebrates, and paternal care rarer still. But when parental care does occur in invertebrates, what does it look like? Since maternal care is more common, let's start with some examples of that first.

One example of maternal care is provided by insects known as "treehoppers." These remarkable creatures have striking, ornate "helmets" that allow them to mimic plant thorns and thereby camouflage themselves from predators. They feed on sap by piercing plant stems with their beaks. After laying her eggs, the mother sits over them, guarding them from predators and parasites. Since newly hatched treehoppers are unable to pierce the plant stem to extract sap for themselves, their mother makes a series of slits in the stem that the nymphs use to suck sap.[4] Clearly, this is a circumstance where parental care is essential for the young to survive, and that helps us to understand why it evolved in this case.

Invertebrate maternal investment reaches its greatest—and most troubling—extreme in the velvet spider, *Stegodyphus lineat*. In this desert-living spider, females produce a single egg sac of about ninety eggs. After the eggs hatch, the mother's internal organs begin to disintegrate, allowing her to regurgitate her liquefied guts for her young to consume. At some point, the mother stops regurgitating, and her young begin consuming her while she is still living. They leave the nest only after she is fully consumed.[5] Natural selection can produce some cruel adaptations indeed.

Natural selection is the mechanism by which evolution allows organisms to solve evolutionary "problems." For example, treehoppers have solved the problem of avoiding predators by evolving helmets that allow them to mimic tree thorns, and they have solved the problem of the nymphs not being able to access sap by evolving maternal care. Sometimes natural selection independently converges on the same solution to the very same problem in distantly related organisms. Birds and bats, which belong to different vertebrate classes, have converged on wings as a solution for flight. A rather astounding example of convergent evolution is the independent evolution of complex camera eyes in vertebrates and cephalopods (e.g., squid, octopus), which are invertebrates. In the realm of parenting, one of the great mammalian innovations is provisioning offspring with milk produced by the mother. Remarkably, a similar adaptation has evolved independently in jumping spiders (*Toxeus magnus*) that live in Southeast Asia. Mother spiders produce milk that spiderlings suck directly from the ventral surface of her abdomen. Like mammalian milk, spider milk contains sugar, fat, and protein, but it has four times the protein of cow's milk. Without their milk, spiderlings die. This is no trivial investment by the mother, as spiderlings continue to suck milk until they are 80% of their adult body size. To give you a sense of the magnitude of this investment, human children reach 80% of adult body weight by about age fourteen. I doubt many mothers would agree to fourteen years of breastfeeding. Nevertheless, the spider mother's efforts pay dividends, as 76% of hatched offspring survive to adulthood.[6]

What about fathers? Despite the fact that invertebrate parental care is disproportionately provided by females, there are exceptions in which either both males and females care for the offspring (biparental care) or males provide exclusive care (male uniparental care).

Dung beetles eat the dung of large animals like cows, giraffes, and elephants. Males and females cooperate in claiming, transporting, and burying balls of dung in their burrow. The female then lays her eggs inside the dung, and the resulting young feed on the dung. Parents also guard the burrow from intruders, and the female even produces antifungal secretions that prevent infection of the burrow. Given the effort that males exert in building the nest, we assume that his investment increases his reproductive success. However, one study showed that the number of eggs the female laid did not depend on the presence of the male. This suggests that males may have ulterior motives. While building the nest, they are also guarding it and the female from invading males that sometimes succeed in replacing them. So what initially looks like devoted parenting effort may be partially, or even primarily, mating effort.[7] The blurring of the distinction between mating and parenting effort will be a recurring theme throughout this book.

A less ambiguous example of male parental care in invertebrates is provided by the giant water bug. These predatory insects live in freshwater ponds, marshes, and streams. They can be as large as 4.5 inches in length and have a nasty habit of biting humans between their toes. Male giant water bugs carry the developing eggs on their back (figure 2.3). Nurturing these eggs requires some skill. Fathers have to keep the eggs moist to prevent them from drying out, yet also periodically raise them above the surface to properly oxygenate them. They also stroke the eggs with their hind legs while below the surface. These behaviors are essential for egg survival, as eggs that are detached from the father fail to develop.[8] Why do males care for the eggs rather than females? Experiments show that female giant water bugs prefer to mate with males who carry eggs.[9] If this was also true in the past, then egg-carrying males would have been selected for by female mate choice.

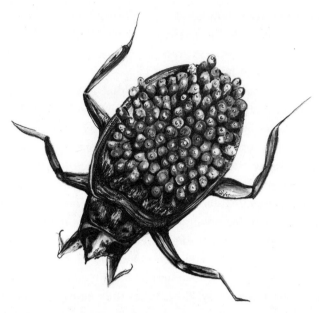

Figure 2.3
Giant water bug father carrying fertilized eggs.

Parental Care in Fish

Returning to our twenty-four-hour-clock, we have only four hours left
to cover the evolution of vertebrate parental care. A monumental event
known as the Cambrian explosion occurred at approximately 21:10 (530
MYA). This was a relatively brief event (about 25 million years) characterized by a tremendous increase in the diversity of animals. Among the animals that appeared at this time were the first vertebrates in the form of fish.
In contrast to most invertebrates that reproduce via internal fertilization,
these early fish likely reproduced via external fertilization. Females would
deposit their eggs, males would fertilize those eggs with their sperm, and
then both partners would be off. There was no parental care, and that is still
the case in the majority of fish species living today. However, in some fish
lineages, males began to guard these fertilized eggs. Paternal care evolved
independently multiple times in fish and more often than did maternal
care, resulting in the situation we see today, where if it occurs at all, parental care in fish is most often provided solely by males.[10]

Why did uniparental male care evolve in some fish? Can it provide us with any clues about when we should expect it more generally? External fertilization seems to be a big part of the explanation. In fact, among fish that do have internal fertilization, uniparental female care is more common. However, male care prevails among fish with external fertilization. This represents the converse of the situation seen in insects, where internal fertilization routinely limits paternity certainty. By contrast, external fertilization can provide male fish with greater paternity certainty, and this increases the benefit of male parental care to their reproductive success. Male and female fish often release their gametes simultaneously, so a male who fertilizes female eggs immediately upon their release and then guards them will have a high probability of paternity. With internal fertilization, however, a female might mate with several males before depositing those already fertilized eggs that if a male cared for he might be misdirecting his paternal investment.

Increased paternity certainty looks like the reason that paternal care is associated with external fertilization in fish, but is it? Female fish after all have maternity certainty, yet they rarely provide care. In other words, paternity certainty looks like a necessary condition for the evolution of paternal care in fish, but in and of itself is insufficient to explain unimale care. So what else is involved? Another factor is the tendency of male fish to establish and defend territories. Those territories often include eggs from many females, which increases the benefit to males of guarding the eggs they have fertilized.[11]

We can imagine that sometime in the Devonian period, approximately 400 MYA (21:52 on our twenty-four-hour clock), a male fish was born with a mutation that made him tend to linger after fertilizing his eggs and establish a territory instead of leaving. He began defending the territory and the eggs within it, and his offspring consequently experienced higher survival. When another female laid her eggs in the same territory, he could simultaneously defend both broods and offset any costs of not pursuing additional mates. His male offspring inherited his tendency to establish a territory and guard their eggs, and those offspring survived well and inherited the same parental tendency . . . and we are off. There is even evidence from a few fish species that females are attracted to males that are already caring for eggs. Even better, he could have his cake and eat it too. That is, once females are selectively mating with males who parent, mating and parenting effort are really one and the same, and there is no trade-off—no temptation to shirk parental responsibilities in favor of seeking additional mates.

In fish, paternal care most commonly involves guarding the fertilized eggs; however, males of some species go further.[12] Males of the three-spined stickleback weave elaborate nests by gluing together plant materials using a special adhesive that is secreted from their kidney (figure 2.4). After building his nest, the male attempts to attract a female mate with a zigzag courtship dance. If successful, she lays her eggs in his nest, and he then fertilizes them. Note that this should allow him paternity certainty, which we have shown is strongly associated with male parental care. The male then guards the eggs but also uses his pectoral fins to direct fresh water toward them in order to oxygenate them. He continues guarding the fry for a few days after they hatch. Remarkably, he will retrieve those that stray from the nest by sucking them into his mouth and then spitting them back into the nest.[13]

If paternity certainty is the key to male parental caregiving in fish, then seahorses are the paradigmatic example. Female seahorses actually lay their

Figure 2.4
Male three-spined stickleback waiting to fertilize eggs that the female lays in the nest he has prepared. Illustration by Alexander Francis Lydon in William Houghton, *British Fresh-Water Fishes* (London: William Mackenzie, 1879).

eggs in a pouch on the male's ventrum, where only he can fertilize the eggs. After doing so, males are effectively pregnant and begin secreting nutrients into the brood pouch that are absorbed by the embryos. The pouch also offers protection to the embryos, oxygenates them, carries away their waste, and controls salinity. In another remarkable example of convergent evolution, the male expels the hatchlings from his pouch via a series of muscular contractions, just like mammalian mothers do. After birth, however, parenting ends.[14]

Parental Care in Amphibians

Around 21:50 (400 MYA), the first vertebrates appear on land. It's after 10:00 p.m. before the earliest amphibians appear, at approximately 22:11 (340 MYA). Most fish that were living at the time continued to evolve as fish, and some of those lineages have managed to survive until the present time. However, one lineage instead evolved into amphibians. These early amphibians likely did not provide parental care, as is the case for most amphibians alive today. Then, a minority of species evolved uniparental care, which was equally likely to be maternal or paternal. Today, parental care is found in 6% to 15% of frog and toad species and 20% of salamanders. Uniparental male and female care are about equally common, and biparental care is rare. Of course, amphibians live both in water and on land. Care tends to be more common in more terrestrial species, presumably because parents are needed to keep the eggs moist on land. All frogs are external fertilizers, but salamanders have both internal and external fertilization, and the mode of fertilization reliably predicts the type of care. As expected, male uniparental care is found with external fertilization, and female uniparental care is found with internal fertilization.[15] As with fish, egg guarding is the most common type of care. However, once again, a few species extend their parental investment well beyond this.

One of these is the Central American strawberry poison dart frog (*Dendrobates pumlico*). The female lays three to five eggs on a plant leaf. The male keeps the eggs hydrated with fluid from his cloaca. After the eggs hatch, both parents transport individual tadpoles on their backs to a water-filled location. But that's not all. Females even provision their young, a form of investment that is rare outside of birds and mammals. The mother visits

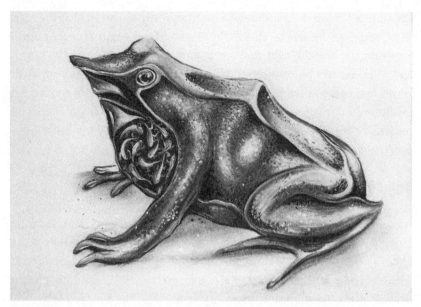

Figure 2.5
Male Darwin's frog brooding tadpoles in his vocal sac.

each of her tadpoles every few days and deposits several unfertilized eggs for their consumption.[16]

Similar to seahorses, there are also a number of frogs in which males shelter and protect the developing young. In Darwin's frog (*Rhinodrema darwinii*), the father guards his eggs for several weeks before swallowing them so the tadpoles can develop in the protected space of his vocal sac (figure 2.5). After about six weeks of metamorphosis inside the sac, they are regurgitated as miniature frogs.[17] In another frog, aptly named the pouched frog (*Assa darlingtoni*), adult males shelter their young tadpoles in a moist pouch on their hip. After two to three months, the young also emerge as miniature frogs.[18]

Parental Care in Reptiles

On our twenty-four-hour clock, the first reptiles appear only about ten minutes after the first amphibians, at approximately 22:21 (310 MYA). Some amphibians continued evolving as amphibians, but one lineage evolved into reptiles. Parental care is particularly rare in reptiles. Most reptiles bury their eggs and do not guard them. Among the small minority of reptiles that do parent, uniparental female care is most common, occurring in 3% of snakes and 1% of lizards. For example, female pythons coil around their eggs to protect them and shiver in order to generate heat to incubate their eggs. Female crocodiles gently carry their newly hatched young in their mouths from their nest to the water. Biparental care evolved, probably from female uniparental care, in some crocodiles. While examples of male uniparental care are known for insects, fish, and amphibians, uniparental male care has never been documented in reptiles.[19] Once again, this lack of male care may be partially attributable to a lack of paternity certainty since reptiles practice internal fertilization.

Parental Care in Birds

Around 23:12 (150 MYA), birds evolve from a branch of dinosaurs known as the theropods. You are already familiar with the theropods because they include animals like *T. rex* and velociraptors. Parental care in birds is remarkable because of how common it is. In each of the other animal taxa we have discussed (invertebrates, fish, amphibians, and reptiles) parental care is uncommon or even rare. In birds, however, it is ubiquitous. Not only is parental care pervasive in birds, that care is usually provided by both parents. In fact, 90% of bird species are biparental.[20]

How can we explain the stark contrast with the closely related reptiles, where parental care is rare and biparental care even more exceptional? One major physiological difference between birds and the other vertebrates we have surveyed is that birds are warm-blooded. This means they have a higher basal metabolic rate in order to produce enough heat to maintain their body temperature. To accommodate this higher metabolism, warm-blooded animals need more calories. Many birds are born in a quite undeveloped state, which biologists refer to as "altricial." Altricial young can't feed themselves and require provisioning. One of the main avian parental duties is therefore provisioning the young and fueling their high metabolism. The other is to incubate the eggs, which in contrast to reptile eggs, need to be maintained at a constant temperature if they are to develop successfully. Biparental care may be necessary to meet these increased caregiving demands of young birds. Indeed, there is clear evidence that male parental care aids survival of the young above and beyond maternal care. This has been demonstrated experimentally by removing the male of breeding pairs and examining the effect on offspring survival. In dark-eyed juncos, for example, 25% of chicks survive to independence when both parents are present, but that number plummits to zero when the father is removed.[21] So male care is likely to be adaptive *if* it can be directed to the male's own offspring—which brings us back to paternity certainty. Like reptiles, birds practice internal fertilization, which limits paternity certainty and is expected to bias males away from caregiving. Then why is it that 90% of male bird species exhibit paternal care? How do male birds solve this problem of limited paternity certainty due to internal fertilization? Monogamous pair-bonding seems to be the solution. That is, if female birds pair off with and mate with only a single male, then males will have paternity certainty despite internal fertilization, and paternal caregiving can evolve. However, bird monogamy is imperfect. Females occasionally sneak away for extra-pair copulations, and males guard their mates to try to prevent this. There is now clear genetic evidence that females sometimes have young sired by extra-pair mates. Nevertheless, most young are fathered by the female's partner, and this probably explains why male caregiving is so common in birds. There is also evidence that males calibrate their provisioning as a function of their paternity. That is, they provide more care to their young when they are in fact the genetic father.[22] Later, we see an interesting parallel to this in our own species.

Perhaps you think you are a pretty good parent. If so, ask yourself this. Would you stand in −40 degree weather around the clock, amid periodic blizzards, and shelter your offspring for months on end, fasting all the while? Sounds pretty miserable, doesn't it? This is the life of an emperor penguin father. Their parental sacrifice is so legendary that they inspired an Oscar-winning documentary, *March of the Penguins*. Emperor penguins are giants, more than three feet tall and weighing close to 100 pounds. They live in Antarctica. Their march begins with all adults walking from the edge of the pack ice inland for thirty to seventy-five miles to their nesting grounds. Males and females pair off with each other, mate, and then the female lays a single egg. The mother carefully transfers the egg to the father and returns to the sea to feed. The father then incubates the egg throughout the grueling Antarctic winter. He balances the egg on top of his feet and surrounds it with loose skin and feathers to keep it warm (figure 2.6). Males form enormous huddles, using one another for heat and as shields

Figure 2.6
Male emperor penguin incubating his chick.

against the bitter-cold wind. After about two months, the egg hatches and the father feeds the chick "crop-milk" from his esophagus. The chick's survival depends on the mother returning shortly after this first meal. Upon her return, the father transfers the chick to her, and she begins feeding the chick regurgitated partially digested food she has obtained. The male will have fasted for approximately four months and lost as much as half his body weight before finally returning to the sea to feed. But afterward, he must return, thirty to seventy-five miles, to feed the chick and spell the mother once again.

I saw *March of the Penguins* in the movie theater with a friend who leaned over and whispered to me, "Thank god they don't have consciousness." I think he was suggesting that if they did, they might realize how grueling their lives were and simply give up. Or would they? If you have read Cormac McCarthy's novel *The Road*, you may identify with the father whose love for his son would not allow him to give up, even in the miserable, postapocalyptic world they inhabited.

The courtship between male and female penguins, as beautifully depicted in the documentary, is particularly striking. One can't help the anthropomorphic feeling that they are falling in love. Perhaps that is too strong, but they are definitely forming an attachment, and a powerful one that will have to last if the chicks will be raised successfully. Both parents need to be able to trust that the other will return from the sea. More generally, monogamy and biparental care depend on mutual commitment and trust. Since successfully raising chicks requires the investment of both parents, each must be sure they can trust the other not to desert. These courtship rituals often involve spectacular bouts of synchronized behavior between partners that are essential for establishing the trust on which these pair-bonds depend.[23]

Although male parental investment is impressive in birds, males are commonly less heavily involved in parenting than females, perhaps because their parentage is less certain overall.[24] Tension between parental commitments and mating pursuits is also evident in some male birds. Male red wing blackbirds are one of the minority of polygynous species in which males defend territories than can accommodate multiple female mates. Males feed and protect their young, but when a new female settles in his territory, males often defer parental care to later broods.[25] Thus, males seem to trade off mating and parenting effort.

Although biparental care is the rule in birds, uniparental care is found in about 10% of bird species. Uniparental care is associated with precocious rather than altricial young. Because precocious young are born in a more developed state and presumably require less care, uniparental care seems to be sufficient and the payoff of biparental care may be limited. Uniparental care is usually provided by the female (4% to 8% of all bird species), but there are some examples of male uniparental care (1% to 2% of all bird species).

One of these is the cassowary, a large, flightless bird native to the forests of Australia and Papua New Guinea. Female cassowaries are larger and more brightly colored than males, and they establish territories within which several males nest. They can be more than 6 feet tall and weigh as much as 130 pounds. Females lay several eggs in each male's nest, but provide no care for the eggs or hatchlings. In contrast, males devotedly incubate their eggs for about fifty days and fiercely defend the young for about nine months after hatching. They also provision their young with fruit and insects.[26] Its not obvious what is responsible for this role reversal, but it may have to do with a male-biased sex ratio (1.5 males per female in one study) that leaves males with limited mating opportunities.[27] With limited female mates available, parenting might be a cassowary father's best option.[28]

In birds, there is another important type of breeding system in which caregiving extends beyond the mother and father to include several adults of either or both sexes. This is known as cooperative breeding. We should take special note of this type of breeding system since it is the one that best characterizes our own species.[29] About 3% of bird species worldwide are cooperative breeders. Among these species, helpers seem to pay dividends since the number of alloparents is positively correlated with offspring survival.[30] The Florida scrub jay is a good example of a cooperative breeding bird. Males and females form monogamous pairs, but some pairs live together with adult offspring from previous seasons who help feed and protect the young. Pairs with helpers fledge more young than pairs lacking help. Furthermore, experimental removal of helpers leads to decreased fledgling survival. Helpers can act as sentinels, allowing for earlier detection of predators. Scrub jays are heavily preyed on by snakes and hawks, and pairs with helpers suffer lower predation rates than pairs without helpers. By provisioning the young, helpers can also free up the father to spend more time acting as a sentinel.[31]

Parental Care in Mammals

Finally, we come to our own vertebrate class, *Mammalia*. The first mammals evolved from small, active carnivorous reptiles known as therapsids. This happened a bit earlier than the emergence of birds, at about 22:55 (200 MYA). Mammals lived among the dinosaurs for millions of years until dinosaurs went extinct about 23:40 (65 MYA).[32] This critical event, just twenty minutes before the end of our twenty-four-hour day, had monumental consequences for mammals, as they were able to evolve into a variety of ecological niches that dinosaurs had previously dominated. This extinction of dinosaurs ushered in the age of mammals.

Like birds, mammals are heavily invested parents. In contrast to insects, fish, amphibians, and reptiles, parental care is found in all mammalian species. However, there are some critical differences between parental care in birds and mammals that have significant consequences. In mammals, embryos develop inside the female. That is, mammalian mothers gestate their young. Internal gestation keeps young mammals safe during development. Birds, in contrast, develop in eggs outside the mother's body. In addition, the defining feature of mammals is that females provision their young with milk made in mammary glands. These twin innovations, internal gestation and lactation, reflect a major escalation in parental investment on the part of female mammals. Male mammals are unable to gestate young or lactate, so this limits their ability to contribute to the well-being of their young. It may also be the case that in general, young mammals need their fathers less than young birds do since mammalian mothers are investing so heavily. The result is quite low rates of paternal caregiving in male mammals. Whereas all female mammals care for their young, paternal caregiving is found in only about 5% of mammalian species.

But which 5% and why? Under what circumstances does paternal caregiving evolve in mammals, and what form does it take? Paternal caregiving is strongly associated with monogamy among mammals, again presumably because monogamy offers increased paternity certainty. The most likely scenario is that monogamy evolved in a number of mammalian lineages from an ancestral condition in which females lived alone and males occupied ranges overlapping several females. In some environments, females need to spread out in order to have enough space to feed themselves. When they spread out far enough, males may have been unable to guard and defend access to multiple females at once, so they reverted to guarding just a single female. By staying with that one female, he could attempt to prevent her from mating with other males. Then, in some of these lineages, paternal care was able to subsequently evolve because males had the prerequisite high paternity certainty provided by monogamy, their care improved offspring survival, and they had few other options for mates given that females were pairing monogamously. This transition apparently happened multiple times, independently in mammals.[33]

Today, mammalian paternal care is most common among rodents, primates, and carnivores, although it is still found in only a minority of species in these three orders.[34] The California mouse (*Peromyscus californicus*) is a nocturnal rodent that lives along the California coast. They practice exclusive monogamy, as verified by genetic testing, and adult males and females form pair-bonds with each other. Their average litter size is two, and they have multiple litters per year. However, they live only between nine and eleven months, so their lifetime reproductive success is between four and five. Before becoming fathers, adult males are not interested in pups and may even kill them. However, they become devoted fathers after their pups are born until the time that they are weaned. Fathers assist in building the nest, retrieve pups who stray from the nest, lick and groom the pups (the rodent equivalent of hugging and kissing their offspring), and huddle with the pups to help them keep warm. Fathers spend as much time with pups as mothers do and even more time licking and grooming them. Father absence severely disrupts their family functioning. When researchers removed fathers from nests in the wild, offspring survival decreased from 1.5 pups to 0.6 pups per litter.[35] How do fathers improve pup survival? When they huddle with their pups and keep them warm, pups spend less energy on thermoregulation so they require less milk to survive. By freeing the mother from huddling, fathers may also alleviate energetic stress on her, allowing her to forage more and produce more milk.[36] This is an important theme in mammals. Even when fathers have a limited ability to feed offspring because offspring depend on maternal milk, they may be able to help offspring indirectly through provisioning the mother or alleviating her energetic burden, and she will in turn provision the offspring.

In humans, childbirth is typically managed by an obstetrician, nurse, or midwife or, in more traditional human societies, experienced female elders. These birth attendants assist with delivery and afterward clear the newborn nostrils to provide an airway, clean the newborn of afterbirth, and in some cases warm the infant.[37] Human fathers are rarely involved with these aspects of birth, but Djungarian hamster (*Phodopus campbelli*) fathers are more ambitious, practicing what is known as male midwifery. They lick and sniff the female anogenital region during labor and delivery of the pup, assist with the delivery by tugging on the pup with forepaws or incisors, clean the pups of afterbirth to help clear the nostrils, carry pups to the nest after birth, and huddle over newborns to keep them warm while the female delivers subsequent pups. To top it off, they eat the placenta, which may function to mask the scent of newborns from potential predators or may simply provide a healthy meal.[38]

The more than six thousand species of mammals are divided into twenty-six orders.[39] One of these are the rodents (Rodentia) discussed above. Another is the meat-eating order Carnivora. Carnivores are further classified into twelve families, and one of these is the canids, with thirty-seven species of dogs and related species. The canids are an outlier among mammals insofar as both paternal caregiving and monogamy are ubiquitous. Young canids are born altricial and have neither the skill nor strength to successfully hunt the large mammals they depend on for their sustenance. They must therefore be provisioned during the relatively long period of development when they are growing and acquiring hunting skills. This is where canid fathers step up. They provision their pups, as well as lactating mothers, by regurgitating partially digested meat. It has been estimated that African wild dogs can carry three days' worth of food in their stomachs to the pups and mothers. In many canids, males are not the only providers. Mothers also provision, and many canids are cooperative breeders, in which alloparents provision too. Provisioning extends well beyond weaning. For example, gray wolf pups are weaned at about six weeks but may be provisioned up until one year of age. In some cases, this alloparental care has been shown to increase the number of young that mothers produce, as well as their survival. Incidentally, there is another mammal whose traditional diet is heavily based on meat and is provisioned well beyond weaning while they develop the skill and strength to become competent hunters themselves. That species is our own.

But back to canids for a few minutes before we turn to our own order, the primates. A prime example of male provisioning is found in the largest of the wild canids, the gray wolf. Wolf packs typically consist of a monogamous mated pair and their young from the previous few breeding seasons, some of whom remain after reaching adult status. Mother wolves do not leave the den for the first few weeks after the birth of their cubs. During this time, they rely on the father and other adults to provide regurgitated meat for them and their young.[40]

Rather than meat, another canid, the bat-eared fox (*Otocyon magalotis*), eats mainly termites. Foraging for termites is time-consuming, so lactating females must spend most of their time foraging, away from the cubs in the den. Males appear to compensate by providing more direct care. Fathers guard, groom, and huddle with cubs. They spend more time in the den than mothers do, and the amount of time they spend there is the best predictor of the number of cubs surviving to weaning age.[41] Thus, paternal care increases male reproductive success in bat-eared foxes.

Parental Care in Primates

As in rodents and canids, paternal caregiving is relatively common in our own mammalian order, the primates. The first primates do not appear until twenty minutes before the end of our twenty-four-hour day, at about 23:40 (60 MYA). Figure 2.7 is a highly simplified phylogenetic tree that illustrates the evolutionary relationship between humans and other primates. There are two suborders of primates: the prosimians and our own suborder, Anthropoidea. Compared with anthropoids, prosimians have smaller brains and rely relatively less on vision and relatively more on olfaction.

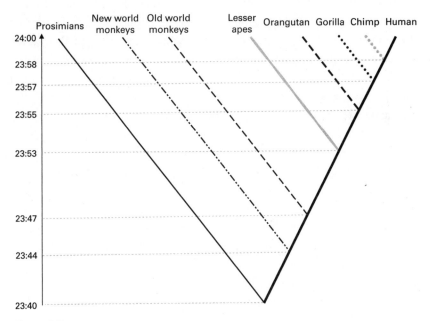

Figure 2.7
Simplified phylogenetic tree for primates, based on twenty-four-hour day analogy; 23:40 is equivalent to 60 million years ago (MYA).

The best-known examples of prosimians are the lemurs from Madagascar. The anthropoids include New World monkeys, Old World monkeys, and apes. New World monkeys, our most distant relatives among the anthropoids, live in South and Central America. We humans are more closely related to Old World monkeys, which live in Africa and Asia. That is, we share a more recent common ancestor with Old World monkeys than we do with New World monkeys. We are most closely related to the apes and are even classified in the same superfamily as them. Among the apes, our most distant relatives are the smaller gibbons and siamangs, known as lesser apes. Our closest living relatives are the great apes, which include, in increasing order of relatedness to us, orangutans, gorillas, and chimpanzees and bonobos. Chimpanzees and bonobos are equally closely related to us and diverged from each other only approximately 2 million years ago when bonobos were isolated south of the Congo River. Our own evolutionary lineage diverged from chimpanzees and bonobos only about 6 to 7 million years ago. On our twenty-four-hour clock, this happened just two minutes before the stroke of midnight at 23:58.

We might expect that paternal caregiving would be most evident in the primate species that are most closely related to us, the great apes. However, this is not the case. This is important to emphasize, because it tells us that we did not simply inherit our paternal potential from our ape ancestors. Rather, that capacity newly evolved during human evolution for reasons that we consider later in this chapter.

The best examples of committed paternal caregiving within the primate order are found in two groups of New World monkeys: titi monkeys and owl monkeys. At this point, it will come as no surprise that both of these species are monogamous and males therefore have high paternity certainty. In fact, a recent study showed that titi monkeys are genetically monogamous, which means they don't just pair off and mostly stay faithful, like birds. They are strict, exclusive monogamists: 100% paternity certainty.[42] This is quite rare in the animal kingdom.[43] The strength of the titi monkey pair-bond is demonstrated by their adorable habit of intertwining tails as they sit together on tree branches (figure 2.8).

Titi monkeys, which live in the Amazon rain forest, are quite small for primates—about the size of a guinea pig (2.5 pounds). Adult male and female pairs perform loud duets, by which they vocally mark their territory from other pairs. From the first week of life, adult males are the primary

Figure 2.8
Pair-bonded titi monkeys with intertwining tails.

caregiver of infant titis. They carry infants on their back almost constantly. Mothers engage in brief bouts of nursing several times a day, but then hand the infant back to the father for him to carry. Infants are carried continually by fathers for the first few weeks. By two months, fathers are still carrying them approximately 60% of the time. By three months, infants are about 40% of adult body weight, and males continue carrying them about 40% of the time. This would be akin to me carrying my 72 pound, ten-year-old son on my back for nearly half the day. No, thank you. By four months, even titi fathers have had enough, and they begin rejecting infant attempts to climb aboard. Nevertheless, in an emergency, as when being chased by predators, even four-month-olds get a ride from their devoted fathers.

Paternal caregiving in titi monkeys is not limited to carrying. Fathers also willingly share food with their infants, and, along with older juveniles, are the infant's playmates.[44] As a consequence of this heavy paternal invest- ment, infant titi monkeys appear to become more strongly attached to their fathers than their mothers as inferred from the relative amount of distress they show upon separation from each.[45]

We don't think that animals consciously strategize about how to maxi- mize their reproductive success. However, natural selection causes them to behave as if they do, and sometimes it is a useful exercise to imagine them doing so. In the case of an adult male titi monkey, where we find strict, exclusive genetic monogamy, there is no point in seeking additional mates because all titi females have paired with only one male, and they are com- pletely faithful. Furthermore, he has absolute paternity certainty so any investment in his mate's offspring will contribute to his own reproductive success. It seems like a no-brainer for titi fathers to help raise their young. But for the sake of argument, let's say that female titis can raise their infants without male help and the infants survive no better when he helps. Even in this case, he may still increase his reproductive success by alleviating the energetic burden on the mother, which should allow her to resume repro- ductive cycling and have another infant sooner. Since his reproductive suc- cess is tied to hers, anything he can do to shorten her interbirth interval will also increase his own reproductive success.

Another group of New World monkeys also shows extensive paternal care, but within the context of a different breeding system. These are the marmoset and tamarin monkeys, collectively known as the callitrichid pri- mates. Marmosets and tamarins are even smaller than titi monkeys, weigh- ing less than 2 pounds. In contrast to biparental titi and owl monkeys, marmosets and tamarins are cooperative breeders.[46] Groups often consist of two or more adult males and two or more adult females, plus younger subadults. Callitrichids have flexible mating systems. In contrast to titi and owl monkeys, they are not exclusively monogamous. Usually only one of the adult females breeds, but she may mate with multiple adult males in her group. The other females often suppress their reproductive cycling (or have it suppressed by the breeding female), presumably because breeding females are known to kill the infants of other female group members. This isn't a conscious choice by these females. Rather, some cue from the dominant breeding female is somehow able to reduce their secretion of reproductive hormones.[47] Instead of reproducing, these females and other group members assist the breeding female with her own reproduction.[48] As in titi and owl monkeys, adult males do most of the infant carrying, but their burden is greater since callitrichids typically give birth to twins, each weighing from 6% to 8% of the mother's body weight. We know that carrying twins taxes males because they are able to feed less and lose significant body weight when caring for them, and they preemptively pack on extra pounds during gestation in preparation for it (figure 2.9).[49] Male care makes a difference since groups with more adult males have more surviving infants.[50] The extensive help that females receive from other group members permits them to reproduce at an unprecedented rate among nonhuman primates. Breeding females are able to produce twin offspring twice per year, and this is supported by a remarkable physiological adaptation to facilitate rapid reproduction. In other primates, female reproductive cycling is suppressed during lactation. Consequently, they are not able to conceive while

Figure 2.9
Tamarin father carrying his twin infants.

intensively nursing. This presumably functions to keep interbirth intervals long enough so that the first infant can be raised to independence before the second comes along. Callitrichids do not have this "lactational amenorrhoea" and can conceive again soon after birth, even while nursing their newborn.[51] This capacity likely evolved because the breeding female can depend on the other group members to help raise her infants. She doesn't need to keep birth intervals long. Quite to the contrary, with all that help, she will maximize her reproductive success by keeping the intervals short.

In addition to carrying the twin infants, adult males groom, protect, and provision the young with food. But why all this paternal investment in a polyandrous breeding system that precludes paternity certainty? The question to ask is, What other alternatives do these males have? As noted, only the dominant female of a group breeds, so there are a limited number of breeding females that males can reproduce with. Males often face the choice of leaving their group and not reproducing or staying in a group where they are not the female's only mate. Callitrichid twins are dizygotic: they can have different fathers. Given that callitrichid females may produce two sets of twins per year, a male who shares access to a breeding female might still be the father of one or two offspring each year. With his reproduction tied to that of the breeding female, helping her raise the offspring may be his best available reproductive strategy. By carrying the twins, he saves her energy for lactation and may enable her to replenish energy reserves faster in order to gestate another litter sooner. His provisioning may also allow her to wean her infants sooner or increase their probability of survival after weaning.

Although it is common for multiple adult males to mate with the breeding female, they may not be equally likely to father her offspring. Within callitrichid groups, some males have much larger testes than others, implying that some are producing much more sperm than others. If multiple males are mating with the female, the one that produces the most sperm may be most likely to fertilize the eggs and father the offspring. The marked variability in testes size within groups suggests that males with smaller testes may be reproductively suppressed, similar to nonbreeding females.[52] These suppressed males may simply be biding their time, staying on in the group and alloparenting to earn their keep, until later when they will potentially have an opportunity to take on a more prominent breeding role.[53]

Another factor that might contribute to male caregiving in callitrichids despite the lack of paternity certainty is that adult male group members may sometimes be brothers or half-brothers. Earlier, I defined the term *biological fitness* as the ability to pass on genes to the next generation and said that we expect organisms to engage in behaviors that increase their fitness. However, we share genes with our relatives. For example, a male shares half of his genes by common descent with his full brother and one-quarter of his genes by common descent with his half-brother. So by helping his brother, he is also increasing his own fitness. When we also consider the effect of a behavior on the fitness of relatives, in addition to one's own fitness, this is called *inclusive fitness*. We expect that evolution has selected for behaviors that increase inclusive fitness, and caring for a brother's young will do so, as long as you aren't sacrificing your own fitness in the process.

Old World monkeys do not show the high level of paternal care found among titi monkeys, owl monkeys, and the callitrichids in the New World. This fits with the very low frequency of monogamy among Old World monkeys.[54] Along our life history strategy continuum, adult male Old World monkeys are generally more oriented toward mating effort than parenting effort. Males aggressively compete for mating opportunities, both directly by defending access to female mates and indirectly by competing for social status or dominance, which often leads to mating opportunities. This reproductive strategy is reflected in their anatomy. Males are equipped to fight. They are considerably larger than females and have large canine teeth they use as weapons to injure their opponent. They are not constantly fighting. They fight to establish their place in the dominance hierarchy. But once they have sorted that out, aggression becomes less frequent because everyone knows their place.

Some Old World monkeys, such as hamadryous baboons, have a breeding system known as single male polygyny, in which a single adult male forms a group with several adult female mates, while many other adult males do not have access to female mates unless they can depose an existing breeding male. Male competition for groups of breeding females in hamadryous baboons is intense. and sexual dimorphism in body size is pronounced. Infanticide by nonbreeding, would-be usurper males is a major source of infant mortality in these species. In fact, infanticide committed by unrelated adult males is an adaptation. Adaptations are traits that have evolved by natural selection because they increase an individual's fitness. To say a trait is adaptive is not to say that it is "moral," "ethical," or "as it should be." Infanticide in species with single male polygyny is a prime example of an adaptation that we humans consider morally reprehensible. Why did it evolve? Female reproductive cycling is typically inhibited by nursing (though not among callitrichids), meaning that these mothers cannot conceive again until they stop lactating. Males that are not the father of their nursing offspring can therefore hasten a female's resumption of reproductive cycling and begin making their own family by killing her infant. And sadly, they do so in species like hanuman langurs and hamadryous baboons.[55] Not surprisingly, then, fathers play a crucial role in physically protecting their own infants from infanticide in these species.[56]

Infanticide is not limited to Old World monkeys with single male polygyny. In fact, it has been documented in every major group of primates, including prosimians, Old World monkeys, New World monkeys, and apes.[57] Nevertheless, it is less common in some primate species where females appear to have evolved effective counterstrategies. While species like hamadryous baboons practice single male polygyny, many other Old World monkey species live in social groups consisting of multiple breeding males and multiple breeding females that do not pair monogamously. Well-known examples include savanna baboons and rhesus macaques (also known as rhesus monkeys). In these species, adult females have evolved a special tactic that can help to prevent infanticide: they mate with multiple—and often most—adult males in the group. How does that prevent infanticide? It confuses paternity, and males are unlikely to kill a female's infants if there is a chance the infant is theirs. So, as the ingenious primatologist Sarah Hrdy realized and popularized, female promiscuity may be an adaptation to prevent male infanticide; she likes to refer to such multimale mating as "assiduously maternal," a mother working to increase the survival chances of her baby, not "promiscuous," a term in our society more often applied to someone who has sex with more individuals than someone else (usually male) thinks she should.[58] Theoretically, males are expected to counter female promiscuity by withholding paternal investment due to low paternity certainty and instead invest their energy in mating with additional, often willing female partners. And this they do in these primate societies. Nevertheless, there are now several examples of Old World monkey species, including savanna baboons, rhesus macaques, and assamese macaques, in which fathers preferentially associate with or physically protect their genetic offspring during aggressive conflicts with other group members.[59]

These observations suggest that males in these species may be less confused about paternity than we think. But how do they discern who their offspring are? Perhaps they use the timing and frequency with which they mated with a female as a cue. Female mammals, including primates, ovulate at a consistent point in their reproductive cycle, when they can conceive. In some species, females provide obvious cues as to when they are ovulating. In baboons, this takes the form of an enlarged, pink sexual swelling at the time of ovulation (figure 2.10). It is quite possible that males find females most attractive during this cycle phase either because of their appearance or some pheromonal cue they are emitting. If a male mates with a female at ovulation, when he finds her most attractive, perhaps he then decides to protect her offspring. This is pure speculation, and I suggest

Figure 2.10
Female baboon with sexual swelling.

it only to illustrate a potential mechanism. A decision like this would be adaptive insofar as he targets his parental investment at offspring that are likely to be his own, but he wouldn't need to know anything about ovulation or conception or have any conscious thoughts about the importance of paternity or reproductive success. Those thoughts are just for scientists. It may also be the case that males are somehow able to detect something in the infant's appearance or odor that identifies them as kin and that it is this similarity between himself and his infant that cues him to care for it. Indeed, there is evidence that chimpanzees are able to infer relatedness between individuals by comparing their facial appearances.[60]

Males in several Old World monkey species protect their infants from both predators and hostile group members. In some cases, it is the infants who seek out their father rather than vice versa, potentially because fathers can provide them with a harassment-free zone where they can forage and engage in other activities undisturbed and more effectively.

Despite evidence that males can direct paternal care to their offspring in many Old World monkey species, there are also species, such as olive baboons and Barbary macaques, where males are known to care for infants that do not belong to them.[61] This seems puzzlingly maladaptive until we realize that the infants' mothers are in turn more likely to mate with those males in the future. In this case, the infant care he offers is actually an effective mating strategy. In olive baboons, males form what have been termed "friendships" with females and offer protection to both her and her infant. In turn, she may mate with him in the future.

Parental Care in Apes

Next, we turn to the apes, with whom we are more closely related than monkeys. Apes differ from monkeys: they are typically larger, have bigger brains, lack tails, and have greater mobility in the shoulder joint that allows them to climb and swing from trees effectively. Apes are divided into greater and lesser apes. Lesser apes, including gibbons and siamangs, are smaller than great apes. They live in the forests of Southeast Asia, where they use their long arms to acrobatically swing between tree branches. Humans and lesser apes share a last common ancestor about 20 million years ago.[62] That means that our two evolutionary lineages diverged from each other at about 23:53 on our twenty-four-hour clock.

In contrast to many of the Old World monkeys, both gibbons and sia-mangs are mostly monogamous. Males and females are sexually monomorphic: they are the same size and look the same. Both gibbon and siamang couples are highly territorial and mark their territories with loud vocalizations. Monogamy and sexual monomorphism would seem to set both up nicely to be good fathers, and this prediction holds true in siamangs. Infant siamangs are initially carried by their mothers, but in the second year of life, the siamang dad takes over this responsibility. Fathers also have been observed grooming and playing with their infants. Many siamangs live in monogamous groups, but others live in polyandrous groups where a single female mates with more than one male. Interestingly, infants are carried more if they live in a monogamous group than a polyandrous group, even though there are more males available to carry them in the latter.[63] Possibly, males from monogamous groups are willing to invest more because they have paternity certainty that polyandrous males lack. When siamang fathers carry infants more, mothers get a break. In turn, mothers who carry their infants less have shorter interbirth intervals. Presumably, male carrying allows her to save energy so that she can better nourish and more quickly wean her infant and then resume reproductive cycling. As in titi monkeys and callitrichids, the father's reproductive success is tied to the mother's. Therefore, fathers are increasing their own reproductive success by decreasing the mother's interbirth interval, and paternal caregiving is adaptive.

Gibbons do not appear to meet our predictions. Unlike siamang fathers, gibbon fathers do not care for their offspring, and that has been something of a mystery given that they are monogamous. However, it may be that gibbons are not as monogamous as we initially thought. One study reported that 12% of all copulations observed were extra-pair copulations, usually with a member of the opposite sex from an adjacent territory.[64] One possible explanation is that paternal care is precluded by their opportunity to potentially obtain extra-pair copulations. That is, a mating effort strategy may be more adaptive than a parenting effort strategy for male gibbons. This isn't a great explanation since many male birds provide care and also pursue extra-pair copulations. But perhaps paternal care is less essential for the well-being of infant gibbons than it is for bird offspring.

Great apes are more closely related to humans than are lesser apes. Among the great apes, our most distant relatives are the orangutans, with which we share a last common ancestor from about 15 MYA (or 23:55).[65] That is, around 15 MYA, there was a speciation event such that an ancestral great ape lineage split into two: one that would become orangutans, and another that would give rise to the African great apes and humans. Orangutans live only in the forests of Borneo and Sumatra. Male orangutans are very far to the mating effort end of the life-history-strategy continuum. Dominant males have large cheek pads on the sides of their face and are referred to as "flanged." These are not found in either immature males or females and have likely evolved because they help males compete more effectively for female mates.[66] This is a classic example of a sexually selected trait. Like all other male sexually selected traits, cheek pads evolved either because they attract females or because they intimidate or help males defeat other males. Based on personal experience, I would bet on the latter explanation in this case.

One Halloween not that long ago, I decided to dress as an orangutan, complete with the mask of a flanged male. This was an adult party, and I found that people were backing away from me all night when I approached them, looking a bit scared. No one seemed particularly attracted to me. So that's not really an experiment, but it is a kind of data.

Dominant male orangutans also have a large throat pouch under their chin that they use to emit booming long calls that serve to mark their territory and attract females. Flanged male orangutans (figure 2.11) are polygynous, their territory encompassing the home range of several adult females, all of whom they mate with.[67] These mature males are intolerant and hostile toward other males. They are mostly solitary creatures that typically interact with females only when mating. In the wild, they provide no parental care. Mothers, however, are heavily invested parents, nursing their young for as long as eight years. This heavy investment has consequences, however, resulting in the longest interbirth interval among great apes and humans: from six to nine years.[68]

After orangutans, our next closest relative among the great apes is the gorilla. Humans and gorillas shared a last common ancestor from about 10 MYA (or 23:57).[69] Like some Old World monkey species, gorillas traditionally practice single male polygyny. A single adult male, among mountain gorillas known as the silverback, lives in a group with several adult females and their young. He is their exclusive mate. Nonbreeding males hang out together in "bachelor bands" and look for opportunities to overthrow

Figure 2.11
Flanged male orangutan.

breeding males and take over their groups. These confrontations are violent, and that is why male gorillas have evolved such impressive size and strength; larger, stronger males were more likely to win these fights. If nonbreeding males are successful at dethroning the resident silverback, they will try to kill his infants, which has the effect of stopping female lactation and hastening their return to reproductive cycling, and the female is more likely to follow the killer afterward, presumably because he is a better bet to defend her next infant than the deposed harem leader was.[70] So the most important function of the silverback is to defend the group from these nonbreeding males. In the process of doing so, he is defending his young against infanticidal intruders, but also defending his mating access to the group's females. That is, he is simultaneously exerting both parenting and mating effort.

Young gorillas seek out the silverback and attempt to maintain proximity to him, presumably to benefit from his protection. While Silverbacks do not routinely provide direct care to them, they are quite tolerant of their proximity, which has been interpreted as a type of low-cost parental effort.[71] In recent years, some gorillas have begun living in multimale groups, possibly as a strategy to protect against human poachers.[72] Within such groups, the highest-ranking males associate most with infants, irrespective of whether they are the father. It turns out that after accounting for the effect of dominance rank on reproductive success, males that associate more with infants end up having more offspring.[73] This has led to the idea that male infant care may attract females and may therefore also be a form of mating effort.

Our closest living primate relatives are the last two African great ape species, chimpanzees and bonobos. Around 6 to 7 MYA (23:58), some ancestral ape species gave rise to two lineages: one that evolved into us and another that, after splitting again at 2 MYA, gave rise to chimpanzees and bonobos. It is worth considering these two species in some detail since they are considered to be the best models we have of the ancestral ape from which humans evolved 6 to 7 MYA. But remember that chimpanzees and bonobos are living species; we did not evolve from them.

On that point, I suffered the following embarrassment as a freshman in college. While sitting in class, I had a moment of profound insight into what I believed was a critical weakness in the theory of evolution. Confidently, I raised my hand and queried my professor: "If humans evolved from chimpanzees, then why have humans changed so much and chimpanzees have stayed the same?" He quickly pointed out that humans did *not* evolve from chimpanzees—that we instead evolved from some other ancestral ape species that we don't know too much about due to an impoverished fossil record from East Africa at the time. He assured me that both humans and chimpanzees have changed in the time since they diverged from their common ancestor. We did not evolve from chimpanzees. This is now a point that I emphasize in all my classes when teaching about human evolution.

Nevertheless, I think we can be confident that humans have changed more than chimpanzees or bonobos over the previous 6 to 7 million years, as only we have evolved bipedalism and language, along with a threefold increase in brain size. There are prominent scholars who believe that chimpanzees are in fact a reasonable model of that common ancestor from whom we evolved.[74] Similar to baboons, chimpanzees live in communities consisting of multiple adult males and multiple adult females. Chimpanzee males spend much of their time aggressively competing for dominance status because dominant animals have preferred access to valued resources such as food and mates.[75] Dominance in chimpanzees is not only based on size and strength, although those definitely help. Chimpanzee social life is more complex. Dominance is also based on alliances and coalitions that males form with other individuals—what the great Dutch primatologist Frans de Waal labeled "chimpanzee politics."[76] Chimpanzee females mate with many males, which may function to confuse paternity and decrease the likelihood of male infanticide. However, males sometimes attempt to form consortships, or short-lived pair-bonds, with females, sometimes by aggressively coercing females into these consortships. Males also aggressively attempt to guard female consorts from other males.[77]

Males are larger than females, as expected in a species with so much aggressive competition. However, males also compete in a more furtive manner. Chimpanzee males have enormous testes that make copious sperm. When females mate with multiple males, as is the case in chimpanzees, males who produce more sperm are more likely to fertilize the egg and to have higher reproductive success. So males are competing covertly, inside the female reproductive tract—hence, the evolution of large testes size in chimpanzees.

Another important preoccupation of male chimpanzees is the collective defense of the group's territory. Males from adjacent territories are hostile to each other, and lethal encounters are well documented. Males work together with other males and the occasional female to patrol and defend their territorial boundaries. When they have a numerical advantage, they may attack chimpanzees from the other group. If they encounter females from the other group who are fertile, males may abduct them.[78]

You are probably getting the sense that chimpanzee social life is not very compatible with paternal nurturance.

Despite female promiscuity, infanticide is still a real threat to chimpanzee infants.[79] Unrelated adult males are the most common perpetrators, presumably as a strategy to accelerate the mother's return to reproductive cycling, although females may also kill another female's offspring. Similar to baboons, their multimale breeding system limits paternity certainty and is therefore expected to constrain the evolution of paternal caregiving. Chimpanzees are certainly not doting fathers. They provide only minimal investment in their offspring, yet there is evidence that they both play and associate more with their own biological infants than with unrelated infants, just as some baboons do.[80] Somehow, they are able to discern who their infants are, at least better than they could by randomly guessing. The tendency of chimpanzee fathers to associate with their infants is most pronounced in early infancy when the risk of infanticide is highest. Chimpanzee fathers thus seem to be motivated to protect their infants.

According to some primatologists, around 2 MYA, a population of apes crossed the Congo River during a drought and became isolated at the south of that river when it re-formed. A remarkable thing then happened to that population: the males appear to have been tamed. Humans have tamed a number of domesticated species, such as dogs and horses, by systematically allowing only the less aggressive individuals to breed, generation after generation. These apes may have been tamed in a different way—by the female members of their own species.

These creatures evolved into modern-day bonobos (figure 2.12). They are a bit smaller than their chimpanzee relatives to the north but not by

Figure 2.12
Adult male bonobo (left) and adult male chimpanzee (right).

much. Males weigh about 100 pounds, females about 75 pounds. According to a prominent hypothesis, food was more plentiful and was found in larger patches to the south of the river, and this allowed female bonobos to congregate and become more social. In contrast, chimpanzee females to the north have to spread out so that each of them has enough food within their individual home range, so there is little opportunity for socializing. Female bonobos were able to form strong bonds with one another and, by cooperating, collectively deter male aggression.[81] The mechanism for this bonding, as is often true in animals, was sex, although few animals use sex as variously, routinely, and enthusiastically as bonobos do. Female bonobos have sex with one another by rubbing their clitori together in a face-to-face position. Bonobo clitori may even be specially adapted for this function, since they are larger and more externalized than in other mammals.[82]

At any rate, these strong female bonds meant that in contrast to male chimpanzees, male bonobos were not able to aggressively coerce females into mating with them. Coalitions of females could gang up against unruly males. As a consequence, female bonobos could choose to preferentially mate with less aggressive male bonobos, and those males' genes were passed to subsequent generations. In this way, they may have "tamed" the males of their species.[83]

Male bonobos can be aggressive, but their aggression is less severe than in chimpanzees. This holds true for male aggression toward other males and females within their group and also for aggression toward members of other groups. Chimpanzees are predictably hostile and even homicidal toward chimpanzees from other groups. In contrast, lethal intergroup aggression has not been observed in bonobos, and intergroup interactions are occasionally amicable. Food sharing between groups has even been observed. Bonobos are less xenophobic than chimps. This reduced severity of aggression is even reflected in their weaponry, with bonobos having smaller canine teeth than chimpanzees.[84] Another striking feature of bonobos is that they commonly resolve conflict and dissipate tension within the group through sexual interactions as opposed to aggression.[85]

Selection for decreased aggression in male bonobos is another example of sexual selection, but this time it takes the form of female choice, the process by which females are able to shape male anatomy and behavior through who they choose to mate with. One classic example of female choice is the extravagant tail of the male peacock. In this case, female peahens seem to be choosing males based on some measure of health or genetic quality, since chicks of fathers with larger tails survive better. That is, it seems that by choosing males with large tails, females are giving their chicks a better chance at survival simply because they have inherited this male's "good genes."[86] In other species, females choose males for different attributes, including signs that they will be good parents.

So if indeed female bonobos selected for less aggressive males, why was this adaptive? How did it increase the reproductive success of these females? One possibility is that it helped to protect their infants from male infanticide, something that both chimpanzee and gorilla mothers have to contend with. In fact, a decreased propensity for aggression might be the reason that male infanticide has never been observed in bonobos.[87]

There is, however, another interesting potential reason why infanticide has not been observed in bonobos. Female chimpanzees, who do have to worry about infanticide, advertise when they are fertile with a prominent sexual swelling on their hind end. The swelling reaches maximal tumescence around the time of ovulation, when they are most likely to conceive. This is when chimpanzee males compete most aggressively for the opportunity to mate with females. Bonobo females also have sexual swellings, but they seem to fool males as to the timing of ovulation. This is because they show maximal tumescence for as long as twenty days out of their cycle.[88] Given this confusion, male bonobos may refrain from infanticide because paternity is so uncertain, and it would be maladaptive for them to kill their own infant.

If bonobo mothers don't need to worry about infanticide, then neither do bonobo fathers. This may be why, in contrast to chimpanzees, there are no reports of bonobo fathers preferentially coming to the aid of or preferentially associating with their young. However, bonobos have not been as intensively studied as chimpanzees owing to their rarity and the difficulty of studying them in the remote Congolese rain forest, so it remains possible that paternal defense of offspring will be observed in the future. More generally, bonobo fathers do not appear to provide any significant care for their offspring. Part of the explanation may be that female bonobos, like female chimpanzees, are sexually promiscuous and males presumably have limited paternity certainty.

But I noted that chimpanzee fathers, despite female promiscuity, seem to be able to target help toward their genetic offspring at better than chance. A chimp male who mates with a female when he finds her most attractive (when she is maximally tumescent) has a reasonable chance of being the father of her offspring. On the other hand, a bonobo male who does so could well be mating with her when she is not ovulating due to her extended period of maximal tumescence. Given this confusion, male bonobos may have more trouble identifying who their offspring are compared with male chimpanzees. Paternal caregiving may be maladaptive under those circumstances, especially when you consider how many other bonobo females are highly attractive, and willing to mate non-monogamously.

Parental Care in Humans

Finally, we arrive at our own species, *Homo sapiens*. The human breeding system diverges markedly from that of our closest ape relatives. One of its striking features is its flexibility and the degree to which family configurations vary across cultures. Humans practice monogamy, serial monogamy, polygyny, and even polyandry. Some form of marriage is a human cross-cultural universal, but marriage can sometimes look quite different from what we imagine. For example, the Mosuo of China are a small, agrarian ethnic group where husband and wife live separately from one another in the households of their respective extended families. At night, husbands may visit their wives in their households. Some of these marriages last a long time, but some are brief, and men and women are free to change partners whenever they like. Among the Mosuo, the mother's brother often has a more prominent role in raising children than does the father.

Despite this flexibility, most human reproductive unions are monogamous.[89] This is in stark contrast to the highly promiscuous mating systems of chimpanzees and bonobos. Furthermore, in most human societies, fathers provision their children, something no other apes are known to do. Typically men in traditional foraging societies produce vastly more calories than they consume, and they distribute the surplus to others, including their children. This provisioning is especially important for children, who won't be producing as many calories as they consume until they are twenty years of age or older. It takes years for human youngsters to develop the strength, skill, and knowledge needed to successfully hunt, and their foraging productivity doesn't peak until well after age twenty.[90]

This pattern of production and consumption contrasts sharply with that of chimpanzees, where juveniles are calorically self-sufficient by age five or so and adults do not generate a surplus. They really can't because they are not able to access high-quality, calorically rich, difficult-to-acquire food resources, such as animal game, nearly as effectively as humans can.[91]

Incidentally, many of the college students I teach are about twenty years old, and when I cover this material, I like to ask them whether they think they are net producers or consumers. This is predictably met by nervous chuckles, on being rudely reminded how far from self-sufficiency they remain. When we consider economic provisioning, it's not just forager children who have a long period of dependency.

The surplus that forager men generate is often shared beyond their immediate family, and may not only be motivated by parental inclinations.[92] In some foraging populations, good hunters have more extramarital affairs.[93] Sharing meat beyond the immediate family may therefore constitute a type of "mating effort." Nevertheless, there is evidence that fathers are often able to target more of the surplus toward their own children, and food that is shared with other families is often reciprocated in the future.[94]

Earlier, I noted the occasional blurring between male mating and parenting effort. When a man provisions his children, it is clearly parenting effort, but it might also be mating effort insofar as it helps him attract and keep his female partner. One ingenious study attempted to unpack these two types of effort by looking at financial contributions that American men made to their children as a function of whether they were biological fathers or stepfathers and whether the child was of their current or a previous partner. The logic was as follows. From a life history theory perspective, money spent on biological children of a current partner includes both mating and parenting effort. Money spent on stepchildren of a current partner was considered purely mating effort since it was not directed at his biological offspring. Money spent on genetic children of a previous partner was considered pure parenting effort since that investment would not help the father attract his current partner. Money spent on stepchildren of previous partners was considered neither parenting nor mating effort. As expected, fathers spent the least money on this last category of children ($156 per year). Fathers spent much more money on genetic children of previous partners ($1,888) and this was pure parenting effort. But they spent about the same amount of money on stepchildren of current partners ($1,861), suggesting that fathers invest just as much in mating effort as they do in parenting effort. Finally, fathers spent the most on biological children of their current partner ($2,570), just as predicted. Comparable results were found when the researchers instead looked at the percent of children who attended college as a measure of paternal investment.

In hunter-gatherer societies, fathers provision children more directly with calories from food that they procure. However, fathers are not the only caloric providers in hunter-gatherer societies. Mothers also generate surpluses, occasionally even larger surpluses than fathers, and grandmothers are an important source of extra calories in some societies as well.[95]

Given that humans are the only hominid species that practices monogamy and exhibits significant paternal investment, these traits most likely evolved anew during the 6 to 7 MYA after we diverged from our great ape relatives, in just the last two minutes of our twenty-four-hour day. But what kind of creature did we evolve from and how, and why were these changes selected for? This is a difficult question to answer, in part because the fossil record from this time period is so scant. Many of us are drawn to the study of human evolution in part because the evidence is so fragmentary. Human evolution is like the world's greatest mystery. We have key pieces of the puzzle but don't know exactly what happened. And we can use those pieces to construct various scenarios of what might have happened. I am going to present one scenario that I believe is plausible in light of the evidence. But keep in mind that alternative scenarios are also plausible.

*   *   *

We might infer speculatively that the common ancestral ape that gave rise to chimpanzees and bonobos had traits found in both living chimpanzees and living bonobos. That is, rather than those traits evolving twice independently, it seems more likely that both species inherited them from a common ancestor. I make the further speculative assumption that these shared chimpanzee and bonobo traits evolved before they separated from humans; however, it could also be that they evolved afterward and that the common ancestor of 6 to 7 million years ago was a different creature (figure 2.13). Nevertheless, some prominent anthropologists have argued that this common ancestor was both anatomically and behaviorally similar to modern chimpanzees.[96]

Based on this assumption, the common ancestor was most likely living in groups composed of multiple males and multiple females, in which individuals mated promiscuously, females advertised when they were

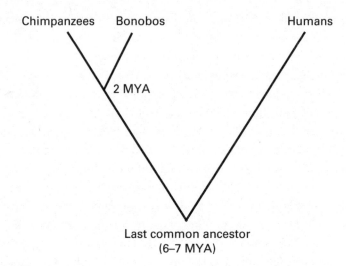

Figure 2.13
Evolutionary relationships between humans, chimpanzees, and bonobos.

ovulating, and fathers, at most, protected their infants from infanticide. They did not likely provision or provide any significant direct care. All of these traits are found in both living chimpanzees and living bonobos. But was the common ancestor that ultimately evolved into humans initially more chimpanzee-like or more bonobo-like? Bonobos seem to have experienced more genetic and morphological change than have chimpanzees since the two lineages split, so perhaps the common ancestor was more chimpanzee-like.[97] If so, then males of this ancestral ape species would also have aggressively competed for dominance and mating opportunities.

So how then could evolution transform a sexually promiscuous, aggressive, nonpaternally investing ape into a species with so many committed fathers and husbands? Scholars have proposed a range of mathematical evolutionary models to try to explain this transition. These suggest the following as one potential scenario. Approximately 2 MYA, Africa became cooler and drier, and the African savannas expanded as the forests shrank. As our ancestors moved into the savanna environment, men began specializing in hunting large mammals that roamed the savanna and women began specializing in gathering plant foods on the savanna, largely because gathering was more compatible with child care than hunting was. In effect, this meant that men began to specialize in procuring fat and protein found in meat, while women specialized in foraging for carbohydrates.[98] Human offspring, with their large and metabolically expensive brains, would fare better if calorically and protein-rich meat were part of their diet. As such, some adult male proto-humans might have enhanced their reproductive success by provisioning their offspring rather than by using that energy to pursue additional mates.[99]

However, there was one big problem: adult male proto-humans couldn't be sure who their offspring were because adult females were typically mating with many males. Of course, adult females would also increase their reproductive success if they could get males to help provision their offspring. And how might they accomplish this? By now, you know the answer: through providing males with increased paternity confidence by selectively mating with them. That is, they began to pair-bond with males. Once females began to partner with individual males, males would know better who their offspring were and could target them with their provisioning and other forms of investment.[100] Some theorists believe this provisioning strategy originally evolved in low-ranking males who did not have plentiful mating opportunities and would therefore not be sacrificing reproductive success by parenting instead of mating.[101]

Imagine a subordinate, adult male proto-human with no female mates. However, let's say he carries a novel mutation in his DNA that alters his brain chemistry in such a way as to increase his interest in juvenile proto-humans. One juvenile, who is not his, approaches him and begs for a scrap of meat. Maybe he likes the way the juvenile playfully ambles over to him and tugs at his ear. While he is not likely to hand it over, in contrast to the other adult males, he allows the juvenile to take a scrap from him. This is known as "tolerated theft" in the literature. The juvenile's mother notices this tolerance and begins to take a liking to this slightly less selfish male. Eventually she mates with him, in addition to the many other males she has already been routinely mating with. Something akin to this happens in baboons. Male "friends" who consistently offer physical protection to females and their offspring sometimes earn the right to mate with those females.

Chimpanzees occasionally hunt monkeys and other smaller prey, and they sometimes share their kill with others. There is evidence that over the long run, female chimpanzees are more likely to mate with males that provision them with meat.[102] So it seems reasonable to think that female proto-humans might mate with male proto-humans that provisioned their young with meat. If indeed human females evolved a preference for men who provisioned them and their offspring with meat, we might expect to find evidence of this in modern foraging populations. Indeed, modern foraging women place a high value on the meat that men provide. Women in foraging societies often strongly encourage their husbands to hunt.[103] In a 1957 ethnographic film documenting four !Kung San men hunting a giraffe, one !Kung women asks her husband to hunt because her "breasts are lacking milk."[104] In another ethnography of the !Kung San, after the death of her husband, a !Kung San woman wonders, "Where will I see the food that will help my children grow? Who is going to help me raise this newborn?"[105]

The female proto-human who mates with our protagonist will have maybe five offspring in her lifetime, but perhaps one will be sired by our provisioning male—and that one offspring may be slightly more likely to survive because he has been provisioned with these extra scraps of meat. The father's reproductive success has increased from nothing to something, allowing him to stay in the evolutionary game and send his genes into the next generation. His offspring, let's say it is male, inherits his tendency to provision. Eventually another fortuitous mutation occurs in one of the male's descendants that makes that male just a tad bit more inclined to allow juveniles to take his food. Females begin finding these males more attractive and mating with them more frequently, giving males more paternity confidence and increasing their incentive to provision. Once set in motion, this process could culminate in a situation where most females are mating selectively with males who deliberately and reliably provision their offspring, much like we find in many modern human societies.

It is worth pointing out that some of these evolutionary models stabilize with a small population of males ("cads") who do not provision, but instead devote all their energy to mating.[106] These would be males who can attract mates simply by virtue of their genetic quality; females are willing to mate with them even if they don't provision their children. (In modern human society, think of some famous athletes and celebrities.) But for most males, provisioning was an adaptive strategy.

If we believe that the last common ancestor was chimp-like, then our male ancestors were also likely transformed in another way; their testes got smaller. I know, I know . . . but please hear me out before you decide to close the book.

Male chimpanzees and male bonobos have enormous testes. As discussed earlier, this is because testes produce sperm and larger testes produce more sperm. Female chimpanzees mate with many different males, often in succession, so the male who produces the most sperm has the best chance of fertilizing her egg cell. It is adaptive for male chimpanzees and bonobos to have very large testes given their mating system. Human male testes are much smaller than those of either chimpanzees or bonobos, and this implies that human females became more selective in terms of mating at some point in human evolution. In response, males didn't need to produce as much sperm, and their testes shrank. Comparison of human and chimpanzee genomes also indicates significant evolutionary change in human testes, since many of the genes identified as having been altered by selection during human evolution are involved with spermatogenesis.[107] The putative reduction in human testes size didn't progress to the point of complete, faithful female genetic monogamy, in which case men would have even smaller testes, but there would have been a significant movement in that direction.

While we are on the topic of genitals, females also likely experienced a profound and consequential change. Instead of advertising when they were ovulating with bright red sexual swellings, they began to disguise it. Human females have no obvious external signals of ovulation and can mate throughout their cycle. Thus, if males want paternity certainty, they can't get away with provisioning their mate and her offspring for only a few days each month. In that case, they might not be around when she ovulates and they might not sire her offspring. Instead, provisioning becomes a full-time male occupation.[108]

This scenario hinges on females having the capacity to choose to mate with low-ranking, provisioning males as opposed to domineering, aggressive alpha males. Did ancestral, proto-human females have that capability, and if not, how did it evolve? While it is debated, some researchers argue that chimpanzee females have little choice and that dominant males are able to coerce females into mating with them.[109] Bonobo females seem to have more choice, likely as a consequence of selection for decreased aggression in male bonobos.[110] But if our common ancestor was more chimp-like, then female choice may have been limited. There is evidence that by around 300,000 years ago, humans had evolved anatomical features that tend to be associated with decreased levels of aggression.[111] These include a shortening of the face and reduction in tooth size, along with reduced brow ridge projection (figure 2.14). This trend toward greater facial feminization continued until recent time,[112] with fascinating implications. Beginning around 300,000 years ago, human males may have started becoming less inclined to use aggression to constrain female choice. It is conceivable that female choice became a powerful evolutionary force only after this reduction in male aggression evolved in humans.

But what might have been responsible for this initial reduction in male aggression around 300,000 years ago? Remember that bonobos went through a similar process. In the bonobo case, it was female bonding and collective aggression against aggressive males that presumably tamed males. However, this does not appear to be a prominent dynamic in human

Figure 2.14
Craniofacial feminization in human evolution. Left: Modern human skull. Right: Hominin skull from before 300,000 years ago (*H. heidelbergensis*). Source: A. Gibbons, "Human Evolution: How We Tamed Ourselves and Became Modern," *Science* 346, no. 6208 (2014), 405–406. DOI:10.1126/science.346.6208.405.

societies. Instead, one intriguing, if chilling, hypothesis is that men "tamed" other men through capital punishment.[113] That is, highly aggressive, domineering men were executed by less dominant men, who were able to use language to plan coordinated attacks on alpha males with minimal risk. If this sounds far-fetched, as it initially did to me, consider reading Richard Wrangham's book, *The Goodness Paradox*, complete with ethnographic examples of executions in foraging societies.[114] It might change your mind.

By provisioning mothers, fathers could provide mothers energy that they could use for lactation. By provisioning their children, they could enable mothers to wean those children earlier, and this would decrease their interbirth interval and increase their lifetime fertility. Indeed, forager interbirth intervals are approximately three to four years, whereas intervals for other great apes are more like five to seven years.[115] Of course, many of us know of siblings who are spaced only a year or two apart–or even less. This is made possible by formula feeding in lieu of intensive breastfeeding. With no lactational amenorrhea, mothers resume reproductive cycling and can conceive much sooner than is possible in natural fertility human populations.

It is interesting, however, how birth intervals often stay long in modern couples who lack support from extended family. My wife and I were one of those couples. We have five years between our two kids. Without local support from family, as we both worked full-time jobs and suffered from parental ignorance, sleep deprivation, and chronic time pressure, not to mention a colicky baby, it took five years before we felt ready to go through it all again. One of my colleagues joked with me, "I see you are adhering to hunter-gatherer birth spacing." I one-upped his nerd humor: "No, more like great ape birth spacing."

Although women have shorter interbirth intervals than female great apes, human infants in natural mortality populations are just as likely, if not more likely, to survive as great ape infants are.[116] That is, despite being weaned early and deprived of the calories they would get from mother's milk, human infants survive as well as great ape infants. This is because they are provisioned by allomothers, including fathers. Across primate species, there is a positive correlation between the amount of allomaternal care infants receive and female fertility.[117] When primate females have more allomaternal help, they reproduce at a faster rate. This comes as no surprise to us when we consider how the cooperative breeding callitrichids discussed earlier are able to raise twin offspring twice per year.

However, there seems to be another crucial benefit to allomaternal care in some mammals. Among carnivores, allomaternal care is positively correlated with brain size. It seems that carnivores are able to use the extra energy provided by allomothers to build larger brains for their offspring. Humans have by far the largest brain size among primates. So allomaternal care may have not only enabled human mothers to increase their fertility and their offspring survival, but it also may have enabled children to develop large, energetically costly brains.[118] In fact, these two products of cooperative breeding, increased fertility and increased brain size, help explain why humans have become such a successful species, with biomass far exceeding that of any other wild mammal.[119] We out-reproduce other apes, and our big brains have provided us with a second informational system, beyond the genetic one, to culturally adapt and inhabit almost every environment on Earth.[120]

One problem with the scenario I have presented is that male provisioning of children might have started well before 300,000 years ago, possibly with *H. erectus*, who appeared closer to 2 MYA. It seems that *H. erectus* were already meat eaters. There is evidence of butchered mammal remains in association with hominin stone tools that date to this time, and the rib cage of *H. erectus* is narrower than what is seen in great apes, which implies they had a smaller gut as one tends to see in animals that eat a lot of meat.[121] There is also some evidence, although debated, for a reduction in sexual dimorphism in body size in *H. erectus*, which tends to coevolve with pair-bonding and paternal care.[122]

Therefore, evolutionary models that do not depend on female choice for the evolution of male parental investment, which may not have been prominent until 300,000 years ago, may have merit. One such model posits that the complementarity of male and female subsistence contributions drove paternal provisioning.[123] That is, human infants needed both the gathered plant foods from their mom, mostly carbohydrates, as well as the hunted animal foods from their dad, mostly protein and fat, to meet the nutritional needs of their large, developing brains. Thus, fathers were provisioning not because they were trying to please mothers or because mothers preferred mating with provisioning males, but because it was necessary for the survival of their offspring.

But if females didn't have the capacity to choose pair-bonding males 2 MYA, how could monogamy evolve so that males have paternity certainty?

Examples from other species show that pair-bonding can evolve from male mate guarding rather than female choice. In this more sobering scenario, proto-human males begin guarding individual females from other potential mates in order to ensure paternity certainty, which allows them to then target their provisioning at their biological children.

If this alternative model is correct and fathers were already provisioning their children by 2 MYA, then what about the craniofacial feminization that initially appears in the fossil record much later, at 300,000 years ago? Did it have any significance for the evolution of paternal care? Indeed, this change may have had behavioral implications beyond potentially signifying a reduction in aggression. Psychology experiments have shown that people think men with less masculine male faces will be better at taking care of children and will provide more resources to their family.[124] This is a plausible prediction since testosterone is known to both masculinize facial anatomy and to bias males toward mating effort and away from parenting effort (see chapter 4). So while this is speculative, it is possible that men were becoming more involved parents at 300,000 years ago. It is intriguing to further speculate that although indirect paternal caregiving was already well established, this may have been the time when females began selecting for direct caregiving behavior in human males: holding, carrying, playing, teaching, babysitting, and so on. Once they had the capacity to choose their mates, women may have selected for men who provided more direct care, and this would have left its signature in the fossil record as increased facial femininity. It might even be the case that men and women were working in tandem to transform men into less aggressive and more nurturing beings, through capital punishment and female choice, respectively. An important caveat here is that although women in these psychology experiments preferred to hypothetically marry men with more feminine facial faces, they preferred more masculine male faces for hypothetical "extra-pair copulations" (i.e., extramarital affairs).[125] If we believe these results generalize beyond the college student participants from this study, then female choice may not be uniformly and consistently aimed at less aggressive, more parental males.

Paternal provisioning is common in human males. But how common is direct caregiving, and would it have contributed to the reproductive success of men and their female partners? Human fathers don't carry infants anywhere near the 90% of daylight hours seen in titi monkeys, but they

do carry or hold infants for a significant amount of time in some societies. Given our interest in the evolution of paternal caregiving, evidence
from foraging societies is most relevant, again because we think their way
of life is in many ways similar to that of our ancestors. Across foraging
societies, the amount of time fathers spend holding infants ranges widely,
from 2% to 22% of daylight hours.[126] This is considerably less than what
mothers do. For example, among the Hadza foragers from northern Tanzania, mothers do 69% of the infant holding and fathers do only 7%. Yet
fathers do more holding than any other allomother.[127] Aka fathers from
Central Africa do more caregiving than fathers in any other known society,
holding their infants 22% of the time (figure 2.15).[128] Yet infants are held
by their mothers 51% of the time. While the figures for fathers alone may
not be impressive, if we pool allomaternal care across fathers and other
caregivers, it becomes significant. Considered in this way, Aka allomothers
carry infants almost half the time and Hadza allomothers almost one-third
of the time. We need to think of fathers as part of a team of allomothers
that provide help to mothers. This is not intuitive to those of us who live
in modern, nuclear family households where there are a limited number
of helpers. If couples, or single parents, in these families are struggling, it

Figure 2.15
Picture of Aka father holding his infant. Courtesy of Barry S. Hewlett.

is with good reason. A recent detailed study of three different traditional, natural fertility societies showed that mothers, on average, have more than ten helpers who provide some degree of physical care for their infant. And the more helpers they have, the less time they themselves spend in direct caregiving.[129] Collectively, direct care from this allomaternal team is likely to have an impact on the mother's energetics, and therefore, plausibly, her reproductive success.

Fathers seem to calibrate their investment based on what other allomothers are providing. Often they do more when others do less, and vice versa. One example comes from the Aka foragers living in Central Africa. In traditional human societies, newly married couples typically live with either the husband's family (patrilocally) or the wife's family (matrilocally). Aka couples may do either. The anthropologist Courtney Meehan showed that when the Aka live matrilocally, where mothers are surrounded by their own mother and their own sisters, fathers hold their infants less compared with when they live patrilocally and mothers lack these kin helpers.[130] Similarly, a study of the Bofi foragers, the northern neighbors of the Aka, found that fathers had more physical contact with infants when older female relatives were not available.[131] Finally, Tsimane fathers from lowland Bolivia provide less direct care to infants and young children if they have older daughters who are available to provide care. They also provide more care when the mother is not present or is otherwise occupied for whatever reason.[132] Thus, it seems that fathers are doing more when they need to due to a dearth of other caregivers.

This tendency resonates to a degree with my own personal experience as a father. When our children were infants, I struggled with sleep deprivation and work–family balance. During this time especially, I celebrated periodic visits by my mother-in-law. In our family, my wife, bless her, is the primary instrumental caregiver, and I am secondary. When grandma was visiting, I could often move to the tertiary caregiving role temporarily, and it was a tremendous relief. My selfishness in moving to the tertiary role at these times, rather than encouraging my wife to do so, is not lost on me.

So, in sum it seems likely that human fathers were selected to provision their young and to provide direct care facultatively. That means they were not selected to be obligate caregivers, but to provide care when, in the cold, hard calculus of natural selection, it would contribute to their reproductive success, either by improving their children's survival or quality or by increasing the fertility of the monogamous partner to whom their reproductive success was linked.

It is essential that we remember that these putative evolved, genetic predispositions are only one of many influences over human paternal behavior. An array of other variables can cause men to deviate from the predictions we make based on evolutionary theory. Many fathers become more or less involved with their children for reasons that have nothing to do with evolution (these reasons are discussed in chapter 6).

As we conclude this wide-ranging survey of paternal caregiving across species and over evolutionary time, how can we summarize what we know about the evolution of paternal caregiving in our species? Although the evidence can be interpreted in multiple ways and no one knows for sure what happened, I have offered one potential scenario in which paternal caregiving was made possible by a transition from sexual promiscuity to pair-bonding that gave men paternity confidence, accompanied by an ecological shift that resulted in men hunting large mammals in order to provide meat that would nourish the prolonged and protracted development of big-brained, energetically costly human children. The parallel with canids is quite striking, where monogamy and cooperative breeding are prominent and males have high paternity confidence and provision large-brained young with hunted meat over their long period of development as they are learning specialized hunting skills. Beyond provisioning, the direct care that fathers and other allomothers collectively provided helped to lessen the mother's energetic burden, allowing her to shorten her interbirth interval and increase her fertility. Paternal caregiving thus played an important role in allowing our large-brained species to evolve, multiply rapidly, and spread across the globe, as the other great apes remained confined to their places of origin in Africa and Asia.

Highlights

1. Across the animal kingdom, paternal caregiving is more common when males have greater paternity certainty.
2. Given the lack of paternal care among our closest living primate relatives, male parental caregiving likely evolved anew during human evolution.

3. Monogamous pair-bonding may have provided ancestral men with the paternity certainty needed for paternal caregiving to evolve.

4. An ecological shift resulted in men hunting and scavenging large mammals on the African savanna. The extra calories provided by fathers and other allomothers may have enabled our specie's dramatic increase in brain size and high fertility rate.

5. Fathers are part of a team of allomothers that typically assist mothers with direct caregiving.

3 Testosterone

In most mammals, males are less nurturing than females. This is not only because they lack something that females have more of (see chapter 4), but also because they possess something that actively interferes with nurturing. That something is the steroid hormone testosterone. Human males have approximately fifteen times as much testosterone as females do.[1]

Testosterone has both costs and benefits, and these were illustrated quite effectively in the fictional 2008 film *The Wrestler*. The main character is a professional wrestler who struggles to compete, or perhaps, more accurately, perform, into middle age. He turns to anabolic steroids to sustain the muscle mass and motivation needed to continue in this physically grueling profession. In the ring, he is magnificent and adored by his fans. But behind the scenes, we see how he suffers with pain from a litany of past injuries, his body broken down, his heart malfunctioning, and we see the lack of love in his life due to his inability to commit to family life. Ultimately, in the film's final scene, we see him seemingly plunging to an early death.

The role of testosterone in men's behavior has become a controversial topic. A number of recent books have addressed the topic, with some arguing that testosterone's influence over men's behavior has been grossly exaggerated, while others have reaffirmed its importance.[2] The topic has been so heavily scrutinized because of concerns that people will attempt to explain away gender differences in behavior as having a biological explanation, which to many people is tantamount to saying those gender differences are fixed or immutable. For example, we don't want to prematurely conclude that men are more aggressive or more sexually promiscuous than women simply because they have higher testosterone levels and then conclude that there is nothing we can do about that. This would wrongly assume that we cannot shape the behavior of boys and men through the manner in which we socialize them. The reality is that biological and social/cultural influences interact in explaining nearly all human behaviors, so we should not ignore either.

With respect to testosterone and men's behavior specifically, part of the confusion in the literature stems from how we interpret evidence showing that correlations between testosterone and certain aspects of men's behavior, such as aggression or sexual behavior, are weak or inconsistent.[3] If men who have higher testosterone levels are not reliably more aggressive than other men, that doesn't necessarily mean that testosterone is not involved with aggression. I'll illustrate with an example.[4] Let's imagine that among professional basketball players, height is not correlated with average number of points scored per game (I don't know if this is true, but it strikes me as plausible—currently, three of the top ten NBA scorers are 6'2", which is 4 inches below the league average height). We would not then want to conclude that height is unimportant for basketball. This is because nearly everyone in professional basketball is taller than average. You are very unlikely to get into the NBA or WNBA if you are shorter than average, but if you are tall enough to make it into those leagues, then height might not matter much for scoring after that. The same could be true for testosterone. Across human societies, men are consistently responsible for more violent crime than women, and men have about fifteen times higher testosterone levels than women.[5] That magnitude of difference dwarfs the three- to four-fold variation in testosterone among typical men.[6] Therefore, it is plausible that having male levels of testosterone makes a man more prone to aggression but variation above that level doesn't matter much.

However, this is not sufficient evidence for a causal relationship. It is instead possible that a third confounding variable explains the association between male levels of testosterone and aggression. For example, if boys are socialized to be more aggressive, that, rather than their high testosterone, may be why they are responsible for more violent crime. It could also be the case that the correlation between testosterone and aggression is attributable to aggression increasing testosterone levels rather than the converse—and there is indeed evidence that the causal arrow can go in this direction.[7] What is really needed to determine if testosterone increases aggression is to experimentally raise (or lower) testosterone levels in human males and observe the consequences on aggressive behavior. It turns out that something like this takes place naturally at puberty when testosterone levels increase dramatically in boys and so do rates of violent crime. But at the same time that testosterone increases, adolescent boys are becoming stronger and hence more capable of committing violent crimes. Maybe it's testosterone's effects on the body rather than the brain that explains the increase in violent crime. Better evidence comes from experiments where adult men are administered testosterone and show increased aggression in laboratory paradigms, but these paradigms involve inflicting financial rather than physical harm on others, and the effects are found only in men with particular personality types (e.g., dominant and impulsive).[8] In sum, it's not easy to conclusively demonstrate that testosterone promotes aggression in men. But even if it does, we need to strongly emphasize that most men are not violent criminals despite having such high testosterone levels. This is because there are so many other important influences over men's behavior.

In this chapter, I do not focus on aggression per se. Instead, I focus on testosterone's more fundamental role in mating and parenting effort. Nevertheless, it's important to keep the nuances and complexities that I have just outlined in mind as we interpret the evidence. I will conclude that in both human and nonhuman males, testosterone motivates the pursuit of mating opportunities and interferes with the nurturing care of offspring.

*   *   *

First, let's consider why males have so much more testosterone than females do. Males, but not females, have a Y chromosome, and that chromosome has a gene known as the SRY gene. The gene secretes a factor that causes the embryonic gonads to develop into testes. Otherwise, they become ovaries. The testes have two principal functions. Most of their mass is dedicated to the production of sperm cells, which takes place inside structures called seminiferous tubules. In between these tubules are cells known as Leydig cells, and their job is to produce testosterone (figure 3.1). Female ovaries do not produce nearly as much testosterone as male testes do.[9]

Testosterone has myriad effects on anatomy, physiology, and behavior by itself turning other genes on and off. It is known in the business as a transcription factor because it has the power to induce other genes to make transcripts that will ultimately be translated into proteins. Testosterone belongs to a class of hormones known as steroid hormones. Other members

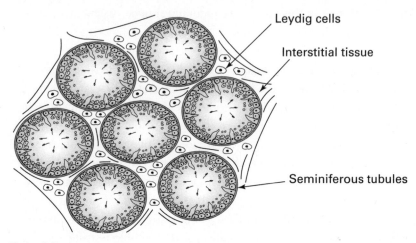

Figure 3.1
Schematic cross-section through the testes, showing the sperm-producing seminiferous tubules surrounded by the testosterone-secreting Leydig cells.

include estrogen and progesterone as well as the stress hormone cortisol.
They are all derived from cholesterol.

The developmental time course of human male testosterone levels provides us with insight into its function (figure 3.2).

The testes begin secreting testosterone early in embryonic development,
around eight weeks after conception. This testosterone masculinizes both

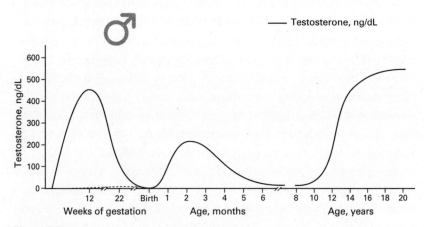

Figure 3.2
Developmental time course of testosterone levels in human males.

the internal reproductive structures and the external genitals. For example, the genital tubercle develops into a penis rather than a clitoris, which is the default in the absence of testosterone. There is also considerable evidence from other animals, as well as some suggestive evidence from humans, that this prenatal testosterone masculinizes the brain insofar as it promotes male-typical mating and levels of aggressive behavior. Following this prenatal surge, levels plummet to close to zero until another smaller surge occurs in the first few months of postnatal life. The function of this secondary surge is less clear, but it may also have to do with changes to the male brain. Then the testes go quiet for quite a long period of development, including both childhood and the juvenile period.[10] During this phase of the human life cycle, the brain grows rapidly, whereas the rest of the body grows slowly, resulting in the familiar phenotype of "big brains on sticks."[11] It looks as if evolution has designed human children to use their big brains to learn how to become a competent adult before actually making that physical transformation. There is a great deal to learn in a cultural species like ours. There are skills to master and norms to learn and internalize. Once all that knowledge is on board, it's time to grow up and adopt an adult role, and that includes seeking and competing for mates. The pubertal surge in testosterone facilitates this transition. In adult males, the most fundamental function of testosterone is to prepare them to succeed in mating: anatomically, physiologically and behaviorally.

Mating is an inherently competitive pursuit, especially for male mammals. Male reproductive success is typically limited by access to female mates, whereas the converse is not true. To illustrate, imagine a female mammal that mates with ten different males over the course of one gestation period. She will give birth to only one offspring, or one litter, as the case may be. In contrast, a male mammal who mates with ten different females could have as many as ten offspring or ten litters. However, if this male has ten offspring with ten different females, this implies that nine other males were not able to find mates and will not have any offspring. Thus, we say there is significant skew, or inequality, in male reproductive success. Since reproductive success is the currency of natural selection, males will evolve to compete for access to female mates, and the intensity of that competition will be in proportion to the amount of reproductive skew.

Male competition can take various forms. In other animals, it most commonly involves direct physical combat. Testosterone prepares males for combat by increasing muscle mass, as well as the number of red blood cells that carry oxygen to those muscles.[12] It also stimulates growth of structures that appear designed to intimidate other males. In human males, for example, pubertal increases in testosterone contribute to growth of the brow ridges and the jaws, as well as deepening of the voice.[13] Testosterone is also responsible for the growth of facial hair. One interesting study found that men from both New Zealand and Samoa rated other men's faces as more aggressive when they had a full beard compared with when they were clean-shaven.[14]

Beyond these anatomical and physiological effects, testosterone also has psychological effects that prepare males for mating and related competition. It stimulates interest in both mating and competition, and it has anxiolytic effects that make it easier to engage in inherently dangerous combat.[15]

Competition for mates does not always involve physical combat. In humans, thankfully, it usually takes other forms. Human males are more likely to compete with one another for social status, which can be achieved through a variety of different means, and testosterone is also involved in status competition.[16] Furthermore, males of many species compete in attracting females, and testosterone sometimes supports growth of the ornaments that have evolved for this purpose.[17]

Competing for mates requires energy, and that energy cannot be used for other life purposes. As introduced in chapter 2, life history theory posits that organisms have a finite amount of energy that must be partitioned among growth, maintenance, parenting effort, and mating effort. Testosterone siphons energy from the former three categories and directs it toward mating.[18] In this chapter, I present evidence in support of this idea from a variety of different species, including humans. The resulting inference will be that paternal caregiving often depends on the suppression of testosterone.

Effects of Testosterone in Birds

We begin with birds. Although they are quite distantly related to humans, we can still learn a great deal from studying them because their mating system is more similar to ours than are the mating systems of most other mammals. Most mammals are nonmonogamous and lack paternal care. Most birds, however, are (relatively) monogamous, and males typically provide parental care.[19]

Many birds are seasonal breeders: they mate during only one part of the year and raise offspring during another. To successfully reproduce, male birds must accomplish several things at the beginning of the breeding season: establish a territory, attract a mate, mate, defend their territory, and prevent other males from mating with their female partner, also known as mate guarding. Testosterone levels increase significantly at the onset of the breeding season, suggesting that it might promote these behaviors.[20] In support of this possibility, one study showed that male European stonechats with higher testosterone levels during this mating phase of the breeding season ultimately fledge more offspring than other males with lower testosterone.[21] However, you've heard the old adage: correlation does not imply causation. Instead, it could be that some other variable is correlated with both testosterone levels and mating success and is causally related to both. Or it could be that mating behaviors increase testosterone levels rather than the converse. So to establish causality, we need to experimentally manipulate testosterone levels and observe if doing so has an impact on mating behaviors. Many such studies have been done in birds, and their findings are quite revealing.

In what are known as testosterone implantation studies, a capsule with testosterone is implanted under the skin early in the breeding season, and this capsule slowly releases testosterone so that levels remain high throughout the entire season. Males must fight and defeat other males in order to establish a territory. In sparrows, testosterone implantation increased territorial aggression so much that normally monogamous males expanded their territory to the point that they could accommodate more than a single female, and they became polygynous, meaning that a male has multiple female mating partners.[22] Take that in for a minute: normally monogamous males become polygynous when given extra testosterone. It looks like testosterone is motivating males to invest more effort in mating.

After establishing a territory, a male must attract a female. Typically, he does this by singing and through bodily ornaments that females find attractive. We know that testosterone supports male birdsong because removing testosterone by castration decreases it and subsequent injections of testosterone reinstate it.[23] Furthermore, testosterone implants have been shown to increase the amount and quality of birdsong in many species.[24] Testosterone also stimulates growth of ornaments such as combs, the colorful fleshy growths or crests that birds sport on their heads. Red grouse, for example, have red combs over their eye, and females prefer males with larger combs as mates. Implanting male red grouse with testosterone increases their comb size, and males with larger combs have greater breeding success.[25]

I mentioned that birds are mostly, but not completely, monogamous. There can be a fair amount of extra-pair paternity in which young are not fathered by the female's partner. Males seek opportunities to mate with females paired with other males in nearby territories, and vice versa. One long-term study of dark-eyed juncos showed that implanting males with testosterone improved their success in these pursuits, increasing the number of extra-pair offspring they sired. Thus, extra-pair mating is another form of mating effort supported by testosterone.[26]

So testosterone clearly helps male birds succeed in mating. However, life history theory predicts that it should also impose costs on males since the energy used for mating must be diverted from some other life enterprise. Specifically, the theory predicts that testosterone should interfere with male survival (maintenance), as well as male parenting.

Let's start with survival. Testosterone implantation has been shown to decrease male survival in both dark-eyed juncos and cowbirds. In cowbirds, thirty-three of eighty-one untreated males survived from one breeding season to the next. However, when a separate group of males were implanted with testosterone, only one of sixteen showed up the following year. Implanted cowbirds have more severe injuries than other birds, so they are presumably getting in more fights, and this might help to explain their increased mortality.[27]

Testosterone may also kill male birds by suppressing immunity and increasing their vulnerability to infection. Indeed, testosterone-implanted red grouse males exhibit reduced immune function, as well as increased levels of parasitic infection. These males also have decreased body weight, suggesting that they have difficulty maintaining their overall health or condition.[28] Similarly, testosterone implantation depletes fat stores in dark-eyed juncos.[29] Yes, testosterone has these effects in human males too. That's great if you happen to live in a society where one of the main challenges we face is, paradoxically, avoiding obesity. However, if you are a dark-eyed junco, that body fat helps you avoid starvation when food is scarce, such as after a snowstorm.

Not only does testosterone make male birds more likely to get in dangerous fights, burn through precious body fat reserves, and disinvest in pathogen defense; it also seems to accelerate senescence. Telomeres are structures found at the ends of chromosomes that shorten each time a cell divides. Eventually they shorten to the point where the cell can no longer divide and becomes senescent. Telomeres shorten with age, and telomere length is predictive of longevity. Therefore, telomere length has been used as a measure of biological age. This is distinct from chronological age because biological age can be affected by a wide range of lifestyle factors. For example, psychological stress has been associated with shorter telomere length.[30] (I like to tell my colleagues that my telomeres shortened dramatically during my three-year term as department chair.). When dark-eyed junco males were treated with testosterone, their telomeres shortened at a faster rate with age than did the telomeres of control males, suggesting that they were senescing more rapidly.[31]

Another reason that organisms senesce is that their metabolism is constantly producing molecules called *free radicals*, which are atoms with unpaired electrons that are highly unstable and reactive. These atoms aggressively try to steal electrons from other atoms, damaging the latter in the process. As a defense against free radicals, many organisms produce molecules known as antioxidants that are able to scavenge free radicals, rendering them harmless. This defense system is not perfect, however, and free radicals are a major cause of damage that accumulates with age. A study in zebra finches showed that implanting males with testosterone decreased their cells' resistance to free radical attacks and therefore would presumably accelerate aging.[32]

So it is safe to say that testosterone interferes with male birds' investments in maintenance and survival.

In most birds, testosterone levels increase at the beginning of the breeding season when males acquire territories and pair with females. Afterward, levels typically decrease to a lower baseline for the remainder of the breeding season. The exception is polygynous birds, whose levels are more likely to remain elevated throughout the breeding season. Polygynous male birds, in contrast to their monogamous counterparts, show little to no parental

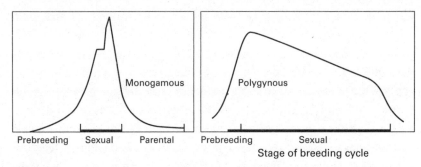

Prebreeding   Sexual   Parental   Prebreeding   Sexual

Stage of breeding cycle

Figure 3.3
Seasonal changes in testosterone levels in monogamous (left) and polygynous (right) birds.

care. Thus, it may be that the breeding season decline in testosterone in monogamous male birds supports their transition to parenting. This would imply that high levels of testosterone normally interfere with paternal caregiving. Once again, we turn to testosterone implantation studies to critically evaluate this hypothesis.

In support of our hypothesis, testosterone-treated dark-eyed juncos were found to feed their young less often than did control males that received implants without testosterone.[33] Similar findings were observed in pied flycatchers and house sparrows.[34] In a study with male black redstarts, rather than implanting males with exogenous testosterone, researchers stimulated birds to produce a short-term increase in testosterone, and this also suppressed feeding behavior.[35] In house sparrows, it was even shown that injecting a drug that blocks the action of testosterone increased paternal feeding rates above those of control males.[36] So it seems quite clear that testosterone interferes with male parental care in birds.

Diminished parenting by T-treated males also had negative consequences for their reproductive success. Offspring of T-implanted house sparrow males were more likely to starve, and T-implanted pied flycatchers produced fewer fledglings. The normal suppression of testosterone in male birds during the parenting phase therefore seems a clear adaptation to support paternal feeding that will increase offspring survival. That is, suppression of testosterone increases male reproductive success.

Before leaving the topic of testosterone effects on life history strategies in birds, let us consider the fascinating case of the white-throated sparrow. In this species, males come in two color morphs: one morph has white stripes on their heads, and the other has tan stripes. The two male morphs adopt quite distinct life history strategies. White-striped males allocate more energy to mating and less to parenting. They display more aggression than tan-striped males when other males intrude on their territory. They also feed their offspring at lower rates than tan-striped males, consistent with the notion of a trade-off between mating and parenting effort. White-striped males also have larger testes than tan-striped males. Large testes evolve for sperm competition, and these are likely an adaptation to support the penchant of white-striped males to engage in more extra-pair copulation compared with tan-striped males. Tan-striped males then are better parents and better partners insofar as they are less likely to stray. The flip side is that they guard their mates more intensively, likely because they have all their eggs in one basket so to speak. That is, since they are more monogamous, it is adaptive for them to attempt to ensure paternity certainty through limiting access of other males to their mate.[37] As we expect by now, white-striped males have higher testosterone levels than tan-striped males. Furthermore, among white-striped males, those with higher testosterone sing more in response to territorial intrusion, a form of territorial aggression, and they feed their offspring less.[38]

Thus, in birds, testosterone biases male life history strategies toward mating and away from parenting. Birds are a great model for the human mating system, but they are not fellow mammals. Does testosterone have similar effects in mammals?

Effects of Testosterone in Rodents

Similar to male birds, testosterone prepares male rodents for mating and competition for mates. Male mice increase their testosterone levels in the presence of a strange female.[39] Similarly, they release testosterone when they smell a novel, receptive female.[40] Testosterone seems to be rising in anticipation of a potential mating opportunity. That increase in turn motivates mating effort since injecting male mice with extra testosterone makes them mount a receptive female faster than normal.[41] In another experiment, male mice were castrated to remove all their testosterone but were then injected with either high or low amounts of testosterone. Researchers then paired one high-testosterone and one low-testosterone male in an arena and observed which one emerged as dominant (basically, who won the fight). With just the two males in the arena, high-T males were no more likely to win than low-T males. However, when researchers added a female to the arena, high-T males more often prevailed.[42] This demonstrates an important principle: testosterone does not always increase aggression, but may do so in specific contexts such as when males are competing for mates.

With some justification, many laypeople associate testosterone with antisocial behavior, so what I am about to tell you will not help its reputation: in animals, testosterone can make males more likely to kill infants. It's not intuitive, but infanticide is really another form of mating effort. When male mammals kill infants, they don't normally kill their own infants. They kill the infants of other males. By doing so, they prevent the mother from nursing those offspring, and nursing is a natural contraceptive in most mammals. As a result, infanticide returns the mother to a reproductive state sooner and provides the killer with a potential reproductive opportunity. Sadly, the most adaptive female response to this troubling male adaptation is for her to start over by mating with the killer of her offspring and hope he can better defend her young.[43] So in this sense, male infanticide is a form of mating effort.

In virgin male house mice, about half of all males are infanticidal and half are parental. However, castration transforms the infanticidal males, and instead of killing pups, they begin licking and grooming them as a parent would. If castrated males are given testosterone replacement, about half again become infanticidal. Remarkably, after mating with a female, all of the males that were not infanticidal become so, up until the time when their own pups are born three weeks later. Thus, they are likely to kill any unrelated pups but not their own. This transition to infanticide after mating has been shown to depend on testosterone released at ejaculation.[44] Testosterone also seems to promote infanticide in rats since implanting males with testosterone increases the number of males exhibiting infanticide.[45] Testosterone is not the only hormone involved with male infanticide. For example, progesterone also increases infanticide in male house mice.[46]

Nevertheless, testosterone appears to support mating effort in rodents. What about maintenance and parenting? Does testosterone interfere with investment in these, as it does in birds?

Castration has been shown to increase longevity in both rats and mice, which implies that testosterone increases mortality.[47] But how exactly does it do this? As in birds, at least part of the explanation is that it increases vulnerability to infection. Rodents are vulnerable to tick infestations. Ticks are more abundant in adult males than in immature males, suggesting that immunosuppressive effects of higher adult levels of testosterone could be involved. Indeed, castrated adult males are less susceptible to tick infestation than castrated males with testosterone replaced. Tick larvae are more likely to attach to the skin of testosterone-treated males, and the ticks on these males had greater engorged weight, implying that they were feeding more effectively on their host.[48]

The role of testosterone in mediating trade-offs between mating and maintenance is illustrated quite dramatically by the marsupial mouse, which is not technically a rodent even though it looks like one. Marsupial mice have a single one- to two-week frenzied mating period during which males engage in 8- to 12-hour-long mating sessions with as many females as possible, in rapid succession. By the end of this mating period, all of the males die. Only castrated males have been known to survive the mating season, again implicating testosterone in male mortality. In fact, testosterone levels increase eight-fold during this mating period, and males become highly aggressive, invading the territories of other males. The intensity of the competition among males produces a dramatic increase in the stress hormone cortisol, which has well established anti-inflammatory and immunosuppressive effects. The testosterone binds to a protein that normally carries cortisol in the blood. In the process, cortisol is freed from that protein, rendering it more available to bind to cortisol receptors on tissues throughout the body.[49] Effectively, testosterone increases the impact of massively elevated cortisol levels. And ultimately, it is cortisol that kills the males by increasing susceptibility to infection and internal bleeding. In essence, male marsupial mice trade their lives for a single frantic orgy of mating.

In addition to compromising survival, we also expect testosterone to interfere with parental caregiving in male rodents. Male rats and marsupial mice, like males in most rodent species, do not provide parental care. So in order to test this prediction, we must turn to other species where males do habitually care for their young.

One such species is the Mongolian gerbil, a small, burrowing rodent that lives on the steppes of China, Mongolia, and Russia. Similar to birds, they are mostly monogamous and biparental. Also like birds, testosterone decreases significantly in fathers after the birth of pups, and this may prepare them to parent.[50] There is tremendous variation in testosterone levels among adult male Mongolian gerbils, more than ten-fold. The intrauterine environment that males experienced turns out to be an important factor in determining how much testosterone they produce as adults. Offspring are typically born as triplets. Males that gestate in between two females (2F males) have about half as much testosterone as adults compared with males that gestate between two other males (2M males). There are some intriguing behavioral differences between 2F and 2M males. 2F males spend more time with pups and are less sexually active than 2M males. 2F males are also less likely to impregnate unfamiliar female hamsters with which they are housed compared with 2M males. Remarkably, there is a subset of 2F males, 20% to 25%, that have even lower testosterone than other 2F males and are apparently completely asexual insofar as they will not mount estrus females. These asexual 2F males are also more likely than other 2F males to assume a brooding posture over pups and to remain with the pups when the mother is away from the nest.[51] These data fit beautifully with the idea that testosterone biases males toward mating and away from parenting. 2F males, with lower testosterone, invest more in parenting and less in mating compared with high-testosterone 2M males. However, once again, the findings are correlational, and we can ask whether experimental manipulation of testosterone has effects in the expected direction. Indeed, compared with males that have normal testosterone levels, castrated male gerbils are more likely to be in contact with their pups, to huddle over their pups, to lick their pups, and to remain with their pups when their mate is away from the nest. Furthermore, they are less likely to mount their mate.[52]

The California mouse is another monogamous, biparental rodent species in which males care for their young. As we have now come to expect, fathers have lower testosterone levels than adult males who are not fathers.[53] However, in stark contrast to Mongolian gerbils and contrary to expectations, castration actually *decreases* paternal caregiving in male California mice. Moreover, replacing testosterone in castrated males reestablishes paternal caregiving.[54] How is this possible given all of the opposing findings from other species already discussed? To understand this counterintuitive finding, we need to make a distinction between hormone levels in the body and hormone levels in the brain. To have an impact on behavior, hormones must bind to their receptors in the brain. What really matters when considering paternal behavior are hormone levels in the brain, but these cannot be easily measured in a living animal. Instead, we normally measure hormone levels in the body and assume those tell us something about what is going on in the brain. For steroid hormones like testosterone, that is a reasonable assumption since they are able to cross the blood-brain barrier and access the brain. However, some brain areas have an enzyme that converts testosterone into another hormone that you are familiar with, estrogen.[55] That enzyme is known as aromatase, and when California mouse males become fathers, their aromatase levels increase in a brain area that is critically involved in paternal caregiving.[56] (I discuss that region, the medial preoptic area [MPOA], in more detail in chapter 5.) Due to the increase in aromatase, MPOA estrogen levels increase and MPOA testosterone levels decrease. Other experiments have shown convincingly that paternal behavior in California mice depends on estrogen in the MPOA, but paradoxically, the source of that estrogen is testosterone, and this is why castrated males show less paternal care.[57] So just like other species, fatherhood in California mice involves a decrease in T, particularly in the brain, where it really matters.

Effects of Testosterone in Red Deer

Males of some mammalian species exhibit dramatic increases in testosterone during the breeding season, so it will come as no surprise that these species lack paternal care. In male red deer, testosterone levels increase more than twenty-fold during the fall rut, which prepares males to compete for access to groups of female mates. Males initially roar at each other to assess who is bigger and stronger, and this can escalate to physical combat if neither one is clearly superior at bellowing. How does testosterone help males in these competitions? First, males with higher testosterone have stronger antlers that are less likely to break during combat. That's important because broken antlers make the deer lose fights, as well as its place in the dominance hierarchy. Second, deer need a strong neck to be good at antler fighting, and testosterone makes male neck muscles grow thicker. Testosterone also makes males more competitive. Castrated males become less aggressive and roar less, and testosterone replacement reinstates both.[58] The rut is extremely physically demanding. Males travel large distances in search of females, mate, scent mark, and aggressively defend access to females, all while eating next to nothing over about six weeks. This results in marked depletion of body fat. They need stamina during this arduous period, and this is supported by a testosterone-fueled increase in red blood cells that allows males to better oxygenate their muscles. Remember how all those Tour de France riders were taking erythropoietin (EPO) to enhance their performance? This is the same principle. Higher testosterone is also associated with higher-quality sperm, known to be linked with male fertility in red deer.[59]

Stags do not maintain these high testosterone levels year round. Instead, after the breeding season ends, testes shrink, testosterone plummets, antlers are shed, and neck muscles atrophy. But why not just keep testosterone high all the time? Because that has costs. In addition to the metabolic costs that lead to body fat depletion, male red deer with higher testosterone have a higher parasite load.[60] Breeding season levels of testosterone are not sustainable if males are going to be healthy enough to partake again in the next year's mating season.

Effects of Testosterone in Nonhuman Primates

In considering our own primate order, let us begin with our closest living relative, the chimpanzee. Remember that chimps have very large testes for their body size, which suggests heavy investment in mating and sperm competition. As life history theory would predict, chimpanzee fathers provide little by way of paternal care, although they do seem to offer their offspring some protection from infanticide.[61]

Chimpanzees live in fission-fusion communities, where subgroups of animals break off from the rest of the group and form smaller parties that travel throughout their territory. These parties are fluid and dynamic, changing in composition over time. When male chimpanzees are traveling in parties that include parous, estrus females, their testosterone levels are elevated compared to when traveling in parties without such females. A parous female is one who has previously given birth, and an estrus female is one who is sexually receptive and near the time of ovulation when she can conceive. In female chimpanzees, this is advertised by a highly conspicuous sexually swelling on her hind end that reaches maximal tumescence around the time of ovulation. Parous females may be more attractive to males because of their demonstrated fecundity, as opposed to nulliparous females who often go through an initial transient period of infertility before giving birth for the first time. When males are in the presence of experienced mothers with pronounced sexual swellings, their testosterone levels rise. At the same time, they become more aggressive in competing with other males to mate with those females.[62] This is only a correlation, but it is consistent with the possibility that the increases in testosterone facilitate increased aggression in this mating situation.

Primates typically form dominance hierarchies in which higher-ranking animals have preferential access to limited resources such as food and mates. Much of male primate mating competition occurs indirectly through competition for rank within these hierarchies. If a male can establish his dominance over other males and achieve high rank, other benefits, including access to female mates, often follow. In fact, high-ranking male chimpanzees sire more offspring.[63] Does testosterone help male chimps achieve higher rank in the dominance hierarchy? Probably. Two studies have found that male dominance rank is correlated with testosterone levels, although another found no correlation.[64] One possible explanation for the inconsistency is the stability of the dominance hierarchy. In a stable hierarchy, each animal knows its position in the hierarchy and has at least temporarily accepted that station. In this situation, there is no need for high testosterone because there is no need to fight. We know well enough by now that high testosterone typically costs animals in terms of maintenance and survival, so it makes sense to keep levels lower when the hierarchy is stable. When the hierarchy is unstable and animals have opportunities to rise and fall in rank, levels of aggression increase, and higher testosterone is likely to be advantageous. This is exactly the pattern seen in another primate species, olive baboons.[65] Dominant males have higher testosterone than other males only when the hierarchy is unstable and they have to defend their rank. Higher testosterone probably increases motivation to compete for status. The other important point to remember when discussing correlations between testosterone and dominance like these is that aggression can also cause changes in testosterone. Especially when animals win competitions, testosterone tends to increase.[66] So the causal arrow likely operates in both directions.

Testosterone has some other interesting associations in male chimpanzees that are consistent with the hypothesis that it increases mating effort. Males with higher testosterone have more muscle mass, and muscle benefits males when fighting to establish dominance.[67] Higher-testosterone males also produce more "pant-hoot" vocalizations, which may function to advertise their social status since high-ranking males produce more pant-hoots than low-ranking males. Not only do high-T males make more of these vocalizations, they make pant-hoots that are acoustically different.[68]

As mentioned in chapter 2, chimpanzees are territorial and males from the same community often work together to collectively patrol the border of the territory. In doing so, they not only attempt to defend their own territory but actively seek opportunities to expand it. Interactions among chimpanzees from different territorial communities are hostile and sometimes lethal. When males expand their territory, that may allow more females to occupy that territory and may also increase the feeding success of individual females within the territory. We can think of chimpanzee territorial aggression as another form of mating effort. Male testosterone levels are elevated when they are patrolling their territorial borders, so this may be yet another way in which testosterone is supporting mating effort.[69]

As in birds and other mammals, testosterone also appears to have costs in male chimpanzees. Illness is the leading cause of death in chimpanzees.[70] Males with higher testosterone levels have more intestinal parasites, suggesting that testosterone may also suppress immunity in chimpanzees and increase vulnerability to disease and associated mortality.[71] The second leading cause of death is aggression, and testosterone seems likely to fuel some of this aggression in chimpanzees. A final piece of evidence that testosterone interferes with survival is the existence of sex differences in chimpanzee mortality. Before puberty, males and females have similar survival rates, but after males start producing adult levels of testosterone, male mortality becomes greater.[72]

Like most other primates, chimpanzee males do not provide significant paternal care, so we can't examine the effects of testosterone on paternal caregiving in chimps. However, there are a number of New World monkey species where males do care for their offspring, and we can ask how testosterone affects male life history strategies in those species.

Marmoset monkeys are one of these paternally investing species. Fathers do most of the infant carrying in this species. They also groom, protect, and share food with their infants. Experiments show that marmoset fathers are more motivated to respond to infants than nonfathers are.[73] This increased paternal motivation has some revealing hormonal correlates.

When captive male marmosets are presented with the scent of an ovulating female, testosterone levels increase. This is similar to chimpanzees and many of the other species we have discussed, but there is an important caveat: this increase occurs in single or paired males who are not fathers but is notably absent in fathers.[74] The lack of a testosterone surge in fathers may help them avoid the temptation to stray from their pair-bond and shirk their parental responsibilities. Moreover, when presented with the scent of their own infant, testosterone decreases within twenty minutes in marmoset fathers.[75] However, this hormonal response is lacking in both fathers and nonfathers exposed to the scent of an unfamiliar infant.[76] The decline in testosterone may well promote infant care since testosterone levels are negatively correlated with infant carrying in male marmosets. Baseline testosterone levels decrease when males are carrying infants at the highest rate around three to four weeks after birth, and males with lower testosterone carry their infants more than males with higher testosterone. do [77]

Despite the apparent benefits of low testosterone for paternal care, some marmoset fathers respond aggressively to males who intrude on their territory, and these males show an increase in testosterone in response to such threats.[78] This may be male mate guarding since extra-pair mating has been observed during intergroup encounters like these. In general, these marmoset studies teach us that testosterone levels can be quite dynamic in males, flexibly increasing and decreasing as a function of the demands of the current context. Infant stimuli drive testosterone levels down in fathers within thirty minutes, yet paternal testosterone increases in response to mating competition in the form of male intruders. We will find that this dynamic, flexible hormonal response is also characteristic of our own species.

Effects of Testosterone in Humans

In at least some circumstances, men respond hormonally to women in much the same way that males of other species respond hormonally to females of their respective species. One study showed that testosterone levels of young men increase within twenty minutes of having a brief interaction with a young woman but not in response to a brief interaction with another man.[79] This response was more pronounced when women judged the men to be high in extraversion and self-disclosure during these interactions. A smaller study by the same research group found larger increases in men's testosterone when the women they were interacting with felt that the man was trying to impress them.[80] Consistent with this finding, male skateboarders showed elevated testosterone when they were being watched by an attractive female versus a male, and they also aborted fewer difficult tricks, which resulted in both more successful tricks and more crash landings.[81] Another study found that male competitors during an ultimate Frisbee tournament experienced larger increases in testosterone when there was a higher ratio of women to men present.[82] Yet another quite remarkable study showed that men who were exposed to the scent of an ovulating woman, through smelling her T-shirt, maintained higher testosterone levels than did men who smelled the T-shirts of women who were not ovulating.[83]

But these studies were all done with college students, most of whom are unmarried nonparents between the ages of 18 and 22. Many of us probably feel that our psychology and our biology have changed a bit since college, so we can reasonably ask whether the physiology works in the same way in partnered fathers. In fact, marmoset fathers, in contrast to nonfather marmosets, did not show an increase in testosterone in response to the scent of an ovulating female. To my knowledge, the equivalent experiment has not been done in human fathers; however, there is evidence that this hormonal response is not automatic or uniform in men. One study of men from the Caribbean island nation of Dominica showed that testosterone levels were elevated when men were interacting with young women who were potential mates, but not when interacting with female relatives or with females who were in relationships with close friends.[84] In an elegant illustration of life history strategy trade-offs, another study found that men who are more interested in babies have a smaller testosterone increase to viewing erotic stimuli compared with men who are less interested in babies.[85] So men's testosterone response to mating stimuli may be a flexible indicator of their motivation to pursue mating opportunities.[86]

Several other studies also suggest a role for testosterone in human male mating effort.[87] For example, men who report more lifetime sexual partners have higher testosterone.[88] In a sample of middle-aged men who had been married an average of twenty-two years, those who had engaged in an extramarital affair had higher testosterone than those who had not, and testosterone levels were positively correlated with the overall number of extramarital affairs men had.[89] Extramarital affairs are a common cause of divorce, so it comes then as no surprise that men with higher testosterone levels are more likely to have experienced divorce at some time in their life.[90] Many of those men remarry, so it may also not be surprising that married men who remarried had higher T than other married men.[91]

Single men tend to have higher testosterone than men in committed relationships, perhaps because they are looking for mates.[92] In fact, one longitudinal study of men from the Philippines showed that young, single men who had higher testosterone at the beginning of the study were more likely to become partnered fathers four years later.[93] That is, high testosterone predicted their future mating success.

After helping motivate a man to find a mate and perhaps enter into a committed relationship, testosterone may also be involved in defending that relationship from real or imagined threats from other men. In other words, it may be involved in what behavioral ecologists call *mate guarding*. In one rather extraordinary study of heterosexual romantic couples on a college campus, each member of the couple came to the lab and viewed a series of profiles of other men who, they were told, attended the same university. Each profile included a photograph and a one-paragraph purported self-description of these potential "rivals." Men were told that their female partner would rate each of these men for attractiveness. Some men viewed ten high-competitive profiles, consisting of men previously rated to be attractive and described as highly competitive. Other men viewed ten low-competitive profiles. In response to viewing the high-competitive profiles only, men's testosterone levels increased when their female partner was near ovulation as compared with less fertile times in her cycle.[94] It was as if men were hormonally preparing for competition at the specific time when their partners were most fertile, when mate guarding would be most needed.

In addition to motivating the pursuit and defense of mates, testosterone may also make men more attractive to women. I know many of my female readers are shaking their heads. Well, let me qualify that statement by saying it's probably true for only some women and only in certain circumstances. In another study of college students, young women rated photographs of young men who had higher testosterone to be more attractive "as a short-term romantic partner" compared with men with lower testosterone. Men with higher testosterone were judged by these women to have more masculine-looking faces, and this presumably made those faces more attractive to women as potential short-term partners. When instead evaluating photos for attractiveness as "long-term romantic partners"—like someone you want to marry and have a family with—there was no correlation with testosterone levels. Instead, it was a perceived "interest in children" that was positively associated with attractiveness ratings. Can you really discern interest in children from looking at someone's photograph? Remarkably, women's ratings of men's interest in infants, based on their photographs, was indeed positively correlated with men's actual self-reported interest in infants. Although men were not smiling in their photos, it seemed that something about the positive affect they expressed communicated their actual and inferred interest in infants.[95] This study adds credence to the suggestion in chapter 2 that ancestral women may have shaped men by selectively mating with those who were willing to help care for and provide for children.

Nevertheless, in short-term mating situations, higher-testosterone men may be viewed as more attractive and have an evolutionary advantage. Similar results have been found for male voices. Men with higher testosterone levels tend to have lower-pitched voices, and women tend to find such voices more attractive, especially when evaluating men as short-term romantic partners.[96] Why might women prefer higher-testosterone men over lower-testosterone men as short-term mates? One theoretical explanation is that men with masculine traits, supported by high testosterone, tend to have better genes because they can better withstand the immunological handicap that testosterone imposes. Testosterone decreases when men get sick or after they mount an immune response to vaccination, presumably because energy is most urgently needed for the immune system to deal with its present challenge and testosterone would draw energy away from that.[97] Men who are able to maintain higher testosterone are therefore expected to be healthier. In fact, some studies find that men with more masculine-looking faces tend to be healthier.[98] There is even some evidence that women find men with stronger immunity to be more attractive. A recent study among university students in South Africa measured men's immune response and found that the strength of this response was positively correlated with how attractive women judged their photograph to be.[99] By preferring and mating with such men, a woman's children would stand to inherit these good genes. In essence, a short-term mating strategy would prioritize genetic inheritance over paternal investment. None of this logic is expected to necessarily be conscious in the minds of women; it merely explains how the preference evolved. However, "good genes" is not the only potential evolutionary explanation for why women have evolved a preference for higher-testosterone men in short-term mating situations. For example, such men might be better able to offer immediate physical protection in situations where that is needed.[100]

As with other primates, another way in which human males compete for mates is indirect, by way of competing for social status, and testosterone is also involved in human male status competition. In one fascinating experiment of young American men, participants were insulted by a confederate who bumped into them and then cursed at them. Men from the American South, traditionally characterized by a "culture of honor," were more likely to think their masculine reputation was threatened by this insult as compared with men from the North. Southerners were also more likely to respond to the insult with aggressive or dominant behavior, and they experienced a larger increase in testosterone in response to the insult.[101] Thus, increases in testosterone may have been fueling an attempt to reestablish status or reputation in Southern men.

Athletic competitions are another form of status competition. In fact, sports may attract some of the most competitive people in society, and we all know that successful athletes can achieve very high status. Testosterone increases as athletes warm up in anticipation of competing and then remains elevated throughout the competition. Some studies also find that testosterone remains elevated longer after winning than after losing.[102] This might function to motivate subsequent competition after a victory that would potentially allow the winner to achieve even higher status. In one experimental study in the laboratory, men played a competitive game against another man but the game was rigged such that the experimenters determined who won and lost. Among losers, some men showed an increase in testosterone across the contest, whereas others showed a decrease. The men were then given the opportunity to again compete against the opponent who just beat them, and those whose testosterone increased were more likely to opt for a rematch.[103] This effect of testosterone on competitive motivation may help to explain some associations between baseline testosterone levels and social status. For example, executives with authority over more subordinates have higher testosterone levels compared with other executives. In this study, it was more precisely the combination of high testosterone and low levels of the stress hormone cortisol that predicted the number of subordinates.[104] Cortisol can interfere with the effects of testosterone, so the pursuit of social status may be dependent on high levels of testosterone combined with low levels of cortisol.[105]

In summary, testosterone seems to motivate men to pursue and defend their social status. However, the path by which men achieve social status can vary across cultures, and even among subcultures or social classes within a society. If aggression is a path to status, then testosterone will be associated with aggression. For example, we all know that hockey players fight a lot. One study showed that hockey players who are more likely to respond to provocation with aggression have higher testosterone levels than other hockey players.[106] Among the !Kung San people of Botswana, men with more scars on their head, resulting from fights, had higher testosterone levels.[107] Thankfully, prosocial behaviors can also raise social status in humans. For example, in many cultures, generosity can lead to social status, and testosterone has also been shown to increase generosity.[108]

So it seems that testosterone motivates the pursuit of mates and social status and that it can make men more attractive to women as short-term mating partners. These pursuits require a lot of energy. Life history theory, as well as the animal research, therefore predicts that the transition to fatherhood among human males will involve a decrease in testosterone to free up energy for parental care. We now know that this is indeed what happens, at least when fathers become involved with their infants.[109] One large longitudinal study of Filipino men was particularly revealing. Over the course of this four-year study, single men who became fathers experienced a decrease in testosterone, whereas men who stayed single did not. Furthermore, new fathers who spent more time on child care had lower testosterone levels than less involved fathers.[110] Note the similarity with the pattern we discussed in birds. Testosterone increases at the beginning of the breeding season when males are seeking mates, but then decreases when males become fathers. It was also the case that male birds with the highest testosterone at the beginning of the breeding season fledged the most offspring, just as single Filipino men with higher testosterone were more likely to become partnered fathers over the course of the study. Another striking parallel is that polygynous men, who presumably engage in less direct caregiving, have higher testosterone than monogamous men, just as polygynous birds sustain high levels of testosterone throughout the breeding season.[111]

This decline in testosterone that men experience when they become a father is believed to somehow support parental caregiving. How might it do so? First, it may curtail their motivation to pursue mates and social status, leaving more energy available for parenting. In addition, we know that testosterone interferes with several psychological processes that are important when tending to infants. For example, testosterone can inhibit empathy. People tend to automatically mimic facial expressions made by others, and that mimicry can be measured using electromyography (EMG), a technique that records subtle activity in muscles of the face. People who automatically mimic to a greater degree tend to be more empathic. One study showed that treating women with testosterone decreased their facial mimicry compared with when they received a placebo, suggesting that it was interfering with empathy.[112] Testosterone can also impair impulse control and frustration tolerance, both of which are crucial when tending to a crying, or, as was the case in my family, a screaming infant.[113] Frustration induced by infant crying is a well-known trigger for infant abuse, and fathers are the most common perpetrators. As an infant, my son cried inconsolably, and I remember feeling tortured by it. *We've tried everything he could possibly need. Why on earth is he still crying? Am I a bad parent? Is he rejecting me? Why can't we fix this? How will I function at work tomorrow after yet another stressful night without sleep?*

So infant crying can definitely be a challenge for fathers, and mothers, and there is reason to suspect that testosterone increases the likelihood that men will lose their temper with their crying infant and in rare cases do something awful that they will regret for the rest of their life. Consistent with this idea, one study found that men with higher testosterone reported less sympathy for a crying infant compared to men with lower testosterone.[114] Hence, the decline in paternal testosterone is likely adaptive insofar as it decreases the likelihood that fathers will harm their infants. Still, that decline is not sufficient to eliminate the sex difference in infant abuse.

Fathers don't have conscious control over their testosterone levels. They can't just say to their testes, "I have a child now, so please stop making so much testosterone." How does it happen that testosterone decreases? What is the signal? One clue is that new fathers who spend more time with their infant have lower testosterone levels, suggesting that cues from the infant may be important. This idea is reinforced by a revealing study that compared testosterone levels between fathers in two neighboring African societies that differ in their normative levels of paternal care. The Hadza people are a traditional foraging population in which men are highly involved in carrying, holding, cleaning, feeding, and pacifying infants, whereas the Datoga are a society in which fathers have very little contact with their infant and infant care is considered women's work. Whereas Hadza fathers had lower testosterone levels than nonfathers, there was no such difference in testosterone between fathers and nonfathers among the Datoga.[115] Presumably the lack of infant contact among Datoga fathers resulted in their maintaining levels found in nonfathers. Further evidence that direct contact with infants suppresses testosterone is provided by a study that examined the effect of father–infant cosleeping. Filipino men who slept together with their infant at night had lower testosterone levels than fathers who slept in a separate room. Cosleeping fathers also experienced a more pronounced decrease in testosterone after becoming fathers compared with other men.[116]

But how does contact with infants signal the testes to decrease production? Testosterone secretion by the testes is ultimately controlled by the brain. A brain area known as the hypothalamus secretes a hormone (GnRH) that causes the anterior pituitary gland to secrete another hormone (LH), and the latter stimulates the testes to make testosterone. What is it about the infant that makes the hypothalamus decrease GnRH secretion? It could be something about their appearance. As the early ethologists pointed out, infant mammals have a unique appearance—for example, big eyes, big foreheads, short, stubby limbs—that is designed to attract parents and elicit parental care (figure 3.4).

Or it could be something about the scent of the infant. In general, primates are more reliant on visual than olfactory cues, but the latter are certainly relevant. Remember how the scent of an ovulating woman was able to increase testosterone levels in men? The scent of women's tears actually has the opposite effect, decreasing men's testosterone.[117] So it seems plausible that the scent of an infant might decrease paternal testosterone, but this has not yet been tested as far as I know.

Some men are more involved and more nurturing fathers than others. When we examine this variation in paternal caregiving among men, we also find interesting correlations with testosterone that make sense in the light of life history theory. Fathers with lower testosterone levels have been shown to look at their infants more, to affectionately touch them more, to provide more instrumental care for them, and to direct more "baby talk" at them.[118]

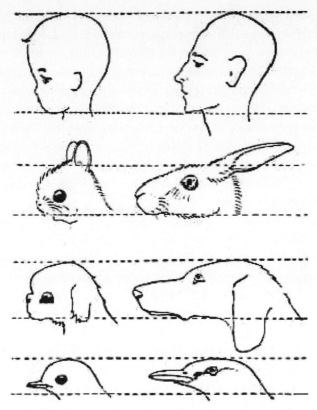

Figure 3.4
Infant animals have a characteristic physical appearance, known as "baby schema," that makes adults find them attractive and want to care for them. Source: K. Lorenz, *Studies in Animal and Human Behaviour*, volume II (Cambridge, MA: Harvard University Press, 1971), 155.

In addition to making men better caregivers, there is evidence that the decline in testosterone makes them better husbands. One study tracked men's salivary testosterone levels across their partner's pregnancy and found that it decreased by an average of 17% before they had ever even seen their infant. This suggests that there might be olfactory cues from the pregnant mother that can decrease paternal testosterone. More to the point, men who experienced a larger decline in testosterone across the pregnancy later reported greater investment, commitment, and satisfaction with their partner three to four months after the baby was born. Not only that, but mothers reported higher postpartum investment and satisfaction with fathers who showed larger declines in testosterone across pregnancy.[119] These effects of testosterone on marital quality may be mediated by its effects on men's behavior since fathers with larger declines in testosterone were more involved in household chores, as well as care of male (but not female) infants.[120] Another fascinating study measured testosterone levels in fathers of nine-month-old infants and showed that their female partners reported higher marital satisfaction and were less likely to be depressed when the father had lower testosterone.[121] The family is a system and children benefit when their parents get along, so decreases in testosterone may benefit infants both directly through the father and indirectly through their parents' relationship.

The magnitude of the decrease in testosterone that new fathers experience is significant but not dramatic. For example, in the longitudinal study of Filipino men already discussed, morning salivary testosterone levels decreased from an average of 207 picograms per millimeter (pg/ml) before fatherhood to 153 pg/ml after the birth of their infant, for a decrease of about 26%. But these levels in new fathers are still way above what is found in women or prepubertal boys. So why don't human fathers decrease their testosterone further, even all the way to zero? We don't know the answer, but we can speculate.

A cynical possibility is that men maintain levels that are sufficient for basic reproductive function in case another mating opportunity comes along. Presumably the father's partner is temporarily infertile due to lactational amenorrhea, but there is always the possibility of an extra-pair mating opportunity. And while we hope men refrain from these opportunities, doing so might have sometimes been maladaptive throughout our evolutionary history, depending on the associated costs. That is, ancestral men who acted on these opportunities may have been selected for under some circumstances.

However, it is also conceivable that maintaining low levels of testosterone, as opposed to none, is somehow helpful for parenting. One possibility is that human paternal behavior, similar to California mouse fathers, depends on estrogen that is derived from testosterone. In fact, one study found higher levels of estrogen in the saliva of fathers compared to non-fathers.[122] But we know very little about the potential role of estrogen in human paternal behavior.

Another interesting possibility is that testosterone supports paternal aggression in defense of infants.[123] This may initially seem far-fetched, but we need to imagine the context in which our physiology evolved. Throughout most of human evolution, infants were sometimes vulnerable to predators or even hostile conspecifics. Remember that infanticide is quite common in many primate species, including our closest living relative, the chimpanzee. Although human fathers have lower baseline testosterone, they exhibit transient increases in testosterone in response to a crying infant.[124] To the extent that testosterone motivates aggression and reduces fear and anxiety, such an increase might help fathers successfully respond to these threats and protect their infants. We don't know if this hormonal response extends into later childhood, but if you have ever seen someone mistreat your child, the notion that testosterone increases to motivate defensive aggression may resonate with you. I have a good friend who confessed to me that if he sees another kid mistreat his daughter, he has to let his wife handle it because he is afraid he might overreact.

Testosterone has effects on brain and behavior during adulthood. However, it may have even more profound effects on behavior through its actions early in development. As noted, testosterone increases in male fetuses during the prenatal period and again for a brief time in male infants. In addition to masculinizing the genitals, this early testosterone affects the brain. One behavior that it influences is rough-and-tumble play. Fathers often specialize in this play, which is good for child development when done correctly. Juvenile male monkeys engage in more rough-and-tumble play than do juvenile female monkeys, and we know that this sex difference is dependent on prenatal testosterone exposure.[125] So there is a possibility that perinatal exposure to testosterone facilitates paternal rough-and-tumble play in humans. I don't think we know yet whether adult levels of testosterone are involved in this play, but it would not surprise me if they were. One recent study is consistent with this possibility. Testosterone levels are highest in the early morning and decrease over the course of the day. However, this diurnal decline in testosterone is blunted, meaning afternoon levels stay higher among fathers who engage in lots of rough-and-tumble play.[126]

Testosterone might also support certain types of provisioning behaviors. In small-scale, nonindustrial societies, fathers provision their children through hunting and other subsistence activities. As just mentioned, testosterone levels typically decrease over the course of the day, but a study of !Kung San hunter-gatherers showed that this diurnal decrease is not found when men are hunting. Instead, they maintain high levels of testosterone throughout the day.[127] What function might this serve? Physical exercise is known to transiently increase testosterone levels, and this may support muscle growth needed to meet the demands of more of the same type of physical activity in the future.[128] Exercise-induced increases in testosterone may also increase hematocrit levels to support greater oxygen-carrying capacity in the blood, which will fuel muscular activity. So increases in testosterone that accompany hunting may help to ensure that men stay in proper physical condition to continue hunting in the future. Another study of the Tsimane people from South America showed that men who participate in a successful hunt experience a further surge in testosterone that is not seen in unsuccessful hunters. The authors suggest that this surge in testosterone is rewarding and may help motivate future hunting attempts.[129]

Testosterone may also be involved in other types of subsistence activities. Among the Bondongo fisher-farmers from the Republic of Congo, subsistence activities are often dangerous and involve risks of drowning while fishing, falling while climbing trees to harvest palm wine, and getting burned while clearing garden plots with fire.[130] Bondongo men who were ranked by their peers as better providers had higher testosterone levels compared with other men. Testosterone has known anxiolytic effects, which may help facilitate participation in dangerous or risky activities such as these. The Bondongo people value paternal provisioning to a much greater extent than direct caregiving, so it may be adaptive for Bondongo fathers to maintain higher levels of testosterone to support their provisioning.

In addition to hunting, the Tsimane men chop down trees to clear land for small-scale horticulture. An hour of tree chopping was found to increase testosterone levels in Tsimane men by nearly 50%.[131] Collectively, these studies raise the intriguing prospect that human fathers decrease their baseline testosterone to circumvent the associated maintenance and direct caregiving costs, but retain the ability to rapidly and transiently increase testosterone when needed for offspring protection and provisioning, or even possibly for the pursuit of new mating opportunities.

A theory known as the *paternal provisioning hypothesis* argues that the muscle mass required for paternal provisioning has become less dependent on testosterone levels in humans. In other species, testosterone is strongly associated with muscle mass, but this correlation appears to be weaker among men. According to this theory, there has been selection to allow fathers to preserve muscle needed for provisioning, despite their declining testosterone that facilitates more direct forms of caregiving. Evidence in support of this theory comes from men living in rural Poland where fathers had lower levels of testosterone than nonfathers, as expected. However, fathers also worked harder in the fields and had greater strength and muscle mass compared with nonfathers.[132] While this study examined baseline testosterone levels, it could be that transient, exercise-induced increases in testosterone are more important than baseline levels for building and maintaining muscle mass needed for male provisioning.

I speculated above that it might be adaptive for fathers to increase their testosterone when their infant is in danger and crying. But often infants cry when their lives are not in danger, and this is what parents today primarily experience. In these situations, testosterone has been observed to either increase or decrease. Decreases in testosterone predict sensitive and nurturing paternal responses to crying, whereas increases may instead predict aggravation.[133] One interesting study exposed men to an infant simulator that either cried frequently or infrequently. While men exposed to the infrequent crying experienced a decrease in testosterone, those exposed to frequent crying showed an increase and also provided less sensitive care to the simulator.[134] These findings are generally consistent with the idea that lower testosterone levels are more conducive to sensitive infant care.

We have discussed the effects of testosterone on mating and parenting effort in human males, but does testosterone also curtail investment in maintenance and survival as predicted by life history theory? Stored body fat in animals, including humans, is an investment in their future survival.[135] That is, if food is available unpredictably, body fat stored during times of plenty can be drawn on during future shortages. As we have already noted, testosterone burns body fat and mobilizes that energy so it can be used immediately for mating effort. However, in doing so, it jeopardizes future survival, at least in the environments similar to those in which we evolved where food shortages were common. This may be why men who are undernourished have low testosterone levels.[136] The body simply can't afford to divert any of its limited energy stores away from maintenance to mating effort—or else it may starve. Men in Western societies tend to gain body fat when they become fathers, and this increase is mediated by a decrease in testosterone.[137] So in a sense, when men become fathers, they begin to invest more in their own future survival, so they can be around to raise their children. Of course, this increased weight gain paradoxically has negative health consequences in our modern environment, but that is an evolutionary anomaly.

For every 100 baby girls born in the world, 105 baby boys are born. By age 110, 90% of all living people are women.[138] In every country in the world, women currently have a greater life expectancy at birth than men do.[139] This sex difference averages about five years globally.[140] For example, a girl born in the United States today can expect to live about eighty-two years, whereas a boy can expect to live only about seventy-seven years. This of course means that human males suffer higher mortality rates throughout their lifetime. A similar but more pronounced sex difference in longevity is found among wild mammal species, which, together with the cross-cultural consistency of the effect, implies a biological explanation.[141]

The sex difference in mortality increases dramatically at puberty, and this is of course the time when boys begin secreting testosterone at adult levels (figure 3.5). One of the most pronounced sex differences in mortality is for accidents.[142] Think back to our skateboarders. Testosterone made them attempt tricks that they often could not pull off. What if it has the same effect on a young man driving his car on a freeway with a group of friends? Another striking sex difference in mortality is found for death by homicide. Most homicides involve men killing other men, and the fights most commonly involve one man insulting another and an ensuing competition for, or defense of, social status.[143] It appears that young men are sometimes willing to risk their lives in order to increase or defend social status, and testosterone may drive this behavior.

Not only are men more likely than women to die from accidents and homicide, they are also more likely to die from parasitic and infectious disease.[144] This is consistent with other evidence we have seen that testosterone suppresses immune function. Additionally, the rate of telomere shortening is greater in men than in women, again implying that men invest less in maintenance compared with women.[145]

But once again these are all just correlations. Men have higher testosterone than women, and men also suffer higher mortality than women, especially once they begin secreting high levels of testosterone after puberty.

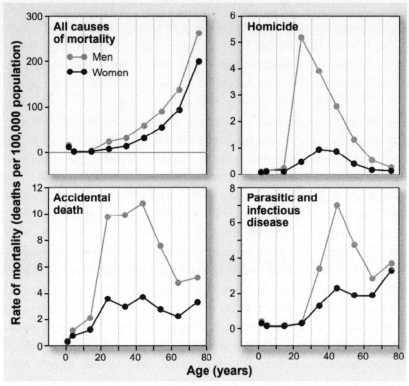

Figure 3.5
Human males suffer higher mortality rates than human females, particularly after puberty when males begin secreting high levels of testosterone. Sex differences in mortality are largest for homicides, accidents, and infectious disease, all of which are linked with higher levels of testosterone. Source: I. P. F. Owens, "Sex Differences in Mortality Rate," *Science* 297, no. 5589 (2002): 2008–2009.

Do we have any causal evidence that testosterone interferes with survival and maintenance in human males? Thankfully, we don't experiment on men with castration, but there have been some tragic natural experiments conducted in the past.

In historical times, throughout parts of Asia and the Middle East, boys were sometimes castrated and later employed as guards and servants of the royal court. Eunuchs, as they were called, could be trusted to work among the king's harem. A sample of eighty-one such eunuchs from the imperial court of the Korean Chosun dynasty (1392–1910) was found to outlive their intact male peers by a substantial fourteen years. In fact, three of the eunuchs lived to be one hundred years old, which turns out to be about 130 times the rate found for men today living in developed countries.[146]

We can also ask about the opposite type of natural experiment in which men are given, or rather voluntarily take, excess exogenous testosterone. There is no shortage of men willing to expose themselves to this manipulation. About 6% of men have used anabolic androgenic steroids in their lifetime, usually to enhance their appearance or improve their athletic performance. Anabolic, androgenic steroids are compounds that bind to the androgen receptor and function like testosterone. Several studies have now demonstrated that steroid users have greater mortality than nonusers. For example, men from Danish fitness centers who tested positive for steroids had three times the mortality rate of nonusers. These men were also more likely to be admitted to the hospital for injuries, infections, and other causes than were men in the control group.[147] Again, we don't know that this relationship is causal because it could be, for example, that the kind of man who uses steroids is inherently more prone to risk taking and that explains their increased mortality rather than their steroid use. A causal relationship is better supported by the observation that mortality rates of Finnish power lifters exceeded that of controls only once steroid use became widespread in this community in the 1970s and 1980s.[148] Prior to that, Finnish powerlifters had mortality rates as low as or even lower than controls.[149] So steroid users may engage in more risk taking and may invest less in maintenance. They may even invest less in maintaining their brains since long-term steroid use has been associated with accelerated brain aging.[150] As we age, we lose gray matter in our brains, the place where neurons reside. This happens at a faster rate in long-term steroid users. Laboratory studies have even shown that very high levels of testosterone can increase the death of neurons in a petri dish.[151]

For all of these studies of anabolic steroids, it is important to note that they establish what is referred to as supraphysiological levels of testosterone in men, by which is meant levels that well exceed what would normally be found in men. So, for example, normal levels of testosterone do not kill neurons; only abnormally high levels do. In fact, testosterone replacement therapy for hypogonadal men with very low testosterone may even improve their survival.[152]

Until now, we have been focusing exclusively on testosterone levels, but this only partially explains androgen signaling in the body and brain, which also depends on how many receptors there are to sense the testosterone signal. The receptor that binds testosterone is known as the androgen receptor. We don't have a way to directly measure the number of androgen receptors in the living human brain, but we can get at it indirectly. The gene that codes for the androgen receptor is polymorphic, meaning that different people have different versions of the gene. They vary in the number of copies they have of a sequence of three DNA base pairs (CAG) within the gene. People can have anywhere from nine to thirty-one copies of this repeat. The gene makes more androgen receptors when it has fewer repeats. These receptors are also better at binding to the DNA of other genes and influencing the expression of those genes compared with receptors made from genes with more repeats. So a man with fewer CAG repeats is expected to have greater androgenicity, and we might expect that to be associated with increased mating effort. Indeed, fewer CAG repeats have been associated with a striking array of traits related to mating effort, including greater lean muscle mass, greater strength, higher rates of sperm production, larger testosterone response to social interactions with women, and greater self-reported social status.[153] We might also predict that fewer CAG repeats are associated with less involved or lower-quality parenting. Among Filipino men, those with both fewer CAG repeats and high testosterone levels—men with the highest level of androgenicity—were less likely to be involved fathers, as would be expected. However, men with the lowest level of androgenicity—those with many CAG repeats and low testosterone—were also less likely to be involved. Instead, men with moderate levels of androgenicity—either few CAG repeats and low testosterone or the converse—were the most likely to be involved in child care.[154] It's not clear why these low-androgenic men tend to be less involved in child care, but these findings reinforce the earlier suggestion that testosterone in moderation may support certain aspects of paternal care.

Throughout this chapter, we have been discussing evidence for hormonal and genetic influences over human behavior. Many people are appropriately wary of genetic explanations for human behavior. One concern is that people will misuse the science to excuse unethical or antisocial behaviors, as in, "His genes made him cheat" or similar statements. Another concern is that people will use the science to impose limitations on people, as in, "He can't be a good father because he has so few CAG repeats." These are of course naive and inaccurate statements. It is crucial to emphasize that neither genes nor hormones *determine* behavior. Instead, they *influence* behavior. Human behavior is complex and multifactorial. Genes and hormones are just two of these many influences that bias behavior in various directions. There are many other influences, including social, cultural and developmental influences, that are discussed in chapter 6. As an analogy, imagine that you release a helium-filled balloon from your home and want to predict where it will land. That would depend on a slew of factors such as the balloon size, wind direction, wind speed, temperature, pressure, and humidity and how all those variables interact with each other. If you only knew the wind direction, that would be useful, but it wouldn't give you a precise estimate of where the balloon will land because you haven't considered any of the other factors. Similarly, if we only know a man's testosterone levels or how many CAG repeats he has, we will not be able to reliably predict what kind of father he is. In fact, there will be some men with high testosterone, or few CAG repeats, who are involved and committed fathers and husbands, just as there will be some men with low testosterone, or many CAG repeats, who are not. If you want to know if someone is a good father, don't measure his testosterone or genotype him. Just watch him interact with his kids.

A summary of testosterone's effects on male behavior and physiology is provided in figure 3.6.

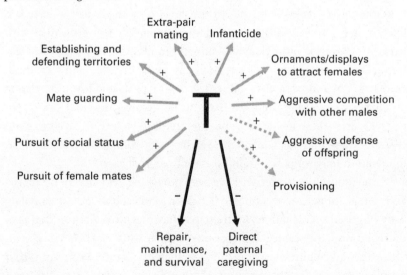

Figure 3.6
Effects of testosterone on male behavior and physiology. Effects with gray arrows are facilitated by testosterone. Effects with black arrows are inhibited by testosterone. Dashed gray lines are parental behaviors and solid gray lines are mating behaviors.

Testosterone's influences on mating and parenting behavior are likely to be mediated by its effects in the brain. So far, I have neglected these neural mechanisms of testosterone, but chapter 5 considers how testosterone and other hormones affect brain structure and function to influence behavior.

Highlights

1. Testosterone biases male animals toward investing more of their energy in mating and less into other important life history goals such as parenting and survival.
2. Human fathers who spend time with their infants experience a significant decline in testosterone that makes them better caregivers and partners.
3. However, human fathers can experience transient increases in testosterone that may support indirect caregiving activities such as provisioning and protection of offspring.

4 Oxytocin

My mom loved to nurture things—not only us but also animals, and especially kittens. And she strongly reinforced those tendencies in me as a boy. I had a menagerie of stuffed animals that I cared for. There was a line-up that I slept with, and it was my job to tuck them in and make them feel safe. They made me feel safe too, as I slept alone in the dark. My favorite stuffed animal was a three-foot-tall Raggedy Andy that I loved to death. After I wore one out and Mom could no longer stitch his limbs back on, I pleaded for my parents to buy me another while still mourning the loss of the former. I remember tucking him in at night—wanting to make sure he was safe and comfortable. I was naturally drawn to this; it was something I craved. My three older brothers, who have all become nurturing fathers themselves, nevertheless used to enjoy kicking my Raggedy Andy around the room to antagonize me. It was quite effective. When I became a parent, I realized that my children were benefiting from being raised by a nurturing father, and I knew that, like my mom, I wanted to reinforce those tendencies in them. My nine-year-old son, Toby, also sleeps with a line-up of stuffed animals. Each night at about 9:00 p.m. as I'm typically responding to the day's emails, my wife predictably calls up to my office on the third floor and says, "Toby wants you." I come down to his room. I could just tuck him in, but I get in bed with him to snuggle for five or ten minutes. I do this for him and for me. As I crawl into the bed, he predictably begins to arrange his stuffed animals in a very particular order. Those that he can't fit in the bed are placed on a large beanbag chair next to the bed. He carefully covers them with his own T-shirts so that the stuffed animals "can smell him," which is supposed to comfort them. He then crouches over those that are in bed with him, looking remarkably similar to a female mammal crouching over her pups when nursing, and then lies gently beside them. I never taught him to do any of this. One night he began crying as he started to think about going to college. Who would take care of his stuffed animals? They would be so lonely!

At this point, you might be wondering if I am worried about any of this. The answer is no. He is pretty much like I was, and I grew up to be a fully functioning adult. Quite the contrary, it gives me confidence that he will raise his children with the love and empathy that they need to be mentally healthy, empathic adults.

Why was my mom so nurturing, and how did that get passed to me and then Toby? I am guessing it involves a little hormone known as oxytocin. I've never had my oxytocin levels measured, but I'll bet they are high, at least when I am with my family. I think my mom's are through the roof.

*   *   *

Most biologists associate oxytocin with motherhood, and appropriately so. The word itself means "rapid birth" in Greek and references its fundamental role in the contraction of the smooth muscles of the uterus during childbirth. In mammals, it also stimulates contractions of smooth muscles that squeeze milk from the milk ducts in breast tissue.[1] Oxytocin is made in the brain, in an ancient structure at its base known as the hypothalamus. The neurons that produce oxytocin extend into the pituitary gland below the hypothalamus, where they release their oxytocin into the bloodstream. From there, the circulatory system carries it throughout the body and it binds to oxytocin receptors, including those in the uterus and breast tissue. Many of these neurons also have branches that project upward and release their oxytocin into the brain. Here, there are also receptors for oxytocin, enabling it to have an influence over behavior.[2] In addition to producing milk, a good mammalian mother needs to be motivated to nurse and otherwise care for and protect her offspring, and oxytocin facilitates these psychological and behavioral components of motherhood.

As you know by now, males also provide parental care in approximately 5% of all mammals. As outlined in chapter 2, paternal caregiving seems to have evolved many times independently among mammals. In many cases, it appears that natural selection enabled paternal caregiving by using physiology that was already in place in mothers of the same species. As such, oxytocin also now promotes paternal care in many of the mammalian species that exhibit it.

This chapter begins by surveying the role of oxytocin in maternal behavior and proceeds to consider its role in caregiving by mammalian fathers, including men. I then discuss the role of oxytocin in pair-bonding, which allows parents to cooperate in raising their offspring.

Oxytocin and Maternal Behavior

Much of the research on the biology of mammalian motherhood has been done on laboratory rats. Mother rats care for their pups by nursing them, retrieving them when they stray from the nest, and licking and grooming them, which is the rat equivalent of maternal affection. But not all female rats are maternal—far from it, in fact. Females experience a dramatic transition when they become mothers. Before the birth, they actively avoid pups or even attack and kill them. After giving birth, however, they become "promiscuously maternal," meaning they will provide care to any pup they can find. Pups become an extremely rewarding stimulus to mothers. In fact, researchers have shown that rat mothers exert tremendous effort, by repetitively pressing a lever, to have pups delivered into their nest.[3]

This transformation is caused by hormonal changes that females experience during pregnancy, including increases in estrogen and oxytocin. This was proven when researchers showed they could elicit maternal behavior in nulliparous (nonmothers) female rats by treating them with estrogen, followed by oxytocin. Estrogen increased the number of oxytocin receptors in the brain. and oxytocin activated those receptors to trigger maternal behavior.[4] Soon after, scientists showed that the opposite experiment, injecting mothers with a drug that blocked oxytocin receptors, was able to prevent the onset of maternal care in female rats that had just given birth.[5]

Like human mothers, rat mothers vary in terms of how affectionate they are. Some lick and groom their pups a lot, others less so. It turns out that more affectionate rat mothers have more oxytocin receptors in a key hub of the parental brain circuitry, the medial preoptic area (MPOA). Remarkably, maternal licking and grooming seem to increase the number of oxytocin receptors that pups have, and these pups go on to become more affectionate parents themselves.[6] This is a classic example of *epigenetic inheritance*, in which a trait, in this case maternal affection, is transmitted from one generation to the next, due not to genetic inheritance but rather to a stable characteristic of the environment: a nurturing mother. It appears that affectionate mothering is being transmitted across generations through upregulation of oxytocin receptors in the brain. I think this is probably why both my mom and Toby are so nurturing.

Oxytocin also seems to support maternal caregiving in humans. More attentive and affectionate mothers tend to have more oxytocin in their blood. During interactions with newborn infants, mothers with more oxytocin spend more time gazing at their infant's face, expressing positive emotions, and affectionately touching their infant. They also engage in more motherese vocalizations (i.e., baby talk) and show better temporal coordination between their own and their infant's positive emotion. That is, they are better synchronized with their infant's emotional expressions.[7] Mothers with postpartum depression tend to have lower oxytocin levels.[8] They also tend to show lower levels of maternal sensitivity, more anger and intrusiveness, and fewer positive interactions with their infant, and this may be due to their low oxytocin levels.[9]

Other studies have examined how maternal oxytocin levels respond to interactions with their infants. Oxytocin increases in mothers who touch and gaze at their infants a lot and not in other mothers.[10] Interestingly, mothers who themselves have a more secure attachment style show increases in oxytocin in response to interactions with their infants, whereas mothers who are insecurely attached show decreases.[11] This is an obvious mechanism by which attachment styles could be passed from one generation to the next. An insecurely attached mother fails to show appropriate bonding behaviors with her infant due to an insufficient oxytocin response, and this lack of maternal sensitivity leads to an insecure attachment in her infant.[12]

Breastfeeding stimulates mothers to release oxytocin, so if oxytocin is important for maternal motivation, we might expect that women who breast-feed would be more sensitive and nurturing mothers, on average, than mothers who do not breast-feed.[13] One very large Australian study found this to be the case. Non-breast-fed children were nearly four times more likely to suffer maternal neglect compared with children who were breast-fed for at least four months. Of course, it is possible that preexisting motivational differences predict breastfeeding rather than the converse, but the study controlled for both pregnancy ambivalence and postpartum depression. Despite these findings, it is important to emphasize that there was no evidence of neglect in 93% of non-breastfeeding mothers, so breast-feeding is clearly not required to form strong, nurturing bonds with children. It probably just facilitates the process.[14]

As with testosterone, oxytocin can have effects throughout the body and brain only by binding to its receptor (figure 4.1). More affectionate rat mothers have more oxytocin receptors in the medial preoptic area of the brain (MPOA). We would love to know if the same is true in human mothers. However, we currently lack the ability to measure oxytocin receptors in the brains of living humans. We can, however, learn something by instead looking at the gene that codes for the oxytocin receptor. This gene is polymorphic in humans, meaning that different individuals have different versions of the gene. By looking at the brains of people after they die, we can measure receptor levels and identify genetic variants that are associated with more or fewer receptors. Then we can ask whether particular genetic

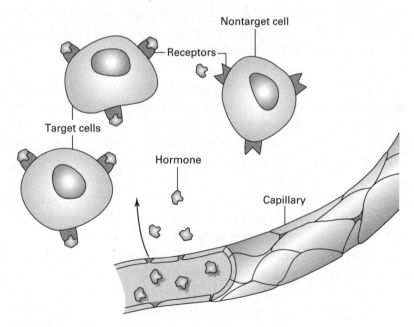

Figure 4.1
Oxytocin has biological effects by binding to oxytocin receptors in the cell membrane.

variants in living people are related to maternal caregiving and, if so, infer that this may have to do with the gene's effects on oxytocin receptor levels.

Genes are sequences of DNA base pairs. The oxytocin receptor gene consists of about 17,000 base pairs. Of these base pairs, 390 are polymorphic, meaning that different people can have different bases at those sites, called single nucleotide polymorphisms (SNPs). The rest of the bases are the same in everyone. SNPs are labeled with numbers. One SNP, rs53576, is of particular interest because it is known to affect expression of the oxytocin receptor in the brain, and it is associated with positive maternal behaviors, including maternal warmth and sensitivity.[15]

Given this substantial evidence that oxytocin supports warm and sensitive maternal behavior, is there also evidence that it is involved in fathers and paternal caregiving?

Oxytocin and Paternal Behavior in Rodents

Mandarin voles are one of the minority of mammals in which males are involved in raising their offspring. Similar to female rats, male mandarin voles experience a profound behavioral transformation when they become fathers, and oxytocin is critically involved. If you present pups to a male who is not a father, he will most likely attack and try to kill them. However, if you do the same to a mandarin vole father, he will approach them and lick and groom them, even if the pups are not his. Fathers have higher levels of oxytocin in their blood compared with nonfathers, and they also have more oxytocin receptors in that critical neural hub for parental behavior, the medial preoptic area of the hypothalamus (MPOA). If you give fathers a drug that blocks their MPOA oxytocin receptors, their licking and grooming of the pups decreases. So it seems that when male mandarin voles become fathers, there is an increase in oxytocin secretion as well as an increase in the number of oxytocin receptors in the MPOA, and the resulting increase in oxytocin signaling facilitates paternal behavior.[16]

In contrast to mandarin vole males, males of another vole species, prairie voles, typically care for pups even before they become fathers. In nature, this often involves juvenile males who help care for their younger siblings while they are still in the nest. In the lab, it is manifest as sexually inexperienced males providing care to unfamiliar pups that they are housed with. When these males are presented with a pup, they exhibit a rapid but transient surge in oxytocin. Levels increase within ten minutes of pup exposure but return to baseline after twenty minutes.[17] The function of this oxytocin surge is unknown, but there are a number of possibilities. First, oxytocin is known to enhance social reward, so it may tag the pup as a rewarding stimulus to the male and thereby motivate him to take care of it.[18] Oxytocin is also known to decrease anxiety.[19] Males experience an increase in heart rate when presented with pups, so pups seem to be an arousing stimulus for males. Oxytocin may help calm their nerves, so to speak, so they are not too stressed out to be a good parent.[20] As we will see, oxytocin may also have these effects in human parents.

Oxytocin and Paternal Behavior in Primates

Oxytocin may also be involved in paternal caregiving among our primate relatives. As discussed in chapter 2, male marmoset monkeys are exemplary fathers who carry their twin offspring on their backs, protect them from predators, share food with them, and groom them. One study showed that the hypothalami of male marmoset fathers produced more oxytocin than did the hypothalami of males who were nonfathers. These findings suggest that similar to male mandarin voles, when marmoset males become fathers, they increase oxytocin production.[21]

These physiological differences may explain behavioral differences between father and nonfather marmosets. Marmoset fathers respond more quickly to infants than do males that are not fathers. When researchers give male marmosets extra oxytocin via a nasal spray, their response accelerates.[22] Another research group injected oxytocin straight into the brains of marmoset fathers and found that it increased their willingness to share food with their infants.[23]

As discussed in chapter 2, humans are best characterized not as biparental—where offspring are raised by a mother and a father—but rather as *cooperative breeders* in which mothers can receive help from a variety of different relatives, including fathers.[24] So if oxytocin supports both maternal and paternal caregiving, we might naturally wonder if it also supports caregiving by other individuals, known as alloparents. Meerkats aren't primates, but they are cooperative breeders, and researchers have experimentally manipulated oxytocin in them and examined the effects. Meerkats live in groups of up to fifty animals, in which a single male-female pair does most of the breeding and other adults assist in their reproduction. These alloparents both babysit and feed pups. After being injected with oxytocin, meerkat alloparents spent more time in close proximity to pups and also shared more food with them compared to when they were injected with a placebo. Not only that, but they spent more time guarding the colony from predators, and excavating tunnels that are used for sleeping or as escape routes from predators.[25] Thus, not only does oxytocin support alloparental caregiving, it also seems to support meerkat contributions to the public good. Indeed, there is now accumulating evidence that the prosocial effects of oxytocin extend beyond parental care to other types of cooperative or altruistic behavior in humans.

Like meerkats, marmosets are also cooperative breeders. When a new infant marmoset is born into a family group, oxytocin levels increase in the mother and the father and in the nonbreeding helpers as well. Early in infancy, parental and alloparental care includes licking the infants, and caregivers with higher oxytocin levels lick the infants more. Later in infancy, especially after weaning, infants depend on allomothers to share food with them, and again caregivers with higher oxytocin levels do so more often.[26]

Oxytocin and Paternal Behavior in Humans

Although oxytocin is generally considered a maternal hormone, and with good reason, once scientists looked at oxytocin levels in fathers, they discovered something surprising: baseline levels of oxytocin in fathers are as high as they are in mothers, except for after a breastfeeding, when levels temporarily surge higher in mothers.[27] Fathers also have higher levels of oxytocin than nonfathers, and one study showed that these levels increase over the first sixth months of fatherhood.[28] The fact that oxytocin levels rise when men become fathers suggests that the hormone may be preparing them for their new role.

Oxytocin levels in mothers and fathers are correlated.[29] So if the mother has high levels of oxytocin, the father is also likely to have high levels. This reinforces the idea of the family as a system, as discussed in chapter 2. There seem to be high-oxytocin families and low-oxytocin families, and the evidence suggests that children from high-oxytocin families are raised differently. One study done in Israel measured oxytocin levels in mothers and fathers and inquired if these were related to the manner in which they interacted with their newborn infants. Mothers with more oxytocin were more affectionate with their infants, and fathers with higher oxytocin levels engaged in more "stimulatory parenting," which included stimulatory touch of the infant, presenting the infant with objects, and moving the infant's body through space.[30] A study done in the United States also found that fathers with higher oxytocin levels touched their infants more and engaged in more play that involved moving their infants around in space, such as throwing them in the air and playfully bouncing or jostling them.[31]

Childbirth and breastfeeding stimulate oxytocin increases in mothers, but what is responsible for triggering increases in fathers, who neither gestate nor lactate? One possibility is suggested by studies showing that paternal oxytocin levels increase after thirty to sixty minutes of skin-to-skin contact with their preterm infants. Not only do levels increase in fathers, they also increase in the infants themselves, consistent with the possibility that oxytocin is promoting father–infant and infant–father bonding. Indeed, fathers who showed larger increases in oxytocin following skin-to-skin contact engaged in more sensitive and responsive care with their infant the next day.[32] Skin-to-skin contact had the additional benefit of decreasing stress hormone levels (i.e., cortisol) in the infant, as well as the father's self-reported levels of anxiety. Another study found that oxytocin levels increased by about 20% after first-time fathers held their newborn infant for the first time.[33] Thus, tactile stimulation may be one mechanism by which infant stimuli can increase paternal oxytocin levels.

Other types of interaction can also stimulate increases in oxytocin. I have mentioned that oxytocin levels increase in mothers after an interaction with their infant, but only if the mother shows lots of affection toward their infant. Paternal affection does not increase oxytocin levels in fathers, but paternal physical stimulation of the infant does. That is, fathers who physically stimulate their infant a lot show increases in oxytocin, whereas other fathers do not.[34] Most fathers in these studies are presumably secondary caregivers. One wonders whether paternal affection might increase oxytocin levels in primary caregiving fathers, as it does in mothers.

Many fathers specialize in rough-and-tumble play with their children, and it may be that such play also raises paternal oxytocin levels. My son was ten when I started writing this book and is now twelve. In contrast to my very affectionate six-year-old daughter, he rarely tells me he loves me or gives me an unsolicited hug. That sometimes hurts my feelings, but he has always, and continues, to love wrestling with me, and he primarily initiates it. The wrestling isn't competitive. He doesn't try to beat me. We're playing. I'm mostly throwing my mass on top of him and smushing him into the couch, and he somehow seems to like this. I read an article recently showing that jujitsu grappling, which involves close contact tactile interaction, increases salivary oxytocin levels, but these levels were not increased as much by punch-kick sparring during jujitsu.[35] So I suspect that when Toby and I wrestle together, our oxytocin levels increase, and that this is his way of bonding with me and that it also helps me bond with him.

These human studies are all correlational, so they can't tell us whether oxytocin actually *causes* changes in paternal behavior. In order to determine this, we need to experimentally manipulate oxytocin levels and see if doing so affects paternal behavior. What happens if we give fathers extra oxytocin, beyond what their own body produces? In contrast to testosterone, oxytocin has trouble getting into the brain if it is injected into the blood. However, it is able to enter the brain when administered as a nasal spray.[36] Therefore, several experiments have examined the effects of intranasal oxytocin on paternal behavior.[37] When Israeli fathers were given intranasal oxytocin, they spent more time touching their infant and engaged in more social reciprocity with the baby.[38] As outlined in chapter 1, social reciprocity is highly beneficial for infants' psychological and social development. But what exactly is social reciprocity? If your baby coos at you and you coo back at her, that's social reciprocity. Or if she smiles at you and you smile back at her, that is also social reciprocity. Oxytocin seems to promote this type of behavior in fathers.

Another study by the same research group even showed that oxytocin administration increased father's head speed and acceleration when they were interacting with their infant.[39] This is quite consistent with the theme that oxytocin is related to stimulating parenting in fathers. A Dutch study showed that the beneficial effects of oxytocin on paternal behavior extend beyond infancy. When treated with oxytocin, fathers encouraged more learning and exploration in their toddler children and showed less impatience and discontent.[40]

So intranasal oxytocin has a variety of positive effects on paternal behavior. Remarkably, it also seems to indirectly affect the behavior of the infants whose fathers are treated with it. That is, when an infant's father received oxytocin, that infant behaved differently too, presumably because he or she was responding to changes in the father's behavior. These infants spend more time looking at their father's face and more time playing with objects. Even more amazing is that oxytocin levels increase in these infants, even though *they were never given oxytocin*—only their father was. The infant begins producing more oxytocin on its own. Oxytocin administration to the father is triggering a positive feedback cycle of oxytocin release and father–infant bonding.[41]

*   *   *

I have presented evidence that oxytocin positively influences paternal behavior, but we can also ask what psychological processes it affects to have those behavioral effects. That is, how does oxytocin influence male psychology? There are at least five different mental processes that oxytocin affects that may help to explain its positive effects on paternal behavior.

First, we know that oxytocin increases the salience of social stimuli so that people pay more attention to those stimuli.[42] For example, we receive much of our social information from looking other people in the eye. Treating people with intranasal oxytocin causes them to spend more time looking at the eye region of faces.[43] Relatedly, in an experimental task where participants can only see the eye region of faces, oxytocin improves their ability to accurately infer what others are thinking or feeling.[44] So oxytocin might bias fathers to attend to their children's eyes, which may in turn allow them to better understand what the child is thinking, feeling or attending to.

The second mental process, empathy, is closely related to the first. People tend to automatically covertly mimic facial expressions that others make, and that mimicry can be measured using electromyography (EMG), which records subtle activity in the face muscles. People who mimic to a greater degree tend to be more empathic. I noted in chapter 3 that testosterone treatment decreases facial mimicry. Oxytocin, it turns out, has the opposite effect. Specifically, it increases facial mimicry of men when they are viewing pictures of infants.[45] Oxytocin has also been shown to increase self-reported empathy to viewing others in pain.[46] These studies suggest that oxytocin may improve paternal behavior by increasing fathers' empathy for their children.

The third psychological domain affected by oxytocin is social reward. Oxytocin acts on the dopamine reward system in the brain to facilitate maternal behavior, presumably by making offspring more rewarding to the parent.[47] In my lab, we have found that fathers who spend more time taking care of their toddler children more strongly activate components of this dopamine reward system in the brain when they see pictures of their child and that oxytocin increases activation in this system.[48] Thus, oxytocin is likely to be increasing the reward value of children to fathers and thereby increasing their motivation to care for them.

Fourth, oxytocin decreases anxiety.[49] Rats tend to hide in corners when they are scared. When they are given oxytocin, they spend more time out in the open. These effects extend to our own species. Many people find public speaking to be stressful. This has been simulated in the laboratory, in an experiment where researchers give participants five minutes to prepare an oral presentation to deliver in front of a stern panel of judges who do not offer any positive feedback. As you might imagine, this task—known as the *Trier Social Stress Task*—reliably increases levels of the stress hormone cortisol. But if men are pretreated with intranasal oxytocin, it attenuates both their cortisol response and the level of anxiety they report in this test.[50] Are these anxiolytic effects of oxytocin important for parenting? Well, imagine first-time parents who arrive home from the hospital without any prior parenting experience, knowing that the well-being of this new infant is entirely dependent on them . . . and then the infant starts crying and they try a few things to soothe him or her but the infant keeps crying. Yes, parenting can involve anxiety, and keeping it under control is essential for parents and their infants.

Staying calm may even be important for veteran, experienced parents in certain contexts. In nature, mammalian mothers are occasionally faced with the prospect of predators attacking their young, and they need to face these predators and attempt to defend their offspring. Oxytocin suppression of anxiety is also involved here.[51] In fact, one study showed that intranasal oxytocin increased protective behavior in depressed human mothers who were confronted by an intrusive female stranger who attempted to approach and engage with their infant.[52]

Finally, oxytocin is known to facilitate learning. Mice who are genetically engineered to not produce any oxytocin appear unable to remember individuals whom they have interacted with previously.[53] In other words, they don't learn the identity of other mice. Oxytocin is also responsible for prairie voles learning to prefer a specific opposite-sex partner as a mate such that they form a selective pair-bond with that individual.[54]

In humans, oxytocin has been found to enhance social reinforcement learning,[55] that is, learning to repeat behaviors that are socially rewarded (e.g., when someone smiles at you) and avoid behaviors that are socially discouraged (e.g., when someone frowns at you). We parents like to think that we are in control of our interactions with our children, but the reality is that our parental behavior is shaped by the positive and negative reinforcement we receive from our children, beginning when they are infants. We tend to repeat actions that make our infants smile or laugh, and we also learn to avoid behaviors that make them cry and repeat behaviors that make them stop crying. They cry until we figure out what to do, thereby reinforcing the behavior that made the crying stop. In effect, infants teach us how to parent through their smiling and crying. No one instinctively knows how to do everything the first time they have a baby. For new parents, the learning curve is steep. In many primate species, for example, firstborn infants have a lower survival rate than subsequent infants.[56] This is probably because mothers learn from their first attempt how to be a better mother the next time around. There is even evidence from some human populations that firstborn infants have an increased mortality risk.[57] It is conceivable that oxytocin facilitates the process of learning how to be a good parent.

Oxytocin and Pair Bonding

In biparental species, such as mandarin voles, adult male and female pairs cooperate to raise their offspring. Males can't do it alone, and their success in reproducing depends on their female partner, and vice versa. So the formation of a pair-bond becomes an essential component of male reproductive success. Throughout mammals, there is a strong association between biparental care and pair-bonding or monogamy.[58] It is here that oxytocin appears to be doing double-duty. That is, in addition to supporting parent-offspring bonding, it also seems to support pair-bonding between adult males and females.

One pair-bonding species that has been studied extensively is the prairie vole. When scientists place a male and female prairie vole together in a cage for twenty-four hours and let them mate, they form an enduring preference to associate with that specific opposite-sex partner rather than another member of the opposite sex. When researchers later give either member of the pair the opportunity to spend time in a chamber that houses their partner versus a chamber that houses a novel opposite-sex animal, they choose the former. This is a pair-bond. Pair-bonding happens faster when mating is involved. And, guess what? Mating stimulates oxytocin release. In 1994, the biologist Sue Carter and colleagues demonstrated that administering oxytocin into the brains of female prairie voles stimulated pair-bonding in females.[59] More recently, researchers have demonstrated that oxytocin is also involved in pair-bonding in male prairie voles. We know this because pair bonding is prevented when males are given drugs that block oxytocin receptors in their brain. Conversely, drugs that stimulate oxytocin release in the brain facilitate the formation of pair-bonds in males, even without any mating. It is also the case that males who naturally have more oxytocin receptors more readily form pair-bonds and that experimentally increasing the number of oxytocin receptors in a particular brain region (you can do this now in animals) promotes pair-bonding.[60] I know what you are thinking—*Do some people just not have enough oxytocin receptors to pair-bond the way they should?* or maybe even, *Can I increase my partner's oxytocin receptors?* We will get back to that.

In addition to oxytocin, mating also stimulates dopamine release in the brain. Dopamine is responsible for the reward that animals, including humans, get from mating. As my colleague Larry Young, one of the world's leading oxytocin researchers put it, dopamine is responsible for the "Wow!" that animals feel after mating. Oxytocin may be linking that "Wow!" feeling to the identity of a particular female partner. As Young puts it, oxytocin changes the reaction from just "Wow!" to "Wow! Who was that?" The male remembers the specific female that was linked with the reward from mating, and she becomes a rewarding stimulus herself. Hence, the pair-bond is formed. I have focused on male pair-bonding since this is a book on fatherhood, but oxytocin and dopamine are also critically involved in female pair-bonding in prairie voles.[61]

There is also some evidence that oxytocin may be involved in human pair-bonding. New lovers have higher oxytocin levels than singles, and couples with higher oxytocin levels are more likely to stay together. Couples with higher oxytocin levels also show more interactive reciprocity during

interactions, which includes behaviors such as positive emotion and affectionate touch.[62] In addition, both men and women with higher oxytocin
levels report better relationship quality.[63]

Better evidence comes from experimental studies. One important study
showed that treating men with intranasal oxytocin increased their assessment of how attractive their female partner was. Moreover, using functional
brain imaging, the researchers were able to show that intranasal oxytocin
increased men's neural response to viewing a picture of their female partner within the brain's nucleus accumbens.[64] You can think of the nucleus
accumbens as a reward area in the brain, so oxytocin was probably increasing the reward these men experienced when viewing their partner. The
nucleus accumbens also happens to be the same brain area where dopamine and oxytocin interact to support pair-bonding in prairie voles.

A successful relationship depends not only on attraction and reward but
also on trust. Partners must trust each other to uphold their end of the
relationship: to be faithful and committed to the cooperative effort of raising the young. Of course, partners do sometimes cheat or desert and our
trust is sometimes misplaced, but you can't win if you don't play the game.
Oxytocin may help you to play despite the risk. One highly cited study
found that intranasal oxytocin increased trust in an economic game. In this
game, known as the *trust game*, player A is given an endowment of money.
They can give some of their money to player B, in which case the value of
the transferred money is tripled. Then player B can give some, all, or none
of the money they have back to player A. When player A was given intranasal oxytocin, they were more likely to transfer all their money to player
B, presumably based on the assumption that B would reciprocate.[65] Thus,
oxytocin appears to be increasing trust in player A. However, some studies
that have repeated different versions of this experiment have not found
oxytocin to increase trust, and scientists are still debating the legitimacy
of this hypothesis.[66] Part of the explanation for these inconsistencies may
be that the effects of oxytocin vary across people as a function of their personality, attachment style, or social skills.[67] People might also vary in their
sensitivity to oxytocin based on the number of receptors they have in their
brains, so it could well be that oxytocin increases trust in only some people.

Jealousy involves the fear of losing a relationship partner and can be
conceptualized as the opposite of trust. Indeed, severe jealousy can jeopardize relationships. If oxytocin increases trust, we might suspect that it

will decrease jealousy. One study asked men and women to imagine scenarios involving partner infidelity and to rate their resulting level of jealousy. For example, one scenario was, "My girlfriend was sitting on her ex-boyfriend's lap watching TV in his apartment." Treatment with intranasal oxytocin decreased reported jealousy in response to these types of imagined scenarios.[68]

Oxytocin may even help couples successfully navigate conflict. One study asked couples to discuss a topic that caused them conflict and found that intranasal oxytocin increased the amount of positive relative to negative communication during these conversations. Positive communication included making eye contact and emotional self-disclosures.[69] This ratio of positive-to-negative couple communication is known to strongly predict positive long-term relationship outcomes. So these findings are also consistent with the idea that oxytocin supports human pair-bonding.

Finally, oxytocin may help to protect pair-bonds from external temptations that could lead to their dissolution. Researchers asked men to walk toward an attractive female experimenter and to stop at a distance where they felt comfortable. Some men were given oxytocin before this exercise, whereas others were given an inactive placebo spray. Among single men, oxytocin had no effect on how closely they approached the female experimenter. Partnered men treated with placebo kept the same distance as single men did. However, partnered men given oxytocin kept a greater distance between themselves and the attractive woman.[70] These findings suggest that oxytocin may help to preserve pair-bonds by decreasing men's tendency to signal romantic interest in other female partners.

In summary, I have suggested that oxytocin may cultivate a strong and healthy pair-bond in men via multiple routes: increasing attraction to the partner, improving verbal and nonverbal communication with the partner, increasing trust, decreasing jealousy, and decreasing the pursuit of extra-pair mates (figure 4.2). This is relevant to fatherhood insofar as a strong pair-bond is likely to facilitate better cooperation between partners in raising children. This raises the questions, "How can one increase their oxytocin levels?" or "What stimulates oxytocin release?" For men, one of the most potent releasers of oxytocin is sexual intercourse.[71] We have all heard that sex tends to be good for relationships. Relationship quality is higher when couples have more and better sex.[72] There may be many reasons for this, but it is quite possible that oxytocin release is partially responsible.

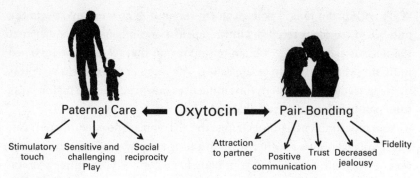

Figure 4.2
Oxytocin effects on paternal caregiving and pair-bonding in human males.

Individual Differences in Oxytocin Signaling (Genetic and Epigenetic Effects)

Based on the discussion thus far, we might expect that men with more oxytocin signaling in their brain would be better fathers and husbands. But what determines how much oxytocin signaling one has, and why do some men have more than others? Oxytocin signaling depends on the amount of oxytocin that is produced, but also on the number of oxytocin receptors that are available to sense that signal. I mentioned a number of factors that can influence oxytocin production in men. For example, it is increased by sexual intercourse and by skin-to-skin contact with infants. Oxytocin levels could also be influenced by experiences one has early in life. For example, one study showed that women who were abused as children had lower levels of oxytocin in their brain as compared with women who were not abused.[73] Again, this could be a mechanism by which abusive or neglectful parenting is transmitted across generations. If my parent abuses or neglects me, my oxytocin levels decrease or my receptors down-regulate, which in turn decreases my motivation to interact with my own child in a sensitive manner.

But what about oxytocin receptors? What influences how many receptors people have? Scientists have learned a great deal about factors that influence oxytocin receptor levels by studying prairie voles. This research shows that both genetic and environmental influences are at play. In terms of genetic influences, we know that one specific single nucleotide polymorphism

(SNP) within the prairie vole oxytocin receptor gene strongly affects the number of oxytocin receptors in the nucleus accumbens. Remember that the nucleus accumbens is a brain reward region that seems to be involved in finding your partner to be appealing. The number of oxytocin receptors in the nucleus accumbens in turn influences how readily male prairie voles form pair-bonds with female mates.[74]

Your environment cannot change the SNP you are born with, but it can affect your genes in another way. To put it simply, genes can be turned on or turned off. Each cell in your body has a copy of your entire genome (except for sex cells). What makes the cells in your heart, for example, look and function differently from the cells in your brain is that different genes are turned on in those two cell types. When a gene is turned on, we say it is *expressed*. Genes can be chemically modified by your body to affect their expression. One common modification is to *methylate* the genes, which means to attach a methyl group (one carbon atom bonded to three hydrogen atoms) to the DNA. This tends to decrease gene expression. So gene expression can be decreased by methylation and increased by the removal of methyl groups. It turns out that the amount of nurturing that prairie voles experience as pups affects methylation of their oxytocin receptor gene, as well as expression of the oxytocin receptor in the nucleus accumbens. Pups that receive more nurturing have less methylation and more oxytocin receptors in the nucleus accumbens. This presumably sets them up to form stronger pair-bonds and to be more nurturing parents as adults.[75]

Similar evidence is now accumulating in humans. By studying the association between oxytocin receptor SNPs and expression of the oxytocin receptor in the brains of people who have died, we now know that there are thirty-one different SNPs in the oxytocin receptor gene that affect the number of oxytocin receptors within the human nucleus accumbens.[76] And the animal research leads us to predict that people with more oxytocin receptors in this brain region are more likely to exhibit strong pair-bonding and high-quality parenting.

There is now also evidence that parental nurturing can affect methylation of the oxytocin receptor gene in human children, which would be expected to affect oxytocin receptor expression in the brain and associated behaviors. One important study measured methylation of the oxytocin receptor gene in five-month-old infants and then again when the infants were eighteen months old. Researchers also recorded a free-play interaction between mother and infant at five months and found that higher quality of maternal engagement at five months predicted a subsequent decrease in methylation of the oxytocin receptor gene in the infant at eighteen months.[77] Thus, early maternal behavior may have lasting effects on oxytocin signaling in their developing child.

Collectively, the research suggests the following hypothesis: boys who are raised in nurturing environments will have a tendency to develop more brain oxytocin receptors and will consequently be more likely to develop into men who are good husbands and fathers. Critically, this does not mean that men who had negative early life experiences and have low oxytocin receptor levels cannot be good fathers because human behavior is complex and multifactorial. It just means that having a nurturing early environment increases the odds of this.

Vasopressin and Paternal Care

Oxytocin and testosterone are not the only hormones that have been implicated in paternal caregiving. Others include prolactin, vasopressin, and even estrogen. Vasopressin deserves a few words before concluding this chapter.

The molecular structure of vasopressin is similar to oxytocin. Both are composed of nine amino acids and they differ from one another at only two of those nine. Consequently, they can even bind to one another's receptors, though not as effectively as they bind to their own. Vasopressin is known to facilitate paternal behavior in many biparental rodent species. For example, injecting it into a specific brain region (the lateral septum) stimulates paternal behavior in male prairie voles.[78] Vasopressin acts in another brain region (the ventral pallidum) to facilitate pair bonding in male prairie voles.[79] Among marmoset monkeys, fathers have more vasopressin receptors in the brain compared with nonfathers.[80]

Human studies of vasopressin effects on paternal behavior have been quite limited in comparison with studies of oxytocin. One study showed that both mothers and fathers with higher levels of vasopressin in their blood provided more stimulatory touch to their four- to six-month-old infants during a face-to-face interaction.[81] Another study administered intranasal vasopressin to expecting fathers and showed that it increased the amount of time they spent looking at baby-related avatars in a virtual reality environment.[82] However, overall we do not yet have a clear picture of whether and how vasopressin influences paternal behavior in humans.

As discussed in chapter 2, paternal behavior seems to have evolved independently in multiple different mammalian species, including our own. It may be that natural selection does not always "evolve" paternal behavior in the same way. That is, it may not always arrive at the same physiological solution to the problem of evolving paternal behavior in previously nonpaternal species. In some cases, it might rely primarily on the oxytocin system, in others vasopressin, and in others both. This is why we ultimately need to study humans to elucidate the biology of human fatherhood.

Oxytocin and Other Types of Social Bonding

The evidence suggests that oxytocin promotes male-female pair bonding as well as father–infant bonding. But oxytocin may also be involved in other types of social bonding. For one, it seems to be involved in the bonds that humans form with dogs.[83] Dogs have effectively evolved to hijack this bonding system and make us form attachments to them and want to take care of them. The writers of the animated children's film *The Boss Baby* had it right.[84]

Oxytocin also seems to be involved in the bonds that people form with other members of their social group, especially when that group is in competition with other groups. Social psychology experiments have shown that intranasal oxytocin increases what is called *parochial altruism*, cooperation that is selectively directed at in-group members.[85] Parochial altruism and in-group bonding are displayed in many contexts, but none so dramatically as during war. In his excellent book *War*, Sebastian Junger describes the extraordinary strength of the bonds among soldiers.[86] Soldiers don't fight and die for their country as much as they fight and die for each other. These powerful feelings may be supported by oxytocin.

In my lab, we gave men intranasal oxytocin and then imaged their brain activity as they cooperated with other men. We found that oxytocin increased activation in brain regions involved in both reward and empathy.[87] Other evidence comes from our closest living primate relative, the chimpanzee. You may remember that chimpanzees engage in something akin to war. Males, and sometimes females, patrol the border of their territories looking for intruders and for the opportunity to attack adjacent communities when they have a numerical advantage. Oxytocin levels in the urine of chimpanzees increase just before and during these intergroup attacks.[88]

How might oxytocin help men in the context of war? First, battles can be deadly and this thought naturally provokes anxiety. Anxiety could be paralyzing to soldiers, but as we know, oxytocin decreases anxiety. So it might make a soldier brave enough to join the fight. Second, given its potential to increase trust, oxytocin might help soldiers believe that their comrades will uphold their responsibilities; that they will not back down; that they will take on their share of the risk. Finally, much as a parent who loves their child or their partner can't stand to see them harmed, the bond that oxytocin promotes among soldiers could help motivate them to fight on behalf of each other and to protect and defend one another. Testosterone is also increasing in this same context, and this may add some competitive motivation to the mix.

Perhaps these hormonal changes can help us to understand why so many people love team sports. I was born and raised in Wisconsin and have been an avid Green Bay Packer fan my entire life. In fall and winter, my Sunday mood rises and falls with the fate of this team. It is profoundly irrational. I don't personally know any of the players, and I have never met any of them. They come from all over the United States. Few are Wisconsin natives. Yet I strongly identify with them and feel emotionally connected to them simply because they wear the green and gold.

For those who haven't experienced it, this is what it feels like to be a team sports fan: It is a Saturday evening in January at legendary Lambeau Field. The Packers quarterback is at the helm. He is in rhythm; directing the offense like a virtuoso conductor, an elite blend of poise, intellect and athleticism. The players are like buffalo, their breath defiant against the bitter cold as it collides with the arctic air. The Packers have a commanding lead in the fourth quarter and are on their way to victory. I am with them. I am a Green Bay Packer, awash in an intoxicating blend of exhilaration, pride and confidence. The offensive line begins to impose its will. You hear the sound of pads cracking in the bitter night. Gaping holes appear in the weary defense. Six yards on first down. Another eight on second. The crowd is triumphant. The noise is deafening. You can almost feel Vince Lombardi in the air. I mutter, "Don't think you can come up to the frozen tundra and beat us in our house." It begins to snow. It is perfection.

I suspect that much of this psychological experience I have is attributable to hormonal changes induced by this vicarious experience of victory in "battle," including increases in both oxytocin and testosterone.

This chapter has focused on the many ways in which oxytocin can influence paternal behavior, but has only minimally considered the parental brain circuitry that oxytocin acts on. The next chapter focuses specifically on the parental brain.

Highlights

1. Oxytocin promotes father–infant bonding, and fathers with more oxytocin tend to physically stimulate their infants more.
2. Oxytocin may cultivate a strong and healthy pair-bond between men and their partners.
3. There is reason to suspect that boys who are raised in nurturing environments will have a tendency to develop more brain oxytocin receptors and will consequently be more likely to develop into good fathers and husbands.

5 Brain

About fifteen years ago, I was reviewing the academic literature in preparation for a new college course I was creating, Human Social Neuroscience. The class would explore what was known about the social brain—the brain circuits and systems involved in human social cognition and behavior. I thought it was essential to include a unit on parental behavior because it is probably the most ancient and fundamental of all mammalian social behaviors, other than sex perhaps. After wading into the literature, I was struck by how much had been explored in mothers and how little in fathers. I knew that positively engaged fathers were important for child development, so I couldn't understand why there was such a dearth of research in this area. Though disappointing, a gap in the literature can also be a tremendous opportunity for a research scientist. It was then that I began thinking about shifting my research in this new direction, wondering if models of maternal brain function based on research conducted in rodents and humans were applicable to human fathers. Over the past fifteen years, my research group has been working in this area. A handful of other research groups have also taken up this question, and this chapter summarizes our collective progress.

In 2014, a group of scientists from Harvard University did something quite astonishing. They were working with male lab mice that were infanticidal; that is, these males attacked and tried to kill any pups placed in front of them. The researchers then shone a light over a specific part of their brain, and as if by magic, the males stopped attacking and instead began caring for the pups by licking and grooming them, a kind of hugging and kissing by rodents. So when a light was shone on their brain, the males transitioned from homicidal aggression to nurturing affection. How is this possible? The researchers had added a gene to a specific population of neurons in a tiny region at the base of the brain known as the medial preoptic area of the hypothalamus (MPOA). The gene that they added is sensitive to light, and this means that light can "turn on" neurons with this specific gene. Thus, by activating neurons in the MPOA, light was able to cause this dramatic transformation in male mice.[1] You can see why neuroscientists are excited about this technique, *optogenetics*. Experiments like these have shown that the MPOA is critically involved in mammalian parental behavior, including paternal behavior specifically. However, it is far from the only brain structure involved in parental behavior. It is embedded within a neural circuit, and it works in concert with other neural circuits. But before I explain those circuits, I provide some background about the mammalian brain that will help you to appreciate what the MPOA is and how it relates to the rest of the brain.

The human brain may be the most complex system that we know of. It consists of 86 billion neurons, each connected to approximately 10,000 other neurons. How can we begin to understand an organ with such dizzying complexity? It is easy to be intimidated and throw up our hands. But some simplified models of the human brain can help us understand its basic organization and function. One of these models is called the *triune brain model* (figure 5.1).[2] According to this model, the mammalian brain consists of three different layers or levels that expanded at different times in evolution. The first layer is called the *reptilian brain* because it is the main component of the brains of living reptiles. Because reptiles evolved before mammals, this is thought to be the oldest layer of the brain. The reptilian brain is involved in highly stereotyped, instinctive behaviors. An example is when a male stickleback fish attacks another male fish that has a red belly. The red color on the belly predictably triggers the attack behavior. Males will even attack an inanimate object with the bottom painted red. Fish and reptiles have lots of these types of behaviors. Their behavior is highly stereotyped and not very flexible.

The reptilian brain is at the core or the center of the brain. The other two layers wrap around it. The immediately adjacent layer is the the *paleomammalian brain*, also known as the *limbic system*. It is involved in feeling and expressing emotion and is believed to have expanded with the evolution of the earliest mammals approximately 200 MYA. The third layer, the *neomammalian brain*, wraps around the paleomammalian brain. This layer expanded with the adaptive radiation of the mammals after the dinosaurs

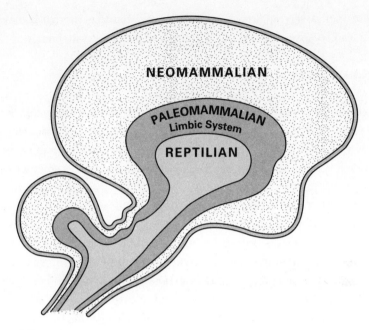

Figure 5.1
The triune brain.

went extinct approximately 65 MYA. The neomammalian brain is involved in thought and reasoning, but it also exerts control over the activity of the lower two levels, allowing us to consciously override instinctive impulses and to suppress emotional responses from the reptilian and paleomammalian layers, respectively (see figure 5.1).

The core parental brain circuitry resides in the two lower levels: the paleomammalian and reptilian brain layers. We know this in part because of a remarkable experiment showing that if the neomammalian brain of a female hamster is removed, she still shows normal maternal behavior. But if you go further and remove portions of the paleomammalian brain, maternal behavior is compromised.[3]

So what are the specific brain structures within the reptilian and paleomammalian brain layers that control parental behavior? Far more research has been done on mothers compared with fathers, so we begin with the former and then ask if what has been learned in mothers also applies to fathers. We will find that there seems to be a generalized parental brain system that largely applies to both males and females, though it may be more readily engaged when the hormones of pregnancy are present.

Maternal Brain Function in Rodents

Rodents are not humans or even primates, but they are mammals like us and so they have the same triune brain layers that we do. Moreover, we know much more about their brains than we do about the human brain because it is possible to employ more revealing methods to study them. For example, we can lesion or inject a chemical into a particular brain area and see if that affects behavior. Or we can measure the amount of a particular neurochemical in a particular brain area—things we can't do in humans. It turns out that we have learned an enormous amount about parental brain function from studying rodents, and much of it applies to humans, so it is worth spending some time considering maternal brain function in rodents.

We start with a crucial structure within the paleomammalian brain: the medial preoptic area of the hypothalamus (MPOA). The MPOA is a group of neurons that sit at the base of the brain (figure 5.2). We know it is critically involved in mammalian parental behavior. If the MPOA in a rat mother is selectively damaged, it will disrupt maternal behavior but leave other behaviors unaffected. We also know that when female rats behave maternally, neurons within their MPOA become more active. If rat mothers are presented with pups, they will care for them, and increased neural activity in the MPOA (and other brain areas) is observed. But if nonmothers are exposed to pups, they will typically avoid or attack the pups, and their MPOA will not show increased activity. That's quite informative because it suggests that the pups will only activate MPOA neurons, which in turn stimulates maternal behavior, if the female has been previously exposed to the hormones of pregnancy.[4]

The MPOA is loaded with receptors that sense these pregnancy hormones. That means it is "designed" by natural selection to listen for and respond to these hormonal signals. For example, estrogen levels increase during the last part of pregnancy in rats, and there are estrogen receptors in the MPOA. We think that when estrogen binds to its receptors in the MPOA, MPOA neurons become more responsive to pups. Part of the evidence for this is that injecting estrogen into the MPOA of a nonmother,

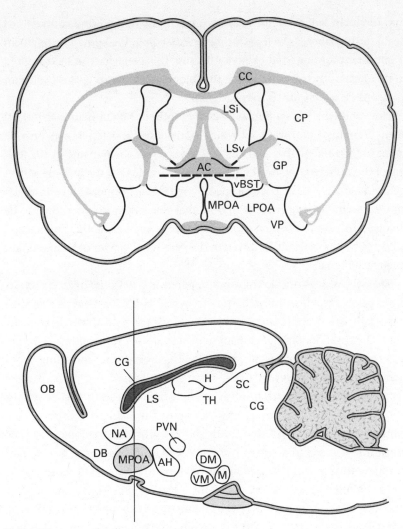

Figure 5.2
Location of the MPOA in the rat brain.

which normally finds pups aversive and will not care for them, will cause her to behave maternally.[5]

Estrogen also does something else very important for the emergence of maternal behavior: it stimulates the production of oxytocin receptors in the MPOA. In effect, estrogen prepares the MPOA to respond to oxytocin.[6]

And oxytocin will arrive in massive quantities when uterine contractions begin during labor, as well as during breastfeeding. We know this oxytocin is important for maternal behavior because this behavior can be disrupted in rat mothers by injecting a drug that blocks oxytocin receptors (called an antagonist), specifically into the MPOA.[7]

How and why does a pup activate its mother's MPOA neurons? In rats, there are neural pathways by which both odors and touch can provide input to the MPOA. This suggests that the scent of the pup and or the feeling of the pup suckling on the nipples may be able to activate these MPOA neurons—if they have been primed by maternal hormones like estrogen and oxytocin.[8] It is important to note that the MPOA is also known to be involved in parental behavior in other species such as rabbits and sheep.[9] Most likely, it is critically involved in the parental behavior of all mammals, including humans.

The MPOA does not do the work of parenting all on its own. It recruits other brain structures to orchestrate parental behavior. One of the most important is a system found within the reptilian brain that relies on the neurochemical dopamine. Dopamine is responsible for feelings of reward and motivation. It makes all kinds of things rewarding, including foods that we love, sex, drugs of abuse, and beautiful faces. But perhaps more critical, it is involved with motivation, by which I mean the desire to seek out a reward—wanting something, as opposed to liking it.[10] Parental care requires motivation, and that motivation is provided by dopamine. The dopamine is made in a circuit called the *mesolimbic dopamine system*, part of the reptilian brain. The circuit has two important nodes. The first is the ventral tegmental area (VTA), where the circuit begins. The second node is the nucleus accumbens, where the circuit ends and dopamine is released. The MPOA is able to tap into the mesolimbic dopamine system because neurons in the MPOA project to the dopamine neurons in the VTA.[11] It is possible to measure dopamine release in the nucleus accumbens. By doing so, researchers learned that dopamine levels increase in the nucleus accumbens of female rats when they behave maternally. We know this dopamine is necessary for maternal behavior because blocking the dopamine receptors that register the dopamine signal disrupts maternal behavior. Damaging dopamine neurons in the VTA also disrupts maternal behavior, so we know the mesolimbic dopamine system is essential for maternal behavior in rats.[12]

Dopamine is strongly implicated in drug addiction and craving.[13] My father-in-law used to joke that his wife, my children's grandmother, was "addicted" to our children. They live about fifteen minutes from us by car. Especially when our kids were infants, my mother-in-law had to come visit them regularly, if only for a few minutes, to get her fix. Dopamine is also involved in food craving.[14] A friend of ours who loves kids once enthusiastically agreed to babysit for us, saying, "I can't wait to gobble them up." And after sending my mom a picture of my daughter, she said, "She looks good enough to eat!" These examples suggest that it is probably no coincidence that a core neural system for parental behavior, the mesolimbic dopamine system, is also implicated in both addiction and food craving.

In chapter 4, we examined the importance of oxytocin for parental care. We therefore might expect that the MPOA, that hub of the parental brain, would somehow link up with the oxytocin system. In fact, it turns out that oxytocin and dopamine are themselves like molecular coparents that work in tandem to support parental motivation. MPOA neurons project to a specific region of the hypothalamus that makes oxytocin, the paraventricular nucleus (PVN).[15] The oxytocin produced by the PVN actually binds to oxytocin receptors back in the MPOA, and that in turn stimulates the mesolimbic dopamine system, which we know increases parental motivation. So oxytocin helps to stimulate dopamine release in response to the pups. In chapter 4, I mentioned that rat mothers with more oxytocin receptors in their MPOA are more affectionate with their pups (e.g., they lick and groom them more). Presumably this is because they more strongly stimulate dopamine release. But the synergy between the two chemicals goes beyond oxytocin stimulating dopamine release. Oxytocin also potentiates the effects of that dopamine. That is, it makes the brain more sensitive and responsive to the dopamine that has been released. This happens because oxytocin and dopamine receptors combine with each other to form what are called heterodimers, and when oxytocin binds to its receptor, it allows dopamine to bind to its own receptor with greater affinity.[16]

Another important function of oxytocin neurons in the PVN is that they project to the posterior pituitary gland, from which they release oxytocin into the bloodstream. Oxytocin is then carried in the blood to the milk ducts, where it stimulates milk ejection during nursing.

So one major job of the MPOA is to stimulate the mesolimbic dopamine system so that mother rats will find their pups rewarding and want to care for them. But that is not enough. It also has to orchestrate specific maternal behaviors, such as licking and grooming, and nursing, and it achieves this through connections with other brain regions that we need not get bogged down with.

Mothers also need to be able to defend their young when they are threatened by predators or hostile adult rats, and this requires something like bravery. Oxytocin helps with the bravery. It is able to suppress activity in a key paleomammalian brain region, the amygdala.[17] The amygdala's job is to detect threats. People with anxiety disorders, who are hypersensitive to threats, have overactive amygdalas. By making the amygdala less sensitive to threat, oxytocin prevents mothers from freezing or fleeing in response to danger. Then MPOA projections to another region in the hypothalamus, the ventromedial nucleus (VMN), make her fight.[18]

Remember that female rats find pups aversive before becoming mothers and that they will attack or avoid pups in this nulliparous state. This is quite common across mammals, and there is actually a brain circuit that supports this antinurturing behavior. Before becoming mothers, female rats dislike the odor of pups. That odor is processed by the olfactory bulb of the brain, which in turn projects to the amygdala, which is involved in processing threatening stimuli. The MPOA projects to this offspring avoidance circuit and may be able to inhibit the circuit to help maternal behavior emerge in mothers after they give birth. Additionally, oxytocin receptors are widespread throughout this avoidance circuit, perhaps allowing oxytocin to dampen it as well, thereby also supporting maternal behavior.[19]

Maternal Brain Function in Humans

Can what we learn from the maternal rat brain be applied to human mothers? As I alluded to above, our ability to study parental brain function in humans is quite limited compared with what we can do in rodents. For example, we obviously don't lesion brain areas like the MPOA in humans and examine the consequences for parenting. Nor can we inject dopamine or oxytocin, or antagonists of dopamine or oxytocin, into specific brain regions. We also cannot optogenetically activate specific neural populations and observe the behavioral consequences. We can't even see which neurons are active when parents are taking care of their children. The only tool we have available to study the parental brain in humans is noninvasive techniques that allow us to image activity across the brain, within "voxels" that include about a million neurons each. Moreover, people can't be moving while we image their brain activity. They have to lie still inside a brain scanner. About the best we can do is present people with stimuli from their children while they are in the scanner and observe which areas of the brain become active.

The most popular human brain imaging technique is functional magnetic resonance imaging (fMRI), which uses a powerful magnet to measure changes in brain blood flow resulting from increases in neural activity (figure 5.3). In one ingenious fMRI study, researchers morphed infant photographs to have either more or less baby schema, the physical characteristics of infants that make them attractive to adults. Basically, the researchers morphed the infant photos to be more or less cute. They asked women (nonmothers) to rate their motivation to take care of each baby that they saw a picture of, and women reported more interest in taking care of the infants with more baby schema (i.e., cuter babies). The researchers also measured brain activity in these women using fMRI and found that as the amount of baby schema increased, women showed more activation within the nucleus accumbens (NA).[20] That is, the same brain area that is involved

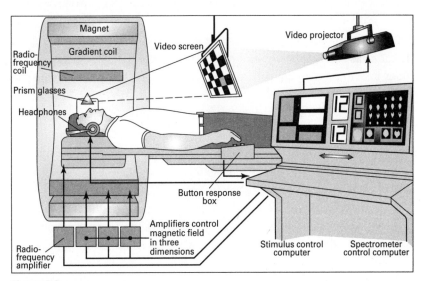

Figure 5.3
The fMRI scanning environment.

in parental motivation in rodents was also tracking how cute the human infants were, as well as women's motivation to take care of them. So cute babies are able to reach into adult brains and activate their reward and motivational system so the adult will want to take care of them. Nice trick.[21]

One noteworthy fact about this study is that it was conducted with women who were not mothers and were viewing pictures of unknown children. So infants may not only be able to seduce their mothers into caring for them; their charms may also work on others. This may be a crucial adaptation in a cooperative breeding species, especially one like ours in which mothers wean infants before they are capable of foraging for themselves independently.[22] It is not hard to imagine scenarios throughout human evolution in which cute babies were able to secure valuable gifts of food from adults at critical times so they could survive or stay healthy enough to fight off pathogens that could kill them later.

One weekend a few years ago when my daughter was two years old, we were at a public fair in our community. She and I were sitting together in the grass playing. There was a man nearby making delectable-looking beignets. He kept looking over at us and smiling, and I could tell he thought Mia was cute. Eventually he came over and gave her a beignet. I thought that was super nice, but it also occurred to me that he did not offer me one. Maybe I didn't look as if I needed one. But I think Mia got one because of her cuteness. After he left, I said, "Great job, Mia. You just got about 100 calories from an unrelated stranger for free!" I couldn't help but think that maybe that ability allowed one of our distant ancestors in Paleolithic times to scrounge up enough calories to survive childhood and continue our lineage.

A number of fMRI studies have examined which regions of mothers' brains respond specifically to viewing photographs of their own children. One study of first-time mothers found stronger activation in the nucleus accumbens when they viewed photographs of their own compared with unknown infants. Remember that the nucleus accumbens is part of the mesolimbic dopamine reward and motivation system, so this result suggests that mothers may find their own infant more rewarding or motivating than an unknown infant. Furthermore, mothers with higher levels of oxytocin in their blood had stronger activation in the nucleus accumbens, suggesting that these mothers may have found their infant particularly rewarding, exactly as the above model based on rodent research would predict.[23]

A recent study measured behavioral synchrony between mothers and their infants in their home and found that those who were more synchronized with their infant showed stronger activation in the nucleus accumbens when viewing videos of their infant.[24] In order to synchronize with her infant, a mother needs to be sensitive to the infant and dynamically adjust her behavior in response to infant cues. This requires motivation, likely provided by activation of the nucleus accumbens. Another notable human study has even managed to directly demonstrate dopamine release in the nucleus accumbens as mothers view videos of their infant, using another neuroimaging technique, positron emission tomography (PET).[25] Researchers found that mothers who displayed greater mother–infant synchrony had more dopamine release in the nucleus accumbens when viewing a video of their infant.[26] What's more, more synchronous mothers had higher levels of oxytocin in their blood. So it seems that oxytocin and dopamine may be working together in the maternal brain to help mothers synchronize their behavior with that of their infant.

There have been enough brain imaging studies of maternal brain function that it is now possible to conduct meta-analyses, studies that synthesize the results over a large number of related studies. While any one study may generate a false-positive or false-negative result just by chance, that should not happen when the results of many studies are pooled and we can look at the overall pattern of findings. A meta-analysis of studies comparing the maternal brain response to viewing photos of their own compared with an unknown child identified the VTA, the other node of the mesolimbic dopamine system, as one of the consistently activated regions.[27] Other studies have imaged maternal brain function as mothers in the MRI scanner listen to infant cries. A number of these studies have also reported activation in the VTA.[28] Although many of us may not find infant cries rewarding, they are certainly motivating: "What do I need to do to make this stop?" or, more empathically, "How can I alleviate my infant's distress?" Given the mesolimbic dopamine system's role in motivation, its engagement by infant crying certainly makes sense.

Further evidence for a role of the mesolimbic dopamine system in maternal motivation comes from fMRI studies of depressed mothers. Mothers with postpartum depression often suffer from anhedonia—an inability to feel pleasure—accompanied by a lack of maternal motivation. And we know that their infants can suffer as a result of the mother's decreased sensitivity and responsiveness. Mothers with postpartum depression have been shown to exhibit reduced activation within the mesolimbic dopamine system when viewing pictures of their infant or when listening to infant cry stimuli.[29]

There is also some tentative evidence that oxytocin interacts with the mesolimbic dopamine system in humans as it does in rats. One study had mothers self-administer intranasal oxytocin and found that it increased their response to pictures of crying infants within the VTA.[30] Oxytocin is also hypothesized to inhibit the parental avoidance system that runs through the amygdala, which would presumably function to decrease the aversiveness of infants. Therefore, it is quite interesting that intranasal oxytocin decreases the amygdala response to infant crying in women, which has been suggested to decrease feelings of anxiety and aversion to the crying infant.[31]

Thus, as in rats, it appears that maternal brain function in humans relies on the mesolimbic dopamine system for caregiving motivation (shown in light gray in figure 5.4). But good parenting depends on more than just motivation. Parents must also be able to understand what their infant needs, and this requires empathy. A great deal of research that has investigated which regions of the human brain are involved with empathy has identified at least three different neural systems that mediate different aspects of empathy.

One of these is involved in the ability to feel what someone else is feeling—what we call emotional empathy. For example, you see your infant in distress—their facial expression, their cry—and you feel that distress yourself. This feeling alerts you that something is wrong, and a parental response may be needed. When I see my six-year-old daughter's face become sad after I scolded her and her lips slowly start to protrude and begin to quiver, and then despite her best efforts to stifle the cry, her belly begins to convulse and she starts sobbing, and I feel like my heart is about to be crushed, that's the emotional empathy system working. Emotional empathy involves a network that includes two regions of the paleomammalian brain: the anterior cingulate cortex and the anterior insula (shown with diagonal gray stripes in figure 5.4).

The second and third empathy systems are part of the neomammalian brain. The second is the mirror simulation network. This network includes "mirror neurons," which have the special property that they fire (produce an

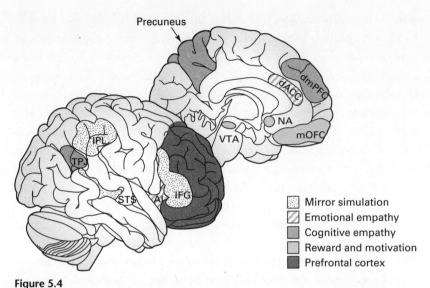

Figure 5.4
Neural systems involved in human parental behavior. VTA = ventral tegmental area;
NA = nucleus accumbens; mOFC = medial orbitofrontal cortex; AI = anterior insula;
dACC = dorsal anterior cingulate cortex; IFG = inferior frontal gyrus; IPL = inferior
parietal lobule; STS = superior temporal sulcus; dmPFC = dorsomedial prefrontal cortex; TPJ = temporoparietal junction.

electrical impulse) both when an individual performs an action and when
that individual observes another individual performing the same action.
Mirror neurons were discovered serendipitously by a group of Italian neuroscientists who were studying neurons involved in reaching and grasping.
They were recording electrical impulses from these neurons and noted that
the impulses, called action potentials, occurred at higher frequency when
the monkey performed a grasping movement, like reaching for a peanut.
But then something surprised them: the same neurons also fired in the
monkey when the human experimenter reached out and grasped the peanut. These neurons were consequently dubbed "mirror neurons" for obvious reasons. They are found in at least two regions of the cerebral cortex:
the inferior frontal gyrus and the inferior parietal cortex (shown in white
with black dots in figure 5.4). Mirror neurons form a natural link between
the sender and the receiver of a gesture and allow the receiver to achieve a
better understanding of the action made by the sender through simulating

it in their own brain.[32] So your infant may smile at you and you are able to feel the emotion behind that smile by covertly simulating the smile in your own brain.

The third neural system is involved in cognitive empathy, also known as theory of mind, which is the ability to understand what someone is thinking or feeling and why. For example, I think my infant is crying because she didn't have a nap yet today and she is tired. Or my infant is crying because she is hungry or because she wants to be held. This component of empathy is mediated by a number of neomammalian brain regions, including the dorsomedial prefrontal cortex, the temporo-parietal junction, and the precuneus (shown in medium gray in figure 5.4).

fMRI studies show that mothers activate all three of these neural systems—emotional empathy, mirror simulation, and theory of mind—when presented with stimuli from their infant. There is also some evidence that more empathic mothers activate these neural systems to a greater extent than less empathic mothers. For example, one study showed that mothers who report more empathy for their infant had stronger activation in the anterior insula when viewing pictures of their infant displaying emotional expressions.[33]

One final brain region plays an essential role in human maternal behavior: the prefrontal cortex (shown in dark gray in figure 5.4). The seat of many of our highest cognitive abilities, such as decision making and planning, it is also involved in keeping our emotions and impulses in check. This is at a premium when an infant won't stop crying or screaming, and perhaps the parent has had a long day and is also sleep deprived from the previous night's crying. Infant crying can be tremendously frustrating at these times, and parents need to keep their emotions under control. The prefrontal cortex is involved in keeping our cool under these circumstances. One study observed mothers interacting with their infants and also imaged maternal brain activity as they listened to their infant crying. Mothers who exhibited greater sensitivity when interacting with their infant had greater prefrontal cortex activation when listening to their infant crying. Furthermore, these mothers had a lower stress hormone response to a brief separation from their infant.[34] These findings suggest that mothers who are able to engage their prefrontal cortex during infant crying are able to stay calmer and deliver more sensitive care to their infant.

Thus far, we have discussed only maternal brain function. Are these models also applicable to fathers?

Paternal Brain Function in Rodents

As with rodent mothers, the MPOA is critically involved in parental caregiving in rodent fathers. Recall that male rats normally avoid pups. However, if they are forced to cohabitate with pups for several days, they seem to habituate to the smell of the pups and begin to find them less aversive until they eventually start to show parental behavior: licking and grooming, retrieving pups to nest, and so on. However, when researchers selectively damaged the MPOA of male rats, they showed deficits in parental behavior in this paradigm.[35] By contrast, implanting estrogen, the primary female sex steroid hormone, into the MPOA of male rats accelerated the onset of their parental behavior.[36]

But what about species in which males naturally and habitually care for their young, such as the California mouse? Is the MPOA involved in paternal behavior in these species? It seems so, since California mouse fathers show increased neuronal activity in the MPOA after being exposed to pups and lesioning the MPOA disrupts their parental behavior.[37] Paternal behavior is also disrupted by lesioning the nucleus accumbens, the target of the mesolimbic dopamine neurons implicated in parental motivation in female rats.[38] So it is likely that the same neural circuit controls parental motivation in both mother and father rodents. This conclusion is further supported by studies in laboratory mice in which specific MPOA neurons are causally implicated in paternal behavior. They become active when males engage in parental behavior, paternal behavior is abolished when these neurons are damaged, and infanticidal males become paternal when these neurons are stimulated optogenetically.[39]

Other research shows that becoming a father actually changes the brain of male rodents. Mandarin vole fathers (figure 5.5) are excellent parents, and their brains are transformed by fatherhood. When males become fathers, oxytocin receptors increase in both the MPOA and the nucleus accumbens.[40] Furthermore, there is an increase in the number of neurons that produce oxytocin within the hypothalamus. Not only that, but dopamine receptors increase in the nucleus accumbens.[41] Presumably this

Figure 5.5
Mandarin vole father, mother, and pups.

means that mandarin vole fathers are making and releasing more oxytocin in their brain and that oxytocin is having more of an impact on the brain circuitry responsible for paternal motivation, where it acts to increase dopamine release and binding.

Quite striking neurological changes have also been described in another biparental species, the California mouse (figure 5.6), as males transition to fatherhood. As described in chapter 3, paternal behavior in California mice depends on estrogen binding to its receptors within the MPOA. However, that estrogen is actually derived from testosterone. That is, testosterone is the precursor to estrogen. The enzyme aromatase is responsible for the conversion of testosterone into estrogen. When California mice become fathers, there is an increase in aromatase levels within the MPOA specifically, which allows testosterone to be converted to estrogen there and to trigger paternal behavior.[42]

Several interesting changes have also been described within California mouse fathers' hippocampus. Most of the neurons in the adult brain are already present at birth, but a small number are produced in adulthood, and the hippocampus is the brain region most involved in this process of adult neurogenesis. The hippocampus is involved in learning and the

Figure 5.6
California mouse father and pup. Reproduced with permission from Dr. Kelly Lambert, https://www.kellylambertlab.com/research.

formation of new memories, so it makes sense that new neurons are born here. Remarkably, the hippocampus begins producing more neurons when California mice become fathers, as if preparing the male to learn how to be a good father. In addition to this increase in hippocampal neurons, those hippocampal neurons undergo changes in their shape—their neuron "spines" actually become denser—which allows them to receive more input from other neurons.[43] These hippocampal changes may help males adapt to their new role as a father. One study showed that new California mouse fathers perform better on a laboratory task designed to measure their foraging abilities compared with nonfathers.[44] Why is it important for new fathers to be efficient foragers? Perhaps it allows them to spend less time away from the nest, making them more available to provide warmth and protection to their pups.

Increased neurogenesis has also been observed among new laboratory mouse fathers, not only in the hippocampus but also in the olfactory bulbs. Interactions with pups trigger this neurogenesis, and the increase in olfactory bulb neurons is what allows the father to learn to recognize the scent of his pups.[45]

A bit closer to home, among our primate relatives, marmoset monkey fathers also experience some noteworthy neurobiological changes. Remember that the neuropeptide vasopressin, which is quite similar to oxytocin, is involved in paternal behavior in some species. Compared with nonfathers, marmoset fathers have more vasopressin receptors in their prefrontal cortex, suggesting that they are likely more responsive to its effects in this brain area.[46]

Paternal Brain Function in Humans

Over the past fifteen years, my lab has been actively engaged in research on paternal brain function in men, specifically asking whether models of parental brain function based on studies done in nonhuman animals and human mothers generalize to human fathers.

In our initial study, we recruited seventy fathers with toddler children (ages 1 or 2) and imaged how their brains responded to viewing pictures of their child. We also asked the child's mother (the father's partner) to complete questionnaires about the father's parenting behavior. When fathers viewed photos of their children, they activated brain regions involved in reward and motivation (the ventral tegmental area or VTA), emotional empathy (thalamus and cingulate cortex), and cognitive empathy (dorsomedial prefrontal cortex). The overall pattern of activation was qualitatively similar to that described for human mothers above. More important, activation in the VTA was positively correlated with how involved the father was in taking care of his toddler, as reported by the mother. That is, if the father was involved in bathing, feeding, changing diapers, and so on, he tended to have stronger VTA activation when viewing a photo of his child.[47] Remember that this area is part of the mesolimbic dopamine system. So it may be that fathers who find their child's appearance more rewarding tend to become more involved in taking care of that child. Or it could mean that as fathers become more involved in caregiving and consequently become more attached to their child, the child becomes a more rewarding or motivating stimulus to them. At any rate, it fits well with the notion that the mesolimbic dopamine system is involved in human parental motivation, whether the parent is a mother or a father.

Oxytocin seems to promote father–infant bonding. The rodent research established that it acts within the mesolimbic dopamine system to increase dopamine release in the nucleus accumbens. We wanted to see if this was the mechanism by which oxytocin facilitates paternal behavior in men. So in a subsequent study, we asked fathers to administer intranasal oxytocin to see if it would alter their neural response to viewing pictures of their toddler child. In fact, oxytocin increased activation in three different brain regions in fathers. The first was the caudate nucleus, a structure that is just above the nucleus accumbens, which is part of the mesolimbic dopamine system. The caudate is not part of the mesolimbic dopamine system, but it does receive dopamine projections from another midbrain region that is also involved in reward and motivation, the substantia nigra. So increased activation in the caudate could mean that oxytocin renders the child a more rewarding stimulus to the father. Oxytocin also increased activation in the anterior cingulate cortex, which is part of the emotional empathy system. Finally, oxytocin increased activation in the visual cortex, which is often found when a participant is focusing more attention on a visual stimulus.[48] Thus, oxytocin may be increasing paternal empathy and rendering the child a more salient and rewarding stimulus.

Studies from other research groups have similarly identified the caudate nucleus as being involved in human paternal brain function. One study found that fathers more strongly activated the caudate nucleus when viewing videos of their own infant compared with an unknown infant.[49] Another study imaged brain activity in fathers as they listened to their own newborn infant crying and found that those who more strongly activated the caudate nucleus reported more positive thoughts about being a parent.[50] Thus, in humans at least, we may have to expand our models of paternal brain function to include the caudate nucleus as part of the parental reward and motivational system.

In addition to the caudate nucleus and the nucleus accumbens, one more region is a major target of midbrain dopamine neurons, the medial orbitofrontal cortex (mOFC). It is located in the cerebral cortex itself, part of the neomammalian brain, situated just above the orbits of the eyes, on the underbelly of the frontal lobes. Like other components of the mesolimbic dopamine system, it consistently activates in response to a wide range of rewarding stimuli.[51] We showed that fathers more strongly activate this region than nonfathers when they are viewing photographs of unknown toddler children, suggesting that fathers may find these children to be a more rewarding stimulus than nonfathers do.[52] Another study of fathers found that the mOFC responded more to pictures of their own child compared with an unknown child, consistent with the expectation that the own child would be a more rewarding stimulus to fathers.[53] In our lab, we also compared the neural response of fathers to viewing pictures of toddler-age daughters versus toddler-age sons. Interestingly, the mOFC of fathers responded more to pictures of smiling daughters than to pictures of smiling sons.[54] There is some evidence that girls and women tend to smile more than boys and men, which, together with our neuroimaging results, raises the intriguing possibility that parents more strongly reinforce smiling in girls because they find it more rewarding, and this reinforcement causes their daughters to smile more.[55]

We can think of children as having been designed by evolution to activate parental brain circuits to make adults want to care for them. One effective method of doing so is simply being cute, but this is not their only tactic. Infants also use crying to shape parental behavior. They cry until we figure out how to meet their needs, and then they mercifully stop, thereby reinforcing whatever we did to solve the problem. Our lab has also investigated how fathers' brains respond to infant crying. We have been particularly interested in this question for a quite sobering but important reason. Although infant crying is normally effective in shaping parental behavior, it is also a known trigger for infant abuse, and fathers are more often the perpetrators than mothers in such cases.[56] Therefore, we would like to know what a father's brain looks like when he is frustrated by infant crying so that future studies can seek ways to attenuate this pattern of brain activation and decrease the likelihood of abuse.

In one of our first studies, we had fathers listen to a newborn infant crying and found that fathers who reported having more empathy for their child had a stronger response to crying within the anterior insula, one of our emotional empathy regions. However, the fathers with the strongest anterior insula response were not necessarily the best fathers. It turned out that fathers with moderate levels of anterior insula activation were more involved in instrumental caregiving than fathers with either low or high activation, a classic inverted-U relationship. We speculatively proposed that fathers with low levels of activation may be lacking in empathy and may disengage from caregiving due to apathy. By contrast, fathers with high activation may be empathically overaroused, a term used to describe the psychological state of a person who shares another's distress to the point that they themselves become overwhelmed and unable to help effectively. So perhaps our empathically overaroused fathers withdrew from instrumental care to manage their stress levels—and perhaps also those of their child.[57] More generally, effective parenting may depend on maintaining an optimal state of empathy and arousal—neither too high nor too low.

In another study, we used fMRI to image how fathers' brains responded to their own infant's cry. We also asked fathers to tell us how aversive they found the cry. Fathers who found the cry more aversive had stronger activation within the anterior cingulate cortex, part of our emotional empathy system. This is also consistent with the idea that too much empathy can lead to excessive negative emotion in the parent and potentially interfere with sensitive caregiving.[58]

As I mentioned above, one study found that intranasal oxytocin decreased the amygdala response to infant crying among women who were not mothers, and this was interpreted to suggest that oxytocin decreased anxiety or aversion to infant crying. The same research group has also examined the effect of intranasal oxytocin on amygdala reactivity to infant crying in first-time fathers and found that it also decreased their amygdala response.[59] These findings are consistent with rodent models of parental brain function that postulate that oxytocin suppresses the parental avoidance system.

fMRI studies of parental brain responses to infant crying have typically involved parents lying in an MRI scanner passively listening to infant crying. However, this is an unnatural situation, since parents rarely passively listen to their infant cry. Instead, they actively respond to the crying in an attempt to meet the infant's needs and stop the crying. So we designed an fMRI study in which fathers had to actively try to sooth infant crying by choosing among a series of soothing options (for example, feeding, holding, swaddling, changing the diaper, using a pacifier) and then exerted effort to implement that choice in a video game format. When the father chose and implemented the correct soothing option, the crying would stop. Compared with passive listening, actively responding to infant crying activated brain regions involved with reward and motivation (ventral tegmental area, nucleus accumbens, and caudate nucleus), emotional empathy (anterior insula and anterior cingulate cortex), and mirror simulation (inferior frontal gyrus). Fathers were also asked to report their level of frustration after each trial of the task. Those who reported more frustration had less activation within the prefrontal cortex as they actively responded to the infant crying. I have noted that the prefrontal cortex is involved in emotion regulation and impulse control and that mothers who have more sensitive interactions with their infants have more active prefrontal cortices. Our results suggest that fathers may need to engage their prefrontal cortex to avoid becoming overly frustrated when caring for a crying infant.[60]

Beyond my own research group, a number of others have conducted important studies that have advanced our understanding of human paternal brain function. An important fMRI study by an Israeli research team directly compared brain function in primary caregiving mothers, secondary caregiving fathers, and primary caregiving fathers. Brain activity was measured as parents viewed videos of their infant. There was considerable similarity among the three groups of parents in terms of which brain areas were activated in response to the videos. For example, all three groups activated regions involved in reward and motivation (ventral tegmental area and mOFC), emotional empathy (insula), and mirror simulation (inferior frontal gyrus). But there were a couple of notable differences among the groups. Primary caregiving mothers had four-fold stronger activation in the amygdala compared with secondary caregiving fathers.[61] This might seem confusing since I have noted that the amygdala is part of the circuit involved in the motivation to avoid offspring. However, the amygdala includes multiple nuclei with different functions. Whereas the avoidance pathway runs through the medial amygdala nucleus, the basolateral nucleus is believed to be a pathway by which olfactory, tactile, and auditory infant stimuli can access the MPOA. So although the researchers could not localize their amygdala activation to a specific nucleus, they may have been detecting activation from the basolateral nucleus.[62]

Interestingly, there was also a brain region that responded more in secondary caregiving fathers as compared with primary caregiving mothers. This is a region in the cerebral cortex known as the superior temporal sulcus (STS) that is involved in visually processing movements of the body and feeds into the mirror neuron system. One potential interpretation of this activation is that secondary caregiving fathers are more attentive to the movements of their child. In fact, STS activation was correlated with the degree of father–infant synchrony observed during a separate parent-infant interaction.[63] Fathers who track and respond to their infant's movements more closely activate the superior temporal sulcus more.

What about primary caregiving fathers? Remarkably, they had amygdala activation like primary caregiving mothers along with STS activation like secondary caregiving fathers—the best of both worlds.[64] One can't help but wonder if some single mothers, who may need to fulfill both traditional maternal and paternal roles, might respond similarly. The augmented amygdala activity of primary caregiving fathers is of particular interest in light of an interesting Dutch study showing that new fathers who were assigned to an intervention that increased their infant carrying time had a stronger amygdala response to infant crying compared with new fathers who went through a control intervention.[65] Perhaps prolonged intimate tactile contact with an infant tunes the basolateral amygdala to be more responsive to infant stimuli in any parent.

Individual brain regions don't function in isolation. They are embedded within circuits or networks. Neuroscientists have recently turned their attention to the behavior of these networks and how they may affect social behavior. Networks may function more efficiently, or at least differently, when activity within the regions that compose the network are synchronized with one another so that they are working together in unison. To take one example, the act of mating synchronizes activity between different regions of the mesolimbic dopamine system in female prairie voles, and this in turn facilitates the formation of a pair-bond with her male partner.[66] As another example, we showed in my lab that intranasal oxytocin synchronized activation within a social behavioral neural network in men who were cooperating with other men.[67]

An ambitious recent fMRI study by the Israeli research group examined synchrony or coherence within brain networks involved in reward and motivation, emotional empathy, and cognitive empathy as parents (mothers and fathers) watched videos of their own and unknown infants. Watching videos of their infant led to increased coherence within these networks, as well as increased coherence among them. Amazingly, the researchers went on to show that the infants of parents who had better synchronized neural networks had better subsequent developmental outcomes in preschool and at age six. For example, children of parents who had greater coherence between emotional and cognitive empathy networks went on to have better emotion regulation abilities in preschool, which in turn predicted better mental health (less depressive tendencies and less anxiety) at age six.[68] This is an extraordinary finding, and it will be important to see it replicated. If it is true, the way that a parent's brain responds to their infant actually predicts the child's psychosocial well-being six years later.

So far, we have discussed evidence that human fathers mostly rely on the same neural systems as human mothers. As with male rodents, there is also emerging evidence that fatherhood actually changes men's brains. We don't currently have the means to look for some of the important molecular changes seen in the brains of rodent fathers. For example, we can't measure changes in oxytocin receptor levels in specific brain regions in humans, although it is quite likely we will be able to do so in the near future once suitable radio-labeled ligands are developed that can be combined with positron emission tomography (PET) imaging. However, we can currently describe macroscopic changes in brain structure that can be visualized with noninvasive structural MRI imaging. It is possible to acquire a very nice structural, or anatomical, MRI scan of a person's entire brain in only about five minutes.

One excellent MRI study showed pronounced changes in the cerebral cortex of women over the course of pregnancy.[69] The cerebral cortex is composed of "gray matter." To understand what gray matter is requires knowing a little bit about neurons. At one end of a neuron (figure 5.7)

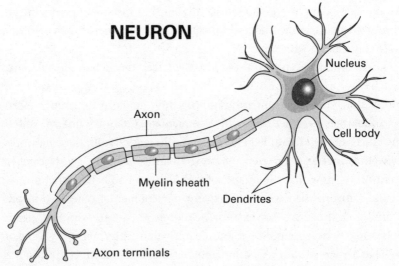

Figure 5.7
Diagram of a neuron.

is a cell body with a nucleus along with dendrites, structures that receive input from other neurons. At the other end is an axon, a cable-like structure that extends away from the cell body and makes connections with other neurons. White matter is composed of these axons, whereas gray matter consists mostly of cell bodies and dendrites.

The study found that as women progress through pregnancy and enter the postnatal period, they experience decreases in gray matter volume across the cerebral cortex. Those decreases happen to be concentrated in regions of the cortex that are involved in cognitive empathy, or theory of mind. So it is interesting to speculate that the mother's brain is getting prepared to empathize with her infant. But shouldn't gray matter be increasing instead? Wouldn't that make more sense? Gray matter actually declines in children beginning very early in development.[70] Late in life, decreases coincide with cognitive decline, probably because there is an actual loss of neurons. But throughout childhood and adolescence, decreases in gray matter are thought to reflect a refinement of neural connections and to be an important component of learning. It turns out that connections between neurons are massively overproduced early in brain development, and these need to be pruned back to sculpt mature neural networks.[71] That pruning is an important part of the learning process. Therefore, we should think of the decline in gray matter that mothers experience as a refinement of their brain wiring.[72]

In addition to the cerebral cortex, gray matter decreases were also seen in the nucleus accumbens and this anatomical change was linked with maternal brain function. Researchers imaged brain activity in these same women with fMRI as they viewed photos of their infants and found that mothers who had a greater decrease in nucleus accumbens gray matter had a stronger nucleus accumbens response to viewing pictures of their infant.[73] So perhaps the structural changes in nucleus accumbens are tuning up the mother's reward and motivational system to be more responsive to her infant.

But what exactly is responsible for this decline in gray matter? One obvious candidate is the hormonal changes that women experience during pregnancy. However, the postnatal scans in this study were acquired at about ten weeks of age, implying that changes could also be due to those ten weeks of postnatal mother–infant interactions. That is, the actual experience of raising an infant could be driving these changes in the brain.

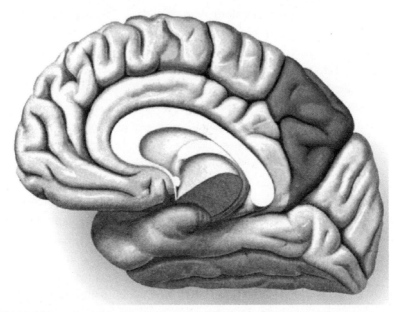

Figure 5.8
Midline drawing of the human brain, illustrating the anatomical position of the
precuneus.

A parallel study in men show that they too experience declines in cortical
gray matter volume across the transition to fatherhood, but those declines
are more subtle and appear to be concentrated in one particular cortical
region within the cognitive empathy system, the precuneus (figure 5.8).[74]

Another study focused specifically on parental changes in gray matter
volume that occurred during the postpartum period, scanning parents in
the early postpartum period (between two and four weeks) and again in the
later postpartum period (between three and four months). Among mothers,
this study reported widespread increases in gray matter and no decreases in
gray matter volume, which is opposite to the above findings. So perhaps
the above gray matter decreases in mothers occur during pregnancy, before
the baby is born, and this is followed by postnatal increases.[75] What then
do the postnatal increases mean? We don't know for sure, but in addition
to synaptic pruning, learning and experience can involve the formation of
new connections and that could result in increases in gray matter volume.

It is quite interesting that the observed increases in new mothers included both the VTA and hypothalamus (inclusive of the MPOA), key regions within the parental reward and motivation system.[76]

Over the same interval postpartum, fathers also showed increases in the hypothalamus. In contrast to mothers, however, fathers also showed postnatal decreases in a number of areas, including, once again, the precuneus.[77] These changes in fathers obviously cannot be due to pregnancy hormones but could plausibly be due to hormonal changes that men experience when they become fathers, such as increases in oxytocin and decreases in testosterone. Alternatively, they could result from father–infant interactions. In principle, it should be possible to determine the correct explanation by testing whether men with more pronounced hormonal changes or men who interact more with their infant experience greater changes in gray matter volume.

Effects of Paternal Care on Offspring Brains

Chapter 1 presented evidence that paternal caregiving can affect the psychological and social development of offspring in both humans and other species where males habitually provide parental care. Given this, paternal caregiving is also expected to affect offspring brain development. Several studies have evaluated this possibility through paternal deprivation experiments in biparental rodent species such as mandarin voles and California mice. In these experiments, fathers are removed from the nest, and offspring are raised exclusively by their mothers. These studies have demonstrated that paternal deprivation indeed has an impact on multiple aspects of offspring brain development.

In mandarin voles, paternal deprivation decreases expression of the oxytocin receptor gene in the nucleus accumbens of offspring.[78] This means that the gene that makes the oxytocin receptor is not as active in this brain region in offspring who are deprived of a father, likely implying that they have fewer oxytocin receptors in that brain area. Female, but not male, offspring also have lower oxytocin levels in their blood. Not to beat a dead horse, but oxytocin acts in the nucleus accumbens to increase parental motivation, so this could explain why paternally deprived mandarin voles also show reduced parental care when they become adults.[79] Not only this, but they show a general lack of motivation to engage in social interactions with other animals that may also be mediated by reduced oxytocin signaling in the nucleus accumbens.[80]

The *Octagon degus* is another biparental rodent species. In this species, paternal deprivation is associated with abnormal development of the prefrontal cortex. Neuronal changes within the prefrontal cortex suggest impaired communication with other brain regions.[81] This is particularly important since we know that prefrontal cortex connections with the amygdala are involved in controlling negative emotions like anxiety and aggression in humans and are therefore critical for mental health.[82]

In chapter 1, I emphasized that children with positively engaged fathers tend to have better emotion regulation skills. This raises the possibility that father–child interactions are shaping the development of the prefrontal cortex in humans, as they do in *O. degus*. But how might this happen exactly? Remember that fathers in modern, developed societies tend to specialize in physical play, particularly rough-and-tumble play. Some intriguing experiments in rats show that depriving juveniles of play fighting leads to both abnormal prefrontal cortex development and abnormal social behavior. Play-deprived rats are more likely to respond aggressively to benign social contact, and they fail to exhibit submissive behavior when encountering dominant males. A special nerve growth factor known as BDNF (brain-derived neurotrophic factor) is normally released into the prefrontal cortex during play, and play is needed for the proper development of prefrontal cortex neurons.[83] So rough-and-tumble play may be one route by which fathers are able to cultivate the development of good emotion regulation and good social skills in their children. In nuclear family settings experienced by many children in high-income nations, fathers may be a key, or even the primary, playmate. In more traditional human societies, other children are often the main source of rough-and-tumble play, and this too is likely to promote prefrontal cortex development.

To my knowledge, no one has yet looked at whether being raised without a father is associated with altered prefrontal cortex structure or function in human children, although we certainly have the ability to conduct such a study (hint: send me grant money). If such an association were found, it would nonetheless be challenging to firmly establish father absence as the cause of prefrontal abnormalities due to many potential confounding variables that are associated with it. Regardless, in a cooperatively breeding species such as ours, we might expect that positive involvement of other allomothers (such as grandmothers and older siblings) might mitigate or eliminate many of the neurobiological consequences of being raised without a father.

What is not clear from these paternal deprivation experiments is whether the observed changes in neural and behavioral development result from the absence of a father in particular rather than from a reduction in the overall amount of parental care. That is, it might be due to having only a single parent rather than to the lack of a father. This possibility was recently addressed in a clever experiment by Professor Karen Bales and colleagues from UC Davis in which some prairie vole pups were raised by their mother and the pup's older sister. Because prairie voles are cooperative breeders, it is not uncommon for juveniles to care for pups in nature. The pups who were raised by the mother–sister combination received as much overall care as pups raised by a mother and father, yet they had fewer oxytocin receptors in their nucleus accumbens, and the male pups grew to have deficits in pair-bonding behavior as adults.[84] These findings suggest that prairie vole fathers care for their pups in a qualitatively different way than mothers or sisters that is uniquely beneficial. Although it remains to be determined what that difference specifically is, it seems to be consequential for development of both prairie vole pup brains and behavior. As the authors of the study note, we don't know the extent to which this finding generalizes to other species, including our own.

Over the past few chapters, we have been examining the biology of fatherhood. The next chapter switches gears and instead focuses on the many important social and cultural influences that powerfully shape paternal behavior.

Highlights

1. Both maternal and paternal behavior appear to depend on a global parental caregiving system.

2. The medial preoptic area is a critical parental brain area that works with oxytocin and midbrain dopamine systems to support parental motivation.

3. Both human mothers and human fathers rely on cortical systems involved in emotional and cognitive empathy, as well as mirror simulation.

4. The prefrontal cortex helps parents keep their negative emotions, such as frustration and anxiety, in check so they can be better parents.

5. Recent studies suggest that, similar to other animal species, when men become fathers they experience changes in their brain that may be preparing them for fatherhood.

6 Norms and Circumstances

My dad was a remarkably patient and kind father. As a boy, I remember going fishing together with him and my three older brothers on the lakes of northern Wisconsin. We were all crammed into a small boat. We boys would eagerly, and frantically, fish all at the same time. Amid the chaos of our tangled lines, snagged lures, and hooked fish, Dad would calmly help each one of us in turn resolve our issues so we could resume fishing and get back to our fun. Dad didn't get to fish much. Even then—I was probably about seven or eight years old—it occurred to me that this was quite altruistic of him and that he seemed to somehow care more about us having fun than his own fun. Throughout our childhood and on into adulthood, he was always available when we needed him. He played sports with us constantly, and coached us in just about everything. And he was unfailingly kind to my mom. So I grew up thinking that this is what a father and husband is and that is how I will be when I grow up and have a family. It was all quite clear to me.

We know from chapter 1 that children benefit from having positively engaged fathers, and yet paternal involvement is highly variable both across and within human societies. Why are some men more involved as fathers than others? How do we explain these differences? Where do they come from? And could answering this question help us develop strategies to increase positive paternal involvement where it is lacking?

As with any other complex human behavior, there are myriad influences on paternal involvement. To organize our discussion of these, I begin by considering influences over the course of boys' childhood that affect their later behavior as fathers. Then I discuss micro, family-level influences that affect adult fathers. Next, I consider more macro, social structural influences over paternal involvement. And finally, I examine the effect of culture on paternal behavior.

We'll start with the developmental influences.

Paternal Role Models

How do we learn to be fathers? There isn't a class about it in school, although there probably should be. Still, a class isn't going to teach you all you need to know. Most of us learn how to be a father in part by observing our own fathers, which allows us to internalize a template or schema of what a father is or should be and what we should aspire to.[1]

I had the privilege of being raised by a warm and nurturing father who was, and is, an excellent role model. But what about kids who don't grow up with a father, or who grow up with one who is not as positively engaged as mine was? Unfortunately, this too tends to be passed from one generation to the next. Men whose fathers were absent when they were children are more likely to become absent fathers themselves, and women whose fathers were absent when they were children are more likely to have children with absent partners.[2] Furthermore, when fathers are present, a better-quality relationship with their son predicts increased father involvement by that son as a parent.[3] Another study showed that men whose fathers were positively engaged with them growing up subsequently gave more praise and physical affection toward their own children.[4]

This means that your relationship with your son likely affects not only him but also his future children by way of your influence on his parenting. In a very real sense, your parenting is affecting your unborn grandchildren, and potentially their children as well. In cases of father absence, what needs to happen then is obvious but challenging: a man must decide to break the cycle of father absence and succeed in doing so.

Barack Obama is an example of such a man. Barack senior was nowhere to be found during his childhood, but Barack junior had this to say in a 2008 Father's Day speech: "So I resolved many years ago that it was my obligation to break the cycle—that if I could be anything in life, I would be a good father to my girls; that if I could give them anything, I would give them that rock—that foundation—on which to build their lives. And that would be the greatest gift I could offer."

He seems to have made good on his promise. Amid the daunting responsibilities of the presidency, he insisted on having dinner with his family five nights a week. He somehow managed to attend every one of his daughter's parent-teacher conferences. He read aloud to his daughter all seven volumes of the Harry Potter series.[5] He even wrote a picture book about America for his daughters.

But breaking the cycle of father absence is hard, and we should encourage and have sympathy for the men who try. When you don't have an internalized template of what a father should be and do, it's hard to know how to go about it. As Leonard Pitts put it in his moving book, *Becoming Dad*, "I have two stepchildren and three biological children, all from the same wife. I fancy myself a decent father to them. But the fact is, there's never a day that passes when I'm not wondering whether I am doing it all wrong."

Or as Obama himself put it, "It is in my capacities as a husband and a father that I entertain the most doubt."

*   *   *

Besides paternal role models, there are other aspects of the early environment that can influence how involved men ultimately become as fathers.

Socioeconomic Status

People living in poverty face many stressors that others are not burdened by. A short list includes substandard housing or homelessness, inadequate nutrition and food insecurity, inadequate child care, lack of access to health care, unsafe neighborhoods, and underresourced schools. With all these challenges to negotiate, it is hardly surprising that low SES parents in urban industrialized countries tend to be less involved with their children and also tend to adopt a harsher and more punitive parental style. [6] It would be wrong to conclude from this that people living in poverty are not good parents. First, many are doing the best that they can under difficult circumstances. Second, some may be trying to prepare their children for a harsh life that they see as inevitable. Finally, an association is only a tendency, and not a 1:1 relationship. Human behavior is complex, and parental behavior is influenced by a multitude of factors. SES is just one of these. There are many highly nurturing parents living in low-SES neighborhoods—some of them heroic—and we should not conclude anything about how good a parent someone is based on their SES. Nevertheless, we may be able to identify relationships that hold at the population level, and these relationships may point toward policies or interventions that could increase positive paternal involvement.

Low SES is also associated with higher rates of single motherhood, not just in the United States but also throughout Europe and North America.[7] Of course, single motherhood implies that fathers are not resident in the home. Although many nonresident fathers are highly involved with their children, they are on average less involved than residential fathers.[8] So if we consider single motherhood an indicator of decreased paternal involvement and investment, then low SES appears to be linked with reduced caregiving by fathers.

Consider one example of a study done in contemporary England.[9] Among a sample of over eight thousand families, lower-SES neighborhoods were associated with younger age at first birth, lower birthweight, shorter duration of breastfeeding and lack of paternal coresidence. When mothers breast-feed their babies for a shorter duration, they are quite literally investing less in their baby's development since producing breast milk requires a lot of energy—so much so that breastfeeding is offered as a weight-loss strategy for postpartum mothers. Not only are mothers investing less in their babies, but fathers seem to be as well, given their low rates of paternal coresidence.

Rates of violent crime are higher in lower-SES neighborhoods, so it is quite relevant that fathers more often live apart from their children in countries with more violent crime.[10] There are also historical correlations over time between violent crime rates and the proportion of children raised outside marriage.[11]

So what is going on here? Why are both low SES and violent crime linked with decreased paternal investment?

One potentially interesting explanation comes from the field of evolutionary developmental psychology, which involves applying basic principles of natural selection to understand the development of human thought and behavior. This explanation is based on life history theory—the basic premise of which is that organisms allocate their limited energy among competing life demands, including growth, maintenance, mating, and parenting. If more energy is invested in one category, less is available to invest in the others. Organisms are predicted to invest their energy among these categories in such a way as to maximize their lifetime reproductive success (i.e., the number of offspring they successfully raise to reproductive maturity). Some species adopt what are called fast life history strategies: they invest heavily in mating but little in growth, maintenance, or parenting. As a result, these species tend to be small, short-lived, reproduce early in life, produce many offspring, and invest minimally in them. A slow life history strategy instead involves heavy investment in growth, maintenance, and parenting and less involvement in mating. These species tend to be large, to be long-lived, and to reproduce late in life, and produce few offspring in which they invest heavily. The classic examples of fast and slow life history strategies among mammals are mice and elephants.

One of the key factors thought to determine whether species adopt fast or slow strategies is their extrinsic mortality risk—basically how dangerous the environment is that they live in. Since mice face very high rates of predation (I once heard the rock pocket mouse referred to as the "Snickers bar" of the desert), they can't expect to live very long. They have a very high extrinsic mortality risk. Their best strategy is to start reproducing before they are killed by a predator and to produce lots of pups and hope some of them don't get eaten. Elephants, in contrast, grow large enough to defend themselves from nearly all predators, so they can wait to reproduce until the time is right and then invest heavily in each offspring, confident that they are likely to survive to adulthood.

Other life history theorists have suggested that this fast versus slow life history pattern may also apply within species, with some individuals within a species adopting faster strategies and others adopting slower strategies. What might cause an individual to adopt a fast or a slow strategy? Again, the risk of mortality has been posited as a key determinant, as well as the inability to control that mortality risk. People living in low-SES neighborhoods face higher rates of mortality, not only due to things like gun violence but also things like pollution and decreased access to quality health care.

These mortality threats are quite salient to people living in low-SES neighborhoods. This became apparent to me in interviews I conducted over the past several years with fathers and grandmothers living in the Atlanta area. One low-SES father I interviewed told me that his father was a homicide victim, and a low-SES grandmother I interviewed relayed the sad story of how she was afraid to take her grandchildren outside to play in the yard for fear of their safety. Life history theory predicts that people will respond to these dangerous conditions by adopting a faster life history strategy, including earlier and more frequent reproduction as well as decreased parental investment. At a psychological level, the adoption of a faster life history strategy is thought to involve a bias toward more immediate gratification over delayed gratification. Critically, these biases are not deemed pathological but rather as an adaptive response to coping and functioning in their high-mortality environment.[12]

Although comparisons within a species are different from those between species, there is some evidence in support of this theory as applied to humans. For example, teenage motherhood is more common in low-SES neighborhoods, and low SES is associated with decreased maternal and paternal investment in some studies like the English one just described.[13] And as I mentioned, paternal involvement is lower in countries with more violent crime. However, here the causal arrow may work in both directions. That is, we know that children raised without a father are at increased risk for behavioral disorders involving increased aggression, so father absence might increase the likelihood that children ultimately engage in violent behavior.

It may also be that children raised in dangerous environments come to adopt a faster life history strategy characterized by father absence. Other evidence is less consistent with the theory, and it continues to be evaluated and refined.[14] Nevertheless, it is useful to consider because it might offer a partial explanation for decreased paternal investment in low-SES neighborhoods. Most important, the theory predicts that life history strategies are responsive to the environment in which people are raised and that reducing poverty and violent crime and improving people's health should increase paternal involvement and investment. The timing of exposure to these environmental cues may be crucial. Indeed, there may be critical or sensitive periods during childhood in which these life history strategies are established, and they may be more difficult to alter later in life.

It has been suggested that exposure to these cues during critical or sensitive periods changes a child's biology to orient them toward a faster or slower life history strategy. One intriguing candidate here is the same oxytocin system discussed at length in chapter 4. You may remember that rat pups raised by highly nurturing mothers develop more oxytocin receptors in their medial preoptic area (MPOA), a brain region critically involved with parental caregiving, and go on to become nurturing parents themselves.

There is now evidence that people who experienced childhood adversity tend to have lower levels of oxytocin, as well as increased methylation of the gene that codes for the oxytocin receptor, also presumed to have the effect of decreasing oxytocin signaling.[15] It is essential to emphasize, however, that the effect size is small. That means that some children raised in difficult circumstances will nonetheless have high oxytocin signaling. Nevertheless, safe and nurturing environments would tend to increase oxytocin signaling in children and help them to become highly motivated caregivers as adults. Conversely, oxytocin signaling would tend to decrease in children exposed to harsh and unpredictable environments, and this may bias them to be less involved caregivers as adults.

Besides life history theory, there are other potential explanations for the link among SES, single motherhood, and decreased paternal involvement. For example, single women in low-SES settings are more likely to become pregnant than single women in higher-SES settings, and this could be due to decreased awareness of or access to contraceptives.[16] When men are not ready to become fathers, they are less likely to commit to the relationship and to raising an unplanned child. This was also a recurring theme in my interviews with fathers in the Atlanta area. A number of nonresident fathers explained to me how they had their first child unintentionally at a young age and with a woman they were not committed to. While their efforts to remain involved in their children's lives were laudable, these early events still imposed limits on the extent and manner in which they were involved. Thus, increased education about and access to contraceptive technologies is also a sensible strategy to increase paternal involvement by making fatherhood a more deliberate choice.

*   *   *

Once boys have grown into men and become fathers, there are microlevel, familial factors that can also have an impact on how involved they become.

Paternity Certainty

Evolutionary theory predicts that male mammals should invest in offspring in proportion to their confidence that they are the biological father. The logic is that males who invest time and energy in raising offspring conceived by other males will be selected against because they are helping those other males send their genes into the next generation at the expense of their own. So we expect that males have inherited an evolved genetic predisposition to invest selectively in their own offspring or to titrate their investment according to their degree of paternity certainty.

Stepfathers raise children with the certainty that they are *not* their father. In small-scale, nonindustrial societies, children are often raised by stepparents due to high rates of parental death and divorce, and there is every reason to believe that stepparenting has always been part of the human experience rather than a recent anomaly.[17] Many stepfathers form close and affectionate relationships with their stepchildren, and this is something to be lauded. Nevertheless, there is unfortunately abundant evidence that on average, stepfathers do not treat their stepchildren as well as they treat their own biological children. First, there is the startling statistic that children are between forty and one hundred times more likely to be abused or murdered if they live in a stepparent household than if they live with both biological parents.[18] Of course, most stepfathers are neither homicidal nor abusive, but on average, fathers invest less in their stepchildren compared to their biological children. For example, fathers have been shown to spend less time and money on stepchildren. Stepfathers are also less likely to provide parental supervision, engage children in interaction, or provide emotional support.[19]

In some cases, fathers are raising children in situations where they are uncertain if they are the biological father of the child, and here again we see evidence of decreased investment with lower paternity certainty. In one study of 1,325 American men surveyed while waiting at a motor vehicle department, low paternity certainty was reported in only 1.5% of births and was associated with higher subsequent divorce rates, less time spent with the child in one-on-one interactions, and less involvement in the child's education.[20] In a smaller study of 170 fathers recruited at London's Heathrow Airport, men who reported greater perceived resemblance to their child, a presumed marker of paternity certainty, also reported spending more time with the child. Additionally, men who reported their partner to be more faithful said they spent more time with their child. Interestingly, men who were less likely to endorse statements such as, "I believe that women find me attractive," were more likely to maintain paternal investment in the context of lower perceived mate fidelity. This suggests that men with higher mate value may be more discriminating in this regard.[21]

Another interesting bit of evidence suggests that men sometimes invest in children as a function of paternity certainty. Among small-scale, nonindustrial societies, men typically pass wealth to the next generation as an inheritance. In cultures where men have high paternity certainty, wealth is typically passed from fathers to sons in what is known as patrilineal inheritance. When paternity certainty is low, however, wealth is more commonly passed from fathers to their sister's son. Presumably this is because men know for certain that their nephew is genetically related to them, whereas they cannot be sure about their wife's son.[22]

In summary, it appears that men often care about biological paternity, and they tend to become more involved and to invest more in children when they are confident that they are the biological father. However, as we will see, there are some important exceptions to this tendency, and that is where we witness the power of social and cultural norms.

Coparental Relationships

Fathers are more positively involved in raising their children when their coparenting relationship with the mother is strong.[23] But when mothers exert excessive control over fathers' parenting, known as maternal gatekeeping, fathers tend to become less involved.[24] In some cases, mothers may even engage in maternal "gate closing," excluding fathers from certain parental activities. Gatekeeping is most common among mothers with very high standards for parenting. As the psychologist Sara Schoppe-Sullivan put it, "The more you care about (being viewed as a good mom), the less likely you are to give up control over that domain." Gatekeeping is also more common among mothers with more traditional attitudes about gender roles.[25] Fathers who receive more encouragement from mothers, however, tend to become more involved in child care.[26] So mothers often seem able to either facilitate or impede paternal involvement, depending on the feedback they provide fathers and their willingness to relinquish control. In the long run, of course, mothers themselves will often benefit if they can get fathers positively involved in child care.

Of course, gatekeeping is sometimes warranted, as in the case of serious paternal incompetence or paternal behaviors that could endanger the child. In fact, evidence suggests that less competent fathers elicit increased gate-closing behavior from mothers.[27] I remember one of the first times that my wife went out of town and left me alone with my son for a weekend. He was a toddler and had just started running. No more than fifteen minutes after she left, he and I were running down the street together. It was cold so I said, "Let's run with our hands in our pockets." That seemed like a responsible paternal suggestion. However, no sooner did Toby put his hands in his pockets than he tripped over a rock and face-planted on the pavement. He started crying, and blood was coming out of his mouth. I thought, *Oh my god! I'm going to have to take him to the hospital and Barbara is never going to feel like it's safe to leave him home with me alone, and she's probably right anyway. Look what has happened!* Thankfully, everything resolved without need for medical intervention, we had a nice weekend together, and I even gained a modicum of parental confidence.

Even when fathers are competent parents, maternal gatekeeping should not be blamed solely on mothers. Fathers need not be passive recipients of maternal gatekeeping. They have agency as well and can call attention to gatekeeping and actively assert their right to be an involved and autonomous parent rather than passively fall into an unfair and unequal pattern.

These studies were conducted within modern developed societies, but strong coparental relationships are also associated with greater paternal involvement across human societies. As Whiting and Whiting summarized, fathers seldom help with infant care in societies where men are aloof from their wives and children and uninvolved in domestic affairs. Instead, men in these societies tend to spend their leisure time with other men and may often eat and sleep with them. Men are more involved with infant care in societies where they eat and sleep with their wife and children and gossip and converse with them in the evening.[28]

Stay-at-Home Fathers

Other aspects of the coparental relationship can also affect paternal involvement. For example, when mothers have better jobs than fathers, some families may opt for fathers to stay at home as primary caregivers, allowing mothers to be the family breadwinner.

In the United States, the number of stay-at-home fathers (SAHFs) has more than doubled in the past twenty years. This has resulted in approximately 20% of all stay-at-home parents now being fathers. At the same time, the number of stay-at-home mothers has been stable.

What is the explanation for this sharp increase? One factor is that women's educational outcomes have steadily improved to the point where considerably more women than men are now receiving college degrees. This has provided women with greater access to high-paying jobs that used to be occupied by men. Moreover, the gender gap in wages has been steadily decreasing. The cumulative effect of these changes is that women are now in a much better position to be the family breadwinner than at any time in the past. Particularly when mothers are better educated and earn more money than their partners, some couples have opted to achieve better work–family balance by having dad stay home with the kids. High rates of unemployment can also contribute to increases in SAHFs. For example, many men became SAHFs because they could not find work after the Great Recession in 2008.[29] Some prominent scholars have argued that the highest-paying and most prestigious jobs are more likely to be occupied by women in the future.[30] If so, the trend for increasing number of SAHFs is likely to continue.

While most men are capable of succeeding in this role and many have valued the experience, being a SAHF is not without challenges. SAHFs often struggle with loneliness and isolation and sometimes feel excluded from SAHM groups. They can also struggle with gender role conflict in which the caregiver role appears to conflict with masculine norms, and this can have a negative impact on their mental health. Finally, it is not always clear that this arrangement suits mothers. One large Australian study found that on average, both mothers and fathers became less satisfied with their relationship when they transitioned from male breadwinner or dual-earner households to female breadwinner households.[31] One encouraging trend is that a number of SAHF groups have emerged to help fathers combat loneliness and isolation.[32]

Single Fathers

While maternal employment opportunities may lead some men to choose to become SAHFs, fathers may also adopt the primary caregiving role as single fathers when competent maternal care is unavailable.

When a child's mother passes away or if that mother is incapable of providing competent care, fathers may receive full custody and take on the primary caregiving role. In the United States, the proportion of children being raised by single fathers has approximately tripled since 1960. Laws have changed in ways that treat parents more equally, facilitating greater paternal custody. Compared with single mothers, single fathers are more likely to have custody of older children and boys. This is probably due to a perception that raising girls and young infants may pose some special challenges for single fathers. Single fathers tend to be more involved in play and spend more time in paid labor compared with single mothers, and they also tend to have a more lenient parenting style that allows the child more room to experiment.[33]

Single fathers often face many of the same challenges as single mothers, including lack of social support, self-doubt, physical and mental fatigue, ongoing conflict with an ex-partner, change of residence, reduced standard of living, and social stigmatization. Many single fathers also face a steep learning curve upon suddenly becoming the primary caregiver and taking on new domestic responsibilities. Potentially as a consequence of these stressors, both single mothers and single fathers are more likely to be depressed or substance dependent compared with mothers and fathers who are married or cohabitating.[34]

Breadwinning

As discussed throughout this book, fathers provide at least two types of paternal investment: direct care, which refers to hands-on caregiving activities in which the child is physically present, and indirect care, which refers to activities like physical protection and economic provisioning. Children clearly benefit from both types of investment. However, in some cases, these two forms of investment compete with one another. That is, fathers become so involved in economic provisioning that they have little to no time for direct caregiving. When faced with the choice of working to feed their children or spending time with them, most men understandably choose the former.

Take the example of a migrant worker forced to live apart from his family in order to provide for them or of a father who must work multiple shifts per day in order to meet his family's basic needs. In societies where men are expected to provide both forms of care, the result can be work–family strain. In the United States, men have reported increasing levels of work–family conflict over the past several decades.[35] An increasing number of women have entered the workforce over that time period, resulting in greater expectations that fathers will help with direct care.[36] In some cases, fathers may be present, but the strain they feel between work and family responsibilities may have a negative impact on the quality of care they deliver. Research in Australia has shown that as fathers experience increased work–family conflict, their mental health, relationship quality, and parenting capabilities all deteriorate and, in addition, have a negative impact on their children's mental health.[37] For some men, work can even become an unhealthy addiction that has a negative impact on their children. Adult children of workaholics have been shown to have higher levels of depression and anxiety compared with other adults.[38] In Japan, addiction to work has manifest as a troubling national phenomenon, with some employees literally working themselves to death. The Japanese call this phenomenon *karoshi*, and it commonly involves death due to stroke, heart attack, or suicide in response to the stress of long work hours.[39] Being a good father then may sometimes involve placing limits on one's commitment to work.

When fathers invest heavily in indirect care at the expense of direct care, this can be a personal decision or it can be a choice that is forced on them by their social and economic circumstances. We now move on to consider other macrolevel, social structural influences on paternal caregiving.

Paternity Leave

Studies in both Europe and the United States show that when fathers take time off work after their baby is born, they become more involved in caring for the child both during the period of leave and later in childhood. For example, one Swedish study showed that fathers who took longer paternity leave spent more time with their children, took more responsibility for child care tasks, and had closer relationships with them.[40] Similarly, an American study showed that fathers who took paternity leave more often read to their two- or three-year-old children and were more engaged with their five-year-old children.[41] In fact, the whole family system may function better when fathers take paternity leave. For example, fathers who take more paternity leave tend to have better relationship quality and coparenting quality a year later.[42] Parenting involves a steep learning curve, especially the first time, so it stands to reason that when a father is immersed in it by paternity leave—the way mothers always have been—he will become better at it.

One important American study showed that paternity leave had a larger benefit among nonresident fathers, and we know that children benefit from increased involvement of their nonresident fathers.[43] It is tempting to conclude that paternity leave causes increased paternal investment, but we need to be careful here. It could instead be the case that men who have strong preexisting beliefs about the importance of being an involved father are both more likely to take leave and more likely to be involved later on, in which case there would be no causal relationship between the two. It is difficult to dismiss this possibility without randomly assigning fathers to paternity leave versus no paternity leave and observing the differences. However, no one has done such a study. Some studies have used advanced statistical techniques to account for the prebirth characteristics of fathers and have still found an association between paternal leave and paternal involvement, favoring the causal interpretation.[44] In reality, the causal arrow likely operates in both directions.

In light of these associations, it is quite surprising that the US government does not offer any paid paternity leave. This contrasts starkly with other high-income countries around the world. Forty-seven other countries offer more than four weeks paid paternity leave. If the United States is a negative outlier in terms of paternity leave, it is a mega-outlier in terms of maternity leave as the only high-income country that does not provide any paid maternity leave. Although the US government does not provide paid parental leave, a minority of US states and US businesses do.[45]

Scandinavian countries have taken the lead in terms of paid paternity leave. In Sweden, for example, parents are provided with 480 days of paid parental leave for mothers and fathers combined, and fathers must use 90 of those days or else forfeit them. We know that this "paternal quota" is effective in motivating fathers to take leave because the proportion of fathers doing so increased dramatically when the quota was initially implemented in Sweden. Prior to the quota, many families would use all of the leave time for mothers. Today, approximately 90% of Swedish fathers take paternity leave.[46]

Still, such policies do not guarantee that fathers will take paternity leave. France, for example, has even more generous paternity leave policies in terms of amount of leave, but French fathers claim paternity leave at much lower rates. Part of the explanation may be that the Swedish government replaces a greater proportion of paternal income compared with the French government.[47] Many fathers still consider "breadwinning" to be their main paternal responsibility, and they are less likely to become involved in direct care if it interferes with their ability to provision. Workplace culture also appears to be an important variable. If men fear that taking leave may lead to negative work evaluations, prevent professional advancement, or, especially, jeopardize their employment, they are less likely to take leave. Similar concerns seem to prevent many Japanese fathers from taking paternity leave. Although Japanese fathers are eligible for a full year of paid leave (at 60% of their wages), only about 6% take it due at least in part to fear of negative repercussions at work.[48]

Compared with Swedish fathers, French fathers may also be less likely to take leave because of stronger commitments to traditional gender norms promoting the idea that caregiving is the mother's responsibility. Of the surveyed French fathers, 46% did not claim their full leave entitlement because they were "not interested."[49] This suggests that paternal leave policies may not be effective in the absence of cultural norms promoting involved fatherhood or when they jeopardize fathers' traditional role as breadwinners.

Sex Ratios

Paternal investment is also associated with the sex ratio of a population. Where women outnumber men, rates of single motherhood, a presumed marker of decreased paternal investment, are higher.[50] This makes sense if we assume that most women in the population wish to become mothers. There simply aren't enough male partners available, so some women will need to raise their children without a father. However, it may be the case that men in such societies also alter their own mating strategy. Remember that life history theory predicts that males will allocate energy to mating versus parenting in such a way as to maximize their reproductive success. Certainly there are many other important influences at play in human reproductive decision making, and I don't want to sound callous about male behavior; I'm just outlining the theory. Life history theory predicts that men will shift toward a mating effort strategy when they are outnumbered by women and therefore have a surplus of mating opportunities.

Simply put, under these circumstances, there may be more reproductive success to be gained by a series of short-term sexual relationships than by a long-term stable relationship.

One example suggestive of this type of life history strategy shift is a study of the Hadza people from Tanzania, a group of nomadic hunter-gatherers from Tanzania that we have discussed earlier in the book. Hadza live in camps of between roughly ten and one hundred individuals that vary in terms of their demographics. The study found that fathers spent less time playing with their children in camps where there were more single women compared with camps with fewer single women.[51] The cynical interpretation is that the availability of potential mating opportunities shifted men's investment away from parenting.

In my own lab, we recently conducted an experiment that had a similar sobering interpretation. We showed that while fathers rated photographs of cute infants as more appealing than did men who were not fathers, exposing fathers to pictures of attractive women beforehand resulted in fathers rating the infants as less appealing, to the point where their assessment no longer differed from that of nonfathers. In contrast, priming fathers with pictures of cute infants did not decrease how appealing they rated photographs of attractive women. [52]

Another insightful study also tested this idea that men's life history strategies shift as a function of mating opportunities by examining changes in male reproductive strategies as a function of the sex ratio across populations within a single society.[53] Across eight Makushi Amerindian populations from Guyana, researchers asked participants to complete a psychological survey known as the sociosexual orientation inventory. People who score high on this measure are oriented toward multiple, short-term relationships and low parental investment, and people who score low prefer longer-term relationships with higher parental investment. As the sex ratio became more biased in favor of women across these eight populations, men's sociosexual scores increased, implying that they became less interested in paternal investment. An alternative way to frame the results is that men become more interested in paternal investment when there is a male-biased sex ratio.

These findings beg the question of what factors affect adult sex ratios. Female-biased sex ratios will result when men go off to war and especially when they are killed and do not come home, as has been the case throughout much of human history. They can also be caused by forced male labor migration or when large numbers of men are imprisoned. College campuses have also become a place with low sex ratios. Current ratios are typically about forty men to sixty women in college. Theoretically, all of these circumstances are expected to be associated with increased mating effort and decreased parenting effort among men. To give an idea of how much sex ratios differ across countries, Qatar has the highest sex ratio at 299 males per 100 females due to an influx of young male immigrants, and Ukraine has among the lowest at 84 males per female due in part to the loss of men during the war with Russia.[54]

Incarceration

Another factor that can obviously limit paternal involvement is incarceration. The United States has the highest rate of incarceration in the world at approximately 700 prisoners per 100,000 residents. For comparison, Canada has a rate of around 100 prisoners per 100,000 residents. Most of these prisoners are men, and many of them are fathers. It wasn't always like this in the United States. In 1960, the rate was only 160 prisoners per 100,000, but it has steadily increased since then. Crime rates have also decreased in the United States over this time period, but it is not clear whether increased incarceration is responsible for this decrease in crime. Much of the increase in incarceration was due to the war on drugs that included mandatory minimum sentences for victimless drug offenses. This hit the African American community particularly hard. Although young Black men do not use or sell drugs more than young White men, they are almost three times more likely to be arrested for using or selling drugs. Once arrested, they also receive longer prison sentences than White men. African Americans are imprisoned at a startling rate of 2,200 per 100,000 residents. One in three Black men will be imprisoned at some point in their lives. This has also severely affected African American children, with about 25% of Black children having a parent who has been imprisoned at some point in their life. Incarcerated fathers often have difficulty finding employment after being released from prison. Coupled with their exposure to prison culture, this may increase the likelihood of future criminal behavior and more time in prison.[55]

Children of incarcerated fathers are seriously disadvantaged. The family loses any paternal income, as well as any direct care provided by the father.

There is also a sense of shame and embarrassment, and some of these children suffer the trauma of witnessing their father's arrest and imprisonment. Unsurprisingly, children of incarcerated fathers perform worse in school and are more likely to drop out of school. They are also more likely to suffer behavioral and mental health problems, including anxiety, depression, and posttraumatic stress disorder. One could argue that this simply reflects heritable antisocial tendencies being passed from father to child, but several studies have attempted to statistically control for this possibility and concluded that incarceration itself has a negative impact on children.[56]

Imagine the following scenario. Reggie is an eight-year-old boy living in the inner city of Atlanta. His twenty-nine-year-old father works as a janitor at a local elementary school. After a long week at work, his dad likes to relax with his friends on Friday night by drinking beer and occasionally smoking marijuana. One night, he is arrested for possession of marijuana. Reggie witnesses his dad being taken away by the police and is scared and confused. Reggie's father is imprisoned. Reggie misses him tremendously, and his family has lost his father's income. As a result, they are forced to move into a more dangerous neighborhood where housing is more affordable. Reggie becomes anxious and depressed. He is unable to focus at school, his grades deteriorate and he eventually drops out of school. His new peers are delinquent, and through his interactions with them and the absence of his father, he comes to view a life of crime as his best way out of poverty.

What if Reggie's father has not been arrested in the first place?

Among imprisoned fathers of Black children, about one-third are imprisoned for nonviolent, victimless drug offenses. If society could find a way to keep nonviolent criminals out of jail and positively engaged with their children, those children would surely benefit, as would society at large.

*   *   *

We've discussed developmental, familial and social structural influences over paternal behavior. Next, we move on to consider the powerful role of social and cultural norms in shaping paternal behavior.

Social Norms versus Genetic Predispositions

When a mating effort strategy leads to greater reproductive success, we expect it to evolve by natural selection. Some human males seem to adopt mating effort strategies that maximize their reproductive success.[57] A number of male athletes and celebrities—Wilt Chamberlain, Charlie Sheen, and Mick Jagger, to name a few—are believed to have had sex with thousands of women. Granted these men are exceptions, but consider the far more common situation throughout history in which men in polygynous societies had multiple wives or wives along with mistresses. Or consider the men who leave their postmenopausal wives and families to start a second family with a younger, more fertile woman. We don't have to approve of them, but there is reason to believe that these male tendencies have evolved by natural selection. As just one example, genetic studies of the human Y chromosome have shown that approximately 16 million men living in Asia today are descended from Genghis Khan or one of his relatives, presumably because these men had the awful habit of killing the men they conquered and taking their wives as concubines or in some cases just raping them and moving on. Their genes consequently spread like wildfire across the generations. Unfortunately, this case is not an anomaly, and this type of male behavior was pervasive over the past ten thousand years.[58] So we expect that men living today may have evolved genetic biases toward polygyny when it is feasible for them. Yet most modern men do not behave like Genghis Khan, and many men who could successfully pursue a "mating effort" strategy choose instead to be committed fathers and husbands, even though that does not maximize their reproductive success. How is this possible? Why are some men not susceptible to these putative evolved predispositions?

A big part of the answer lies in culture. Culture has been defined in many different ways by anthropologists, but one simple definition is the shared attitudes, values, goals, and practices of a population or society. Some anthropologists think of culture as a secondary system of inheritance, in addition to genetics, that guides human behavior. The behavior of other animals is driven primarily by evolved genetic predispositions and what they are able to learn through trial and error, but human behavior is also strongly influenced by cultural norms of behavior or by what anthropologists call cultural models.[59] A cultural model is an internalized mental model that members of a community share and that influences behavior.[60] So if monogamy and devoted paternal caregiving are cultural norms, men may follow these norms even when they conflict with genetic predispositions that may instead favor a mating effort strategy.

Let me provide an example where cultural norms regarding paternal care appear to win out over genetic predispositions. As outlined previously, evolutionary theory predicts that male mammals should invest in offspring in proportion to their confidence that they are the biological father. Above, I provided evidence in support of this prediction among human males. However, here is an exception.

The Himba are a group of people living in northern Namibia who practice pastoralism (i.e., cattle, sheep, and goat herding). Among the Himba, cultural norms permit and protect the existence of informal affairs among married people. Consequently, it is not surprising that rates of extra-pair paternity are a whopping 48%. As a result, many men raise children who are not their biological offspring. For these children, they are the social father even though they are not the biological father. Himba social norms dictate that it is unacceptable to preferentially invest in biological compared to nonbiological children, and men mostly abide by these norms. For example, cattle are the main form of wealth in this society, and fathers give as many cattle to their nonbiological children as to their biological children. Young men must pay a bride-price to the family of their new wife, also typically in livestock, and fathers typically pay this on behalf of their sons. Bride-price payments that fathers make on behalf of nonbiological offspring are as large those they make on behalf of biological offspring.

These forms of investment are very public and highly visible to the community such that others would notice if fathers violated the norm of fully supporting their nonbiological children and reputational damage would likely follow. Nevertheless, there is evidence that fathers may in fact bias investments toward biological children in other domains. For example, children are often fostered to close relatives in this society. Himba fathers are more likely to foster out their nonbiological children, which is tantamount to decreasing investment in them.[61]

Nevertheless, the more general point is that social norms can motivate fathers to provide care for children, even when that conflicts with evolved genetic predispositions to behave in ways that maximize reproductive success. In many parts of the modern world, norms now dictate strict monogamy coupled with considerable paternal investment. Pursuing a mating effort strategy is often not socially acceptable. Recall, for example, the reputational damage that golf phenom Tiger Woods suffered after his string of affairs was made public. Still, there is meaningful cross-cultural variation.

For example, a 2013 poll showed that only a minority of French citizens agreed with the statement, "Married people having an affair is unacceptable." Compare that with 84% in the United States and 94% in Turkey.[62]

So human behavior, including paternal behavior, is influenced by both evolved genetic predispositions and by cultural norms of behavior. But why do humans have this additional system of inheritance that guides our behavior? Some scholars believe that the evolution of cultural norms in our species enabled the suppression of selfish behavior so that people could cooperate more effectively with members of their group, and this in turn allowed them to outcompete other groups of humans who were less cooperative.[63] This process is known as cultural group selection.

When children are exposed to social norms early in development, those norms become internalized and encoded in neural networks in the brain. Consequently, abiding by social norms often just feels right—it is our prepotent bias. But sometimes, especially when they conflict with evolved genetic predispositions, people may consider violating the norms. In doing so, they will often experience guilt or shame, and this may dissuade them from enacting the norm violation. Others may abide by social norms because they are able to anticipate the punishment that would follow for not doing so.[64] Still others may follow social norms because, despite countervailing personal motivation, they reason that it is morally appropriate to do so. Our capacity to do these things—experience guilt and shame, envision the long-term consequences of our behavior, morally reason, and inhibit prepotent genetic predispositions and impulses—depends on our prefrontal cortex.[65]

The prefrontal cortex is the region of the neocortex at the very front of our brain. One way that we can begin to understand the function of the prefrontal cortex is to examine the consequences for people when it is damaged. In 1848, an American railroad foreman named Phineas Gage suffered a traumatic brain injury when an explosion caused a meter-long iron rod to pass completely through his skull and brain. This case is notable in part because the rod specifically damaged his prefrontal cortex and not other brain areas. Remarkably, Gage survived the accident. Even more remarkable, the physician who studied his case in great detail concluded that his intellectual faculties were basically intact. However, his personality was dramatically transformed. He began to exhibit deficits in impulse control, by, for example, using "the grossest profanity" and showing "little deference for his fellows." Previously a highly respected foreman in the railroad company, the company refused to rehire him after the accident. He also became unable to follow through on plans and has been described as possessing a "myopia for the future." That is, he had difficulty anticipating the future consequences of his actions.[66] The deficits described for Gage are very similar to those described in contemporary patients with damage to the same region of the prefrontal cortex.[67]

The human prefrontal cortex dwarfs the prefrontal cortex of other primate species. The human brain as a whole is about three times larger than the brains of our closest living primate relatives, the great apes: chimpanzees, orangutans, and gorillas. However, some parts of the human brain are more than three times larger and others are less than three times as large. Remember from chapter 5 that the brain is composed of both gray matter and white matter. The former includes mainly neuronal cell bodies and the dendrites that receive inputs from other neurons, whereas white matter contains the myelinated axons that connect neurons with each other. In terms of gray matter, the human primary visual cortex—the first place in the cortex where we process visual information—is only 1.9 times larger than chimpanzee visual cortex; however, the human prefrontal cortex is 4.5 times larger than chimpanzee prefrontal cortex. Human prefrontal white matter is even more of an outlier, at over six times larger than that of chimpanzees.[68] The human prefrontal cortex is also more highly convoluted, or gyrified, than the prefrontal cortex of other primates.[69] Together with the dramatic white matter expansion, this suggests that the human prefrontal cortex is more interconnected with other parts of the brain and is likely to be integrating more information compared with the prefrontal cortex of other species. What's clear is that the human prefrontal cortex is special, and that is part of what enables us to benefit from our unique, secondary system of inheritance: our culture.[70] Other primates do have a type of rudimentary culture, but it doesn't build on itself from one generation to the next like ours does, and it does not seem to prescribe norms of social behavior that are followed.[71]

In addition to all the classic parental brain circuitry discussed at length in the previous chapter, being a good father also depends on a well-functioning prefrontal cortex. It allows men to uphold social norms, honor commitments to their wife and children, and refrain from sexual pursuits that could jeopardize their family. But like everything else, people vary in the functioning of their prefrontal cortex and hence their ability to inhibit impulsive, selfish behaviors. An acquaintance of mine who used to be over-weight transformed himself into a fit, athletic physique that he has main-tained for many years now. I asked him how he did it, and he said that he stopped eating breakfast *and* lunch every day, and now only eats a large dinner. I have trouble skipping even one meal, so that always amazed me. I like to kid with him that he has the largest prefrontal cortex of anyone I know.

Are there ways to strengthen your prefrontal cortex? More is known about what weakens it. It probably won't surprise you to learn that alco-hol suppresses the prefrontal cortex since many people make ill-advised decisions under its influence.[72] Child abuse can also interfere with normal development of the prefrontal cortex and thereby contribute to deficits in emotion regulation and impulse control.[73]

Cross-Cultural Variation

The discipline of anthropology is changing rapidly, but one of its tradi-tional pillars has been a commitment to describing the diversity of human cultures. A corollary is that we cannot and should not make generalizations about human nature based solely on studying people living in modern, Western, high-income societies. To properly understand any human phe-nomenon, we need to describe it across a wide range of cultures to deter-mine how variable or uniform it is. In 2010, the Harvard anthropologist Joseph Henrich and colleagues pointed out that over 90% of all participants in psychology experiments come from countries that house just 12% of the world's population—hardly a representative sample of the species.[74] More-over, as Henrich emphasizes in his recent book, *The Weirdest People in the World*, educated Westerners are quite unusual in many respects due to the unique historical circumstances that have shaped them.[75] Henrich and col-leagues devised a clever acronym to describe these people: WEIRD, which stands for Western, Educated, Industrialized, Rich, and Democratic. That acronym probably applies to most of the people reading this book. We need more data from non-WEIRD societies in disciplines like psychology, but this has always been a focus in anthropology. Some modern-day anthro-pologists have called attention to the fact that most studies of non-WEIRD populations have been done by WEIRD anthropologists, which could lead to biases in the way non-WEIRD people are described or depicted or, worse yet, WEIRD people could have imposed aspects of their own culture on people living in other parts of the world. These scholars have made the important point that we should cultivate the development of indigenous scholars who can study their own people and perhaps offer a different perspective.[76] But we also can learn things about our own culture from outsiders who may be able to see things that we take for granted. This point was driven home to me by a clever essay I read in graduate school, "Body Ritual among the Nacirema."[77] The Nacirema are a "North American group living in the territory between the Canadian Cree, the Yaqui and Tarahumare of Mexico, and the Carib and Arawak of the Antilles" who have household shrines with chests built into the walls that hold charms and magical potions that individuals use in private and secret rituals aimed at averting debility and disease. The genius of the article is that Nacirema is *American* spelled backward, and it is only upon this realization that most American readers appreciate it is their own culture being described, and that those "household shrines" are our bathrooms. So we understand humanity best by examining cultures from both within and without, while always trying to be aware of our biases and positionality. As I survey fatherhood across human societies and cultures, readers should be aware of some of the concerns and limitations that have recently been raised about anthropological research.

The cross-cultural evidence that we have suggests that fathers are quite consistently involved in provisioning across human societies, whereas the extent to which they are involved in direct, hands-on caregiving is far more variable.

Prior to the Industrial Revolution, most human societies practiced one or more of the following modes of subsistence: hunting and gathering, pastoralism (herding), horticulture, or agriculture. Some human societies still practice these modes of subsistence, and many such small-scale societies were described by anthropologists before they transitioned to industrial or postindustrial societies.

Anthropologists typically live with a group of people for a couple of years, taking notes, making audiovisual recordings and sometimes quantitative measures, and then write a detailed description of that culture, which is known as an ethnography. Anthropologists have now compiled a large and very valuable database of ethnographies, and they can mine this database for all kinds of information about human societies. One well-known sample of ethnographies, the Standard Cross-Cultural Sample, includes 186 human societies from all around the world. Using this database, it was shown that among nonindustrial societies, fathers were generally closest to their infants in hunting and gathering societies and most distant in pastoralist or herding societies.[78]

Hunting and gathering societies are of particular interest because our ancestors lived as hunters and gatherers for the vast majority of human evolution. Although the lifestyle of modern-day hunter-gatherers has undoubtedly changed over time, there are also likely elements that are similar to those our ancestors experienced. And so hunter-gatherer fathers may tell of something about fathers throughout much of human evolution. Hunter-gatherer fathers commonly hold infants but not nearly as often mothers do. The percentage of all infant holding done by fathers typically ranges from about 2% in the !Kung San from Botswana to about 7% among the Hadza from Tanzania. For comparison, Hadza mothers hold infants 69% of the time.[79]

One group of hunter-gatherers is an inspiring outlier in terms of father–infant closeness and paternal involvement. These are the Aka people from Central Africa. Aka fathers do more caregiving than fathers in any other known society.[80] The anthropologist Barry Hewlett wrote a masterful ethnography about Aka fatherhood aptly titled *Intimate Fathers*. The Aka case is worth considering in some detail, not because Aka fatherhood is particularly representative of our evolutionary past—in fact, there is good reason to think that local historical influences substantially changed Aka family life—but because they illustrate the range of human potential with respect to fatherhood. They provide an illustration of just how involved fathers can be in the right context.

The Aka people live as hunter-gatherers, subsisting mostly on hunted animal game (duikers, hogs, monkeys, and elephants), supplemented by trade with the neighboring society of farmers (the Ngandu people). On average, Aka mothers give birth to six children in their lifetime at intervals of about three and a half years. Infants are nursed until the mother becomes pregnant with her next child. Although fertility is high, so is infant mortality, at 20%. Juvenile mortality is also high: only 55% of children survive to age fifteen.[81] That means a parent can expect to lose two or three children in their lifetime. Most men marry monogamously, but polygyny is permitted, and 17% of men have more than one wife. Aka often hunt with nets:

typically men drive game toward women who wait with the net to trap and kill the animals. Couples work together on the net hunt, and this coopera-tion fosters a sense of intimacy between husbands and wives.

Aka infants are always being held by someone (no cribs or strollers in Aka society), and holding involves skin-to-skin contact, which we know stimulates oxytocin release. Aka fathers hold their infants more than other hunter-gatherer fathers, at 22% of the time versus about 5% for the aver-age hunter-gatherer.[82] This raises the possibility that Aka fathers have high oxytocin levels, which stimulates father–infant bonding. Aka also generally live in close proximity to others with high rates of physical touch, also hypothesized to increase oxytocin levels.[83] Aka fathers often hold infants when mothers are busy collecting firewood, preparing meals, or net hunt-ing. While holding them, they are more likely than mothers or other care-givers to hug and kiss them. Not only are Aka fathers in close contact with their infants for much of the daytime, couples typically co-sleep with their infant, providing an additional nine hours or so of close contact at night. In contrast to fathers in many other societies, Aka fathers do not seem to be their children's playmates.[84] Instead, older children fill this role.

Hewlett lists a number of factors that may help explain why Aka fathers are so involved in caregiving. These include variables such as lack of war-fare or violence, high female contribution to subsistence, and high pater-nity certainty. Additionally, Aka have few accumulable resources that men can strive to monopolize, so more time and energy are available for caregiv-ing. But none of these factors really distinguish the Aka from other hunt-ing and gathering populations and therefore cannot explain their uniquely high levels of paternal involvement. In the end, Hewlett concludes that the net hunt is crucial because it requires couples to spend a great deal of time cooperating with one another and fosters a strong bond between them. They are effectively a team and depend on each other to succeed in the net hunt. Thus, fathers are not only available but also very willing to provide care because of their affection for the child's mother. In contrast, among societies where husbands and wives spend less time together and the marital bond is presumably weaker, husbands tend to do less by way of child care.[85]

Not only do Aka fathers have great affection for their partners, they are also strongly attached to their infants. They are warm and affectionate care-givers who enjoy spending time with their infants. Hewlett attributes this not so much to the manner in which they engage with them, but simply because they spend so much time with them and in so many different contexts. In many modern, developed societies, we emphasize the importance of being engaged and highly responsive to your children as often as possible probably because, especially as fathers, we have limited time with them. But the Aka research suggests another possible pathway to secure father–child attachment—simply being around them so much that you know them intimately.

Finally, Hewlett considers why some Aka fathers are more involved than other Aka fathers. He concludes that more highly involved fathers tend to be monogamous (rather than polygynous) and generally of relatively lower social status, status being measured by the number of brothers a man has. Women generally prefer to marry men with more brothers since brothers work together as an economic unit and a larger unit is more reliable in provisioning. Lower-status men without brothers seem to attract and keep female partners by instead demonstrating their willingness to help more with infant care. From a life history theory perspective, paternal caregiving may be the best reproductive strategy available to lower-status men who have little chance of raising their status or becoming polygynous.

*   *   *

At the other end of the continuum of father–infant closeness among nonindustrial societies are many pastoralist or herding societies. In these societies, fathers were often feared and respected disciplinarians. One example was the Rwala Bedouin people from the north Arabian Desert, who were described in 1913 by the ethnographer Alois Musil. The Bedouin, camel herders who were frequently at war, practiced polygyny. Mary Katz and Melvin Konner summarized the role of fathers in Bedouin culture.[86] Fathers were distant from their children, both physically and emotionally. They spent most of their time in a separate room from the rest of the family, and children under age seven talked with them only occasionally. After age seven, fathers taught their sons how to shoot and go on raids. Fathers used severe corporal punishment (e.g., cutting or sticking them with a dagger) to discipline their sons.

A more contemporary example of distant fathers among pastoralist societies is provided by the Kipsigis people from East Africa, who traditionally herded cattle and grew crops. A comparative study of US and Kipsigis fathers found that while middle-upper class US fathers provided 13% to 17% of direct care for their infants and young children, Kipsigis fathers *never* fed, dressed, bathed or carried their infant outside the house over the first four years of the child's life. They instead viewed their role as that of economic provider and disciplinarian. This lack of involvement in caregiving was supported by Kipsigis cultural models that the infant can be damaged by the strength of the father's gaze and that the father's masculinity can be compromised by the dirtiness of the infant.[87] This is quite interesting in light of the evidence discussed in chapter 3 that close contact with infants decreases levels of testosterone, a hormone that people often associate with masculinity.

In pastoralist societies like these, livestock provide food but are also the primary source of wealth, and this wealth is controlled by men. In order to marry, young men must pay the bride's family in cattle, and fathers typically provide their sons with cattle so they can do so. We saw this in the Himba case. Thus, pastoralist fathers, despite low involvement in direct forms of caregiving, are highly involved in indirect forms of paternal care.

In addition to the mode of subsistence, other variables have been identified that may help to explain why fathers become more involved in some societies and less involved in others.[88] Fathers tend to be less involved and more distant where warfare is common and they engage in military pursuits. In chapter 1, we discussed evidence that father absence is associated with increased aggression in sons. Therefore, the social norm that fathers maintain physical and emotional distance from their sons may be a cultural mechanism by which these societies produce warriors.[89] Alternatively, or additionally, fathers may keep emotional distance from their children in case they are killed in battle, so that the children will not suffer as much from their death.

Polygyny, in which a man has multiple wives, is also associated with less paternal involvement. Most people reading this book live in societies where polygyny is not allowed, but the majority of human societies in the ethnographic record permitted polygyny, even if most men did not achieve it. Presumably polygynous men spend most of their energy on mating effort and provisioning—acquiring the status and resources to attract multiple female mates and economically support their children—rather than direct paternal care.

Cross-culturally, then, greater paternal involvement in direct caregiving is associated with monogamy and the absence of warfare. Fathers also tend to be more engaged in societies where mothers are more involved in provisioning through subsistence activities. For example, hunter-gatherer mothers provide a significant proportion of calories for their group—in some cases, the majority—and fathers tend to be more involved in these societies when men and women make roughly equal subsistence contributions.[90]

Finally, the availability of alloparents is an important factor that influences paternal involvement. As discussed in chapter 2, fathers seem to calibrate their investment based on what other allomothers are providing. Often they do more when others do less, and vice versa. Among the Aka, fathers become far more involved in infant care when they live with their own family (patrilocally) where the mother's kin are not available to help, compared to when they live with the mother's family (matrilocally).[91] Similarly, among the Bofi foragers, northern neighbors of the Aka, fathers have more physical contact with infants when older female relatives are not available.[92] As another example, Tsimane fathers from lowland Bolivia provide less direct care to infants and young children if they have older daughters available to provide care.[93] Thus, it seems that fathers are often doing more when they need to due to a dearth of other helpers.

Although these predictors are derived from studies of mostly small-scale, nonindustrial societies, they can also help us to understand why fathers are becoming more involved with child care in modern, high-income societies. In such societies, marriage is monogamous, warfare is uncommon, mothers have become increasingly involved in provisioning the family by working outside the home, and career aspirations often require couples to move away from their extended family with the result that alloparents are in short supply. All of these factors are expected to increase the amount of paternal involvement.

Cross-cultural variation in paternal behavior is not limited to the quantity of paternal involvement. In some societies, the nature of paternal involvement is quite different from what most of us assume. In several lowland South American cultures, it is normative for women to have sexual relations with one or more men other than their husband. Normally, theory would predict this is a formula for paternal disinvestment. That is, since males lack paternity certainty, they should be reticent to invest in raising their wife's children. However, these cultures tend to be matrilocal so that the patriarchy is weak. Perhaps, as a consequence, the cultural models that have evolved seem to benefit mothers by encouraging multiple men to invest in her offspring. Each of the men who had sex with the mother is considered to be a "father" to the child, and this is supported by a belief in "partible paternity," in which the sperm of each man is believed to contribute to the growth of the fetus. In effect, this cultural system makes men believe that they are all biological fathers, and all of them are therefore inclined to invest in the child. Mothers and children seem to benefit from this ideology, since data show that children with more than one father are more likely to survive to age fifteen, presumably due to extra food provided by additional fathers.[94] Nevertheless, this cultural ideology may still be in conflict with evolved genetic predispositions, since some authors have pointed out that male jealousy is not absent in these societies.[95]

In nonindustrial societies, fathers play an important role in educating their children, especially in teaching boys the skills they need to become successful men in their societies. However, fathers in modern nation-states have significantly diminished roles as educators since much of children's education is left largely to the schools, and few of us groom our children for our own chosen vocation.[96] Twenty thousand years ago, we would have been teaching our sons how to hunt, fish, find honey, and make tools. Today, they are being taught a standard curriculum in our schools, and quite often, our children teach *us* how to use the latest online apps.

This is not to say that modern fathers are not important educators of their children. Indeed, in chapter 1, I discussed their role in teaching children norms, values, and emotion regulation skills. And many modern fathers still teach their children culturally valued skills such as how to play various sports. For a few years, I coached my son in baseball. There was one moment I remember in particular. You might think it was after winning a big game or after my son made a heroic play, or even the opposite. Instead, it was a simple moment when our team was scrimmaging during a practice. The sun was setting behind us, and I felt its warmth on my shoulders. The well-manicured field was beautiful in the evening light. The coaches— there were five of us—looked on from the dugout and the infield. The boys knew what to do. We had been working with them for weeks, and now they all stood in their positions, sure of their responsibilities. Everything was happening as it should. No one was saying anything. No one needed to. There was just beautiful silence. The boys were playing, but time had slowed down. No one wanted to be anywhere else. It was truly a *Field of Dreams* moment. As I looked out at this picturesque scene, it occurred to me how rare it is in our society to witness a group of adult men all collectively passing on a specific tradition to a group of boys. Part of the satisfaction and tranquility of the moment for us coaches was the awareness that we had succeeded in transmitting a valued part of our culture to our sons: *This is a game that we loved as boys and that taught us so much, and now you will be able to experience the same thing as we did and in a way, we will live on through you.* So fathers today still socialize and educate their children, but probably not to the same degree as in the past.

* * *

Next, we move from nonindustrial, small-scale societies to a consideration of paternal caregiving among modern nation-states. But I do so with some trepidation because none of these nations should be considered homogeneous societies in which generalizations apply to all subgroups or all individuals. Rather, nations consist of people varying in class, religion, ethnicity, caste, and subculture. Nevertheless, researchers have offered cautious generalizations about fathers in many of these countries, with the understanding that they do not apply to everyone.

Traditional and Modern Fathers

A survey of fatherhood across modern nation-states reveals that the qualities of the ideal, *traditional* father are remarkably consistent across nations. Rather than warm and nurturing, traditional fathers in many nations were expected to be stern, authoritarian, and emotionally distant providers.

Traditional Chinese fathers were heavily influenced by the ancient Chinese philosophy of Confucianism, which emphasized the father as the undisputed head of the family. Fathers were encouraged to be stern educators and role models for their children, not playmates or caregivers.[97]

Despite being influenced primarily by Hinduism rather than Confucianism, traditional Indian fathers bear many similarities to traditional Chinese fathers. The father was the undisputed head of the household. They were breadwinners and authority figures but not caregivers. Despite often harboring deep affection for their children, traditional Indian fathers maintained emotional distance and did not openly express warmth to their children. Remarkably, grandfathers routinely expressed love and affection toward their grandchildren. Thus, it is said that affection often skips a generation in Indian families.[98]

Similar themes are echoed in the Brazilian context. Portuguese colonists brought with them to Brazil a patriarchal model of fatherhood that Brazilian natives subsequently adopted. Fathers were the unquestioned heads and masters of their families. They were also financial providers and family decision makers on moral and economic issues.[99]

Prior to the Soviet period, Russian peasant fathers spent minimal time with children under seven years of age. Beginning when their sons were seven years old, fathers provided the boys with moral and vocational training, including farming and trade skills. Upper-class fathers were generally strict and emotionally detached from their children, a theme we have heard repeatedly. During the Soviet period, this theme continued. Many children grew up without fathers, meaning that many boys had no role model for how to become a father. When present in the family, fathers tended to have low responsibility for child care. Grandparents often spent more time caring for children than fathers did.

Another tendency that is highly consistent across nations is that modern fathers are moving away from the traditional ideal toward one characterized by greater warmth and involvement. In the United States, for example, fathers have more than tripled their amount of time spent on child care per week, from 1965 to 2016.[100] Many scholars have posited a shift in ideology among Western fathers toward more caring or nurturing masculinities that emphasize caregiving, father–child relationship quality, positive emotions, and gender equality rather than male dominance.[101] To a greater or lesser extent, similar trends have been noted in China, India, and Egypt, and this appears to be a near global phenomenon.[102]

Sociopolitical and Economic Influences

In modern nation-states, economic conditions or sociopolitical factors can often pose challenges for paternal involvement. For example, to fulfill their role as providers, many fathers around the world have been forced into migrant labor that prevents them from living with their children. In China, many rural fathers must migrate to urban areas for work, often with their wives, leaving their children behind to be raised by grandparents. These have been labeled *skipped-generation* families.[103] Similarly, many modern Egyptian fathers have been forced to migrate to oil-producing Gulf States, which has made father absence an endemic and growing problem in the country. As in Western societies, studies show that father absence in Egypt is associated with increased aggression and delinquency as well as lower levels of academic achievement.[104] As another example, the apartheid policy in South Africa from 1960 to 1994 required Black families to live in rural areas of the country when the main source of work was in the cities. This forced Black South African fathers to migrate to urban areas as laborers, making it nearly impossible for them to both financially support and live with their children. Although apartheid has ended, male labor migration has continued. In a 2001 study, only 12% of children were living with their fathers, but half received financial support from their father in a given year.[105]

In Russia, after the fall of the Soviet Union, the transition to capitalism and a market economy placed men under increased pressure to provide for their families. The capitalist economy demanded longer work hours for many fathers, and others had to migrate away from their families to find work and fulfill their roles as breadwinners. So even when Russian fathers have wanted to be involved, there have been considerable barriers to doing so.[106]

After the collapse of the Soviet Union in the mid-1990s, Russian men were suffering shocking rates of mortality from alcoholism, violence, accidents, and cardiovascular disease, likely exacerbated by the psychological stress of transitioning to a market economy. The average life expectancy of Russian men at that time was a dismal fifty-seven years.[107] Alcoholism in Russia is fueled in part by a heavy drinking culture that is an important component of male friendships. It probably goes without saying that alcoholism and sensitive paternal caregiving don't go hand-in-hand. Additionally, theory predicts that high-mortality environments like these will lead to faster life history strategies and less paternal investment, and this may be a partial explanation for low rates of paternal involvement in contemporary Russia. Consistent with the notion of fast life history strategies, rates of teen pregnancy were among the highest for developed countries in the 1990s. Nevertheless, the Russian government subsequently implemented a series of alcohol and tobacco control policies that may be responsible for recent dramatic decreases in male mortality and increases in male life expectancy. More recently, life expectancy at birth for Russian men has increased to sixty-eight years, and rates of teen pregnancy have decreased in parallel[108] This may bode well for paternal investment in future generations. Sadly, these gains may be countered by increased male mortality due to the ongoing war in Ukraine, which will undoubtedly lead to many more children being raised without fathers in both countries.

In the United States and many other high-income nations, there has been a marked rise in the proportion of mothers working outside the home, to the point where the majority of American mothers with children under age eighteen are now working a full-time job outside the home.[109] As a consequence of shared breadwinning responsibilities, the current American cultural model of fathering is that the care of children should also be shared between men and women. Fathers are viewed as coparents, as partners in the parenting process, and couples negotiate how they will raise their children together. Indeed, many American fathers devote substantial time to child care, including feeding, bathing, dressing, and changing diapers. Despite this, mothers still spend more time with children than fathers on average. American fathers also help with homework and are particularly involved in play, especially the rough-and-tumble sort.[110] Many American fathers mentor or coach their children in sports, which may function to prepare them for life in a competitive society.

In my interviews with 120 American fathers from the Atlanta area, it was clear that many of them still believe that they are the person most responsible for the financial well-being of their family, and many reported feeling significant stress and pressure surrounding this responsibility, possibly due to new cultural models involving more direct care by fathers. Many also reported considerable work–family strain as they strive to fulfill both caregiver and provisioning roles. American fathers have reported increases in work–family strain over the past few decades , and more than half of them say that balancing work and family is a challenge.[111] Similar struggles with work–family balance have been noted for Australian fathers and is likely to be the case wherever we see men becoming more involved in direct caregiving while simultaneously upholding their breadwinning role.[112]

Distributed Fathering

In some countries, fathers deviate from the nuclear family model that is common in many parts of the world. Despite the commonplace physical absence of fathers in South Africa, often due to labor migration, many children are raised by men other than their father and many fathers have responsibility for children other than their own. Men are expected to participate in this type of "distributed" fathering, as illustrated in the following essay written by a South African boy.[113]

> I'm 17 years old. I never knew my father, and Mother died when I was four years old. I was raised by my sister and her husband. What amazes me a lot is that their children are younger than me. I call them mother and father because they are my parents. Father, he means a lot to me. He took me in his house when I had no home. He sends me to school and attends parental meetings as it requires. He reminds me to do my school work and to do well at school. When I'm sad he makes jokes just to see me happy. He washes our clothes and sometimes cooks for us. He likes to talk to us (me and my younger sister) about life and the future. He always says he only wants us to have the equal to his two sons. He wants us to grow and be educated and responsible. One day he told me that he wants to see me graduating from the University of Pretoria. He doesn't have much but the little he has he shares with us and his children. There is no difference between us and his sons. He gets angry just like every parent when I misbehave and puts me back to my place as a child and the oldest in the house. Sometimes I wonder what I have done to deserve a father like him. Most men of his age are still running around like boys in the streets but he chose to be the father of two orphans, me and my younger sister, who is very naughty at the age of 14. Even after these years that I've been silly and naughty he's been there for me. He made me realise that it is not only your biological parent that can raise you, send you to school and love you unconditionally.
>
> I call him my father.

Closely related to distributive fathering is social fathering, in which men who are not the biological father provide children with significant nurturance, moral guidance, and emotional support. This is prevalent in the US African American community, where social fathers include older male extended family members, godfathers, ministers, and coaches, for example. [114]

Changes in Marriage and Divorce

In many high-income nations, more people are now having children without being married, and more married couples are divorcing or separating. This trend is particularly pronounced in the United States, which as a result has the highest rate of single parenting in the world. In 2018, almost one-fourth of US children under the age of eighteen lived in a household with a single parent and no other adults present other than adult children.[115] Most of these are single mother households, meaning that an increasing number of American fathers are not living with their children. Similar trends are emerging in other high-income nations.[116]

*   *   *

Beyond culture, a final important factor that can shape paternal behavior is educating new fathers about the importance of paternal involvement.

Fatherhood Programs and Interventions

Several fatherhood programs have been developed that are aimed at increasing the quantity and quality of paternal involvement. These programs can be effective, but some are more effective than others.[117] Programs involving video feedback seem to be particularly effective in improving positive parenting. In these programs, researchers film parent–child interactions and then later review the video with the parent. One such program is video-feedback intervention to promote positive parenting and sensitive discipline (VIPP-SD). This program has two goals. The first is to increase parental sensitivity during parent–child interactions since parental sensitivity predicts secure parent–child attachment, which in turn predicts higher social competence and fewer behavioral problems in children. Sensitive parents accurately perceive and interpret their child's signals and respond promptly and adequately to those signals. The second goal is to help parents implement effective discipline strategies in which positive behaviors are rewarded and rules and limits are explained and enforced. The intervention typically consists of only six visits. Across a large number of well-designed studies, VIPP-SD has been shown to consistently increase both parental sensitivity and child attachment security. Most of this work has been done with mothers, but researchers are now beginning to evaluate the program in fathers as well.[118]

A major challenge for fatherhood interventions is developing programs that fathers will voluntarily participate in and that are engaging enough to sustain their involvement.[119] But a few have been successful in doing so and seem to also be effective in improving father and child outcomes. One such program, Dads Tuning In to Kids, is aimed at helping fathers cultivate emotional competence in their preschool children. Fathers are taught to notice, respond, empathize, and reflect emotions in their child. In effect, they become emotion coaches. This seems a wise strategy given extensive evidence that positive paternal involvement is strongly associated with improved emotion regulation skills in their children. The program consists of seven weekly two-hour sessions and a two-hour booster session eight weeks after program completion. Activities include watching videos of emotion coaching and emotion dismissing, practice exercises, role-play, and group discussions.

One study randomized 162 Australian fathers into either a group that received the intervention or a control group that did not. Remarkably, only one father withdrew from the program. Six months after the intervention, fathers who completed the program reported greater use of empathy and expressive encouragement with their child as compared with fathers in the control group. They also reported greater reductions in their child's emotional and behavioral difficulties as compared with control fathers.[120] Although encouraging, these results may be vulnerable to expectancy bias in which research participants bias their answers in the direction of an expected effect. Clearly, fathers know that they are expected to show improvement in the domains of empathy and expressive encouragement, and they may consciously or unconsciously bias their answers in that direction. Future studies should further evaluate the program by having researchers observe actual father–child interactions in order to determine if there are changes in observed paternal behavior rather than relying on the father's own self-report.

Another program has done just that. Fathers Supporting Success in Preschool is a community parent education program that uses shared book reading as a vehicle for fathers to practice parenting skills. The program consists of eight weekly, ninety-minute group-based sessions. Fathers view videotaped vignettes of exaggerated parenting errors during father–child reading interactions and then break into small groups to identify the errors and brainstorm strategies to correct them. In a homework component, fathers practice dialogic reading, in which the parent uses prompts and feedback to allow the child to become an active storyteller.

One study evaluated the program in 126 low-income, mostly Hispanic fathers recruited from urban communities in the United States. Fathers were randomly assigned to the intervention or a control condition in which they did not receive the intervention. The intervention was successful in engaging fathers, with fathers attending 79% of all sessions. Fathers receiving the intervention showed improvements in not only their self-reported parenting behavior but also experimenter-observed parenting. Specifically, they were less critical of their children and showed more praise and affection toward them. Fathers randomized to the intervention also reported fewer behavioral problems in their children. Furthermore, children of fathers who received the intervention exhibited greater improvements in language development. One advantage of this program is that the education takes place mostly by way of discussions among the fathers rather than a top-down approach in which a highly trained expert instructs or trains the participants. This may increase the acceptability of the intervention to fathers and facilitate their participation. In addition, facilitators need not be highly trained professionals, which expands the pool of potential facilitators, as well as the potential scope of the program.[121]

There is reason to think that fatherhood interventions that begin in the prenatal period may be equally or even more effective than those that start after the child is born. One study found that expectant fathers who showed more sensitivity when interacting with a life-like infant simulator were then more sensitive with their own infant postnatally.[122] Some programs are now starting to target expecting fathers. One group educational intervention targets couples who are first-time parents and aims to enhance fathers' knowledge, skills, and commitment to the fatherhood role; to increase mothers' support and expectations for fathers' involvement; and to foster coparental teamwork in the couple. It consists of eight sessions that span the second trimester of pregnancy through five months postpartum and includes mini-lectures, group discussion, videotapes, demonstrations of skills, role playing, and the use of new parent role models. One study randomized 165 couples who were first-time parents to the intervention or a control group and then observed fathers in a play session with their infant. The attrition rate was only 15%. The intervention was shown to increase paternal warmth, positive affect, and father–infant synchrony while decreasing paternal intrusiveness. It also increased fathers' accessibility on work days.[123]

A Dutch research team has developed the intriguing Prenatal Video Feedback Intervention to Promote Positive Parenting, which involves three prenatal sessions in which fathers interact with their fetus using ultrasound imaging as researchers attempt to cultivate paternal sensitive responding and empathy, "Your baby stopped moving when you started talking. Maybe she is listening. At this age, she is certainly capable of hearing and recognising your voice."[124] This intervention has recently been shown to increase paternal sensitivity.[125]

*   *   *

In this chapter, we have reviewed a large number of environmental factors that influence paternal involvement. These are summarized in figure 6.1, along with the biological influences discussed in chapters 3 to 5. Fathers are expected to be more positively involved when they:

Grow up in safe and healthy environments

Have a positive paternal role model

Intentionally have children who are desired

Are more certain of their paternity

Are offered and accept paid paternal leave

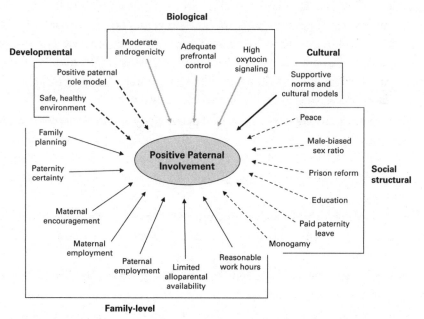

Figure 6.1

Environmental and biological influences over paternal involvement.

Are encouraged and supported by the mother to become involved in
child care

Are employed and work reasonable hours

Are not imprisoned

Outnumber women in the population

Marry monogamously

Have a partner who is also employed

Have few available alloparents

Are not involved in war

Have cultural models that promote involved fatherhood

These observations lead to some straightforward recommendations for policies aimed at increasing positive paternal involvement, and this seems an appropriate way to conclude this chapter.

1. Strive to make low-income neighborhoods safer by decreasing gun violence, for example, and healthier, by increasing access to high-quality, affordable health care.

2. Sadly, it is quite common to hear of fathers' tendency to withdraw from families when they are unable to successfully fulfill their provider role. I think most men want to do the right thing. They want to have a job and fulfill their responsibility to economically provision their child. Providing more jobs and more job training to fathers in low-income neighborhoods should help increase paternal involvement.

3. In countries where it is lacking, paid paternal leave is an obvious recommendation. However, fathers will be more likely to take it when their income is fully replaced and when corporations encourage leave taking and assure men there will not be negative career consequences for doing so.

4. Consider reducing or eliminating prison sentences for nonviolent drug-offending fathers who demonstrate a commitment to raising their children.

5. Avoid war. Societies at war tend to have distant fathers, and many children lose their fathers in battle.

6. Provide education at many different levels, a crucial recommendation. We can start by educating boys and young men about the importance

of fathers for child development. This should be done in high school before boys are likely to become fathers. I recently gave a talk about fatherhood to several classes at an Atlanta public high school with a high proportion of low-income students. I wasn't sure if the material would have an impact on high school students, but I passed out a confidential survey afterward, and one student had this to say: "I thought the presentation was interesting because I am interested in fatherhood and plan to become a father at a certain point in time. I feel more confident now in these facts and now know what to expect." In tandem with education about fatherhood, we should educate adolescent males and females about family planning and the importance of being intentional about having a family. Educating girls and young women in general is also crucial, as it is well known to delay childbearing and decrease the likelihood of teen pregnancy, which we know is linked with higher rates of paternal absence. Finally, we should strive to improve the quality of paternal caregiving with fatherhood programs and interventions such as those outlined above.

Investing in fathers is an investment in the future mental health and well-being of our children and, by extension, in a harmonious and well-functioning society. For both practical and humanitarian reasons, it is an investment that makes sense and is likely to pay substantial dividends.

Frederick Douglass said, "It is easier to build strong children than to repair broken men." But let's not give up on broken men. Many of them are fathers, and we need good fathers to help build those strong children.

7 Conclusion

If I had ten minutes to talk with a room full of young men who wanted to become fathers someday, what would I say to them about fatherhood?

I think I would start by explaining that humans are mammals, and that although most mammalian fathers are not involved in raising their young, something special happened during human evolution that endowed men with the tendency and capacity to do so. Paternal care probably evolved in humans because it provided ancestral children with extra calories that were used to help build a gigantic primate brain and that also enabled mothers to wean their infants early and reproduce at a faster rate. But not only did evolution endow men with the capacity to provision their children; it also gave them the potential to carry, hold, feed, clean, nurture, and play with them. Humans are cooperative breeders, and fathers are often part of a team of caregivers that assist mothers in raising their children. Fathers often get more involved with childcare when other allomothers are scarce. I would emphasize that their caregiving becomes especially critical when couples live in isolated nuclear families that are separated from extended family and therefore lack alloparents.

Then I would tell them that there is a lot of evidence that fathers are important for children's development, especially in Western, high-income societies like ours that tend to have a lot of those isolated nuclear families. I would emphasize that their children will tend to do best if they as fathers are warm, nurturing, sensitive, and responsive and if they set and enforce appropriate limits. Children with such fathers tend to have better mental health, fewer behavioral problems, do better in school, and have more friends. Good fathers help children learn how to cope with negative emotions like anger and anxiety and help prepare them for the world outside the home by confronting them with novelty and unpredictability in a way that teaches them to embrace rather than fear and avoid such experiences.

Next, I would them tell them that evolution prepared them for caregiving by tweaking their physiology and using brain circuitry that probably originally evolved to support maternal caregiving.

A lot of men like the idea of having high testosterone, but it is a double-edged sword. Testosterone can help motivate the pursuit of mates and social status, and it can support certain aspects of paternal caregiving such as provisioning when it requires intense aerobic or muscular exercise. However, it interferes with sensitive and nurturing care of infants, and it siphons energy from maintenance, suppressing the immune system and accelerating aging. So, it makes sense that men experience a decline in testosterone when they become involved fathers. It is not something to be demoralized by. It's a natural and important part of the human life cycle.

Then I would tell them about oxytocin. They may think of it as a maternal hormone, but it also appears to support father–infant bonding and it can be stimulated by skin-to-skin contact with infants. The same oxytocin that helps them bond with their infant also helps them bond with their partner. When they are warm and nurturing toward their children, they may also be shaping the development of their children's oxytocin system so that they will in turn be inclined to be warm and nurturing toward their own partners and children when they are older.

I would also tell them that their oxytocin will act on specific brain circuits that are involved in reward and motivation. In so doing, it will help make them find their infant rewarding and want to take care of him or her. It will also work on other brain circuits to help them empathize with their infant. I would emphasize that their comparatively massive prefrontal cortex will help them control their frustration when their infant is crying inconsolably and will also help them honor their commitments to their family in the face of other temptations that may lead them astray.

I would encourage them to be intentional about starting a family and to do so only once they feel they have found a compatible partner, and the two of them are ready and committed to be positively engaged parents. I would also encourage them to take paternity leave if they can do so without seriously jeopardizing their ability to provide for their family economically. Not only will they help reduce their partner's stress levels at a challenging time, paternity leave will also help form a bond with the infant that will provide a foundation for future bonding and involvement as the infant gets older. I would also tell them that although their financial contributions are likely to be important, they should try to avoid being consumed by work to the point where their involvement in hands-on direct care is negligible. Nevertheless, when circumstances require this in order to provide for their family, they should know that while not ideal, this is an honorable and admirable sacrifice.

Finally, I would tell them that while they can do a lot for their children—it isn't just a one-way street—and their children can also do a lot for them.

But what exactly do children do for their fathers?

My PhD mentor likes to say, "Reproductive success doesn't come easy." It sounds silly when you first hear it, but then you realize how true it is. Raising children is challenging and demands a great deal from parents. In my interviews with fathers from the Atlanta area, the challenges most often mentioned included financial stress related to their role as providers, work–family conflict, loss of intimacy in their relationship with their partner, and sleep loss during infancy.

But most fathers I talked with said that the rewards more than compensated for the challenges. What are some of those rewards?

To begin, having children is associated with increased longevity. One Swedish study showed that sixty-year-old men with at least one child had two years greater life expectancy than childless men. Women also benefited, but not quite as much (one and a half years increased longevity). Since the beneficial effect of children increased as men got older than sixty, one possible explanation is that older parents receive support from their adult children—social, emotional, instrumental—that improves their health and survival.[1]

Beyond this, having children may even lead to physiological changes that slow the rate of aging. One remarkable study recently showed that having children was associated with both improved performance on a memory task and a "younger" brain for one's age in late adulthood. People experience predictable changes in brain anatomy with age, but some people's brains age faster or slower than others. On average, both men and women with two or three children had a lower "brain age" compared with childless people.[2] Since this happened in both mothers and fathers, it is not likely attributable to hormonal changes that mothers experience during pregnancy. Instead, perhaps the mental stimulation that accompanies parenting is responsible for these neuroprotective effects. There are likely to be other forms of mental stimulation that slow brain aging as well, but parenting seems to be one of them.

This finding is counterintuitive to many of us who find parenting occasionally stressful and correctly assume that stress generally ages the brain. For example, it's hard to believe that getting two hours of sleep each night due to a colicky infant is adding years to your life. The key to understanding this may be that although parenting can be challenging and stressful at times, it also usually adds meaning and purpose to your life.[3] What could meaning and purpose have to do with aging? More than we might think. For example, it turns out that purpose in life has been shown to protect against cognitive decline among late-middle-aged adults.[4]

But in my interviews with fathers, they didn't talk about health benefits or longevity. Instead, they talked mostly about how their children made them feel.[5] Many fathers described how gratifying it was to feel loved by their children. When your child runs to greet you at the door or offers you an unsolicited "I love you, Daddy"—well, there aren't too many things better than that. A thirty-seven-year-old physician and father of three children whom I interviewed put it this way: "When you open up that door, running up to you and giving you a hug, like, there's no—there's nothing that I've encountered in life that really can replicate that." Another father confided, "I'm not the best dad in the world, but my children still love me and that feels good."

Other fathers were smitten by their children's charms. A fifty-six-year-old father of two children who worked as a CPA had this to say about his five-year-old daughter: "When she smiles, you know, this is the whole world to me." I have a five-year-old daughter, too, and I have often thought that she is absolutely perfectly designed by natural selection to make me love her. The way she looks, the way she acts, the way she talks, the things she says: I find them all adorable, and the scientist in me doubts it would be possible for evolution to have designed a human being I would be more motivated to take care of. Her power over me is astounding. For example, she loves playing with her Barbie Dreamhouse, and I would probably list that dead last in terms of my favorite activities. Yet when she asks me to play it with her, I am powerless to say no. Indeed, I have dutifully committed to memory the names of all of Barbie's sisters.

Many fathers viewed their children as an extension of themselves that allowed them to experience pleasure through their child. Earlier, I mentioned that my dad would take my brothers and me fishing, but he was rarely able to fish himself because he was busy helping us with all our issues. At the time, I was concerned that my dad was not having fun. But now I know better: he was having fun *through* us.

Seeing your child as an extension of yourself also gives you a kind of immortality—the sense that you will live on into the future through your child. One thirty-one-year-old father of two said, "Having [my kids] was like me being rebirthed. Like I was reborn twice." My daughter recently asked me why we have to die, Why we don't just live forever? Tough question. I decided to tell her that we don't really die, we just put our genes into new bodies, and that I had already done this twice, and that my two new bodies were her and her brother. I'm sure she was thinking, *I'll ask someone more sensible like Mom next time.* Even so, there is good reason why we feel we live on through our children.

Some fathers even felt that having children gave them an opportunity to start over in life and correct mistakes they had made in the past. One PhD student and father of two young children told me, "You do get to kind of live vicariously through, I mean, kids have a clean slate. . . . They can literally be an astronaut, be the president, you have just got to guide them. It's just such a wonderful thing. You look at all the mistakes that you've made. . . . They can really grow up to be great, great people." Another father put it more explicitly in saying that he found it rewarding to "help my children make better choices than I did." Some men even view fatherhood as a path to greater achievement through their children: "I want my guys to be better and do bigger things that I did, and . . . do better and accomplish more goals that I didn't accomplish because that would make me feel pride as a father." And indeed, such achievements were cited by many fathers as being particularly gratifying. For example, a thirty-seven-year-old father of three had this to say: "Seeing 'em accomplish something that they were working hard at and really wanted. Those are the some of the best feelings, best rewards you will ever see."

Many fathers also enjoyed teaching their children skills, knowledge and, values. We feel we have learned things that we can pass on to our children that will help them in life. A twenty-nine-year-old father of three said, "I wanna give him all the keys to everything that I had to figure out through life, through all these different situations . . . so I felt like I had a lot to teach them."

Another reward many men mentioned was simply witnessing the apparent miracle of child development—how a single-celled zygote could ultimately develop into a human being who can walk, talk, think, and love. Watching their children progress through developmental milestones was particularly satisfying. A forty-four-year-old father of two told me, "Just watch them grow; it's a most beautiful thing you can see. When they take their first step. They cut they first tooth. They say they first word. Aye, man, it just make your heart big." Another forty-year-old father of two said, "To me it was really fascinating to see these young babies . . . suddenly they can open their eyes and look around and focus, and the next day they can smile, and the next day they can sit up on their own. . . . It's really amazing just to watch through the first eighteen months of life."

But by far the most pervasive and consistent theme that emerged when I asked men about the rewards of fatherhood was that it added purpose and meaning to their lives. A PhD student with an infant son said to me, "Being a dad . . . gave me, like, another sense of purpose. . . . If all things fall apart . . . but I can provide and keep him happy, I'll be okay." Similarly, a father of a three-month-old infant said, "Before my wife was pregnant, I was having this ongoing philosophical debate about how life HAS no meaning in the grand scheme of things on this one planet, this one universe. . . . But after I had [my daughter], I was, like, I mean everything to this kid and she means everything to me, and it's a solid meaning. You can't really toss it out the window. It's crazy. Never thought something could mean so much."

Sometimes, when we find meaning and purpose in life, it can inspire us to improve on ourselves. One thirty-five-year-old father I interviewed told me, "I feel like I became a better person now that I'm a dad, . . . Definitely think about, you know, your actions before you take them and um, something about how they would impact [his infant daughter]." Another told me that he used to party too much and that he had a number of legal problems. But becoming a father "gave me a reason to stop with all the bull and get my life together." He went on, "I had some warrants . . . and I . . . finally went to court and got everything situated. . . . So now, I'm back, legit, and I got my license back and everything now."

A few fathers cried during our interviews, and some told me how ther-apeutic it was for them to talk about their experience as a father. Those are the ones that really stuck with me.There was this one man who had immigrated from China many years ago, and now had a fifteen-year old daughter and a nine-year old son. Here in the United States, he worked as a research scientist. He told me that for many years, he was not very happy in America, but then, something remarkable happened: his children were born. I could tell he was starting to have trouble talking, and tears began to form in his eyes. He paused for a few seconds and then finally looked at me and said, "I'm not in any religion, but one word from my friend . . . he said that having child is a prayer . . . from . . ." And he couldn't finish the sentence. So I offered, "From God?" He nodded.

Notes

Introduction

1. "THIS Is Dad of the Year," *Daily Mail*, March 10, 2014, https://www.dailymail.co
.uk/news/article-2577520/The-devoted-Chinese-father-carries-son-18-MILES-school
-day-Sichuan-province.html.

2. J. C. Mitani, D. P. Watts, and S. J. Amsler, "Lethal Intergroup Aggression Leads
to Territorial Expansion in Wild Chimpanzees," *Current Biology* 20, no. 12 (2010),
https://doi.org/10.1016/j.cub.2010.04.021; R. W. Wrangham and L. Glowacki,
"Intergroup Aggression in Chimpanzees and War in Nomadic Hunter-Gatherers
Evaluating the Chimpanzee Model," *Human Nature* 23, no. 1 (2012), https://doi.org
/10.1007/s12110-012-9132-1; F. B. M. de Waal, "A Century of Getting to Know the
Chimpanzee," *Nature* 437, no. 7055 (2005), https://doi.org/10.1038/nature03999.

3. A. E. Lowe, C. Hobaiter, and N. E. Newton-Fisher, "Countering Infanticide: Chimpanzee Mothers Are Sensitive to the Relative Risks Posed by Males on Differing Rank
Trajectories," *American Journal of Physical Anthropology* 168, no. 1 (2019), https://doi
.org/10.1002/ajpa.23723.

4. Martin N. Muller, Richard W. Wrangham, and David R. Pilbeam, eds., *Chimpanzees and Human Evolution* (Cambridge, MA: Belknap Press of Harvard University
Press, 2017).

5. Sarah Blaffer Hrdy, *Mothers and Others* (Cambridge, MA: Harvard University Press,
2009).

6. A. P. Moller, "The Evolution of Monogamy: Mating Relationships, Parental
Care, and Sexual Selection," in *Monogamy: Mating Strategies and Partnerships in Birds,
Humans, and Other Mammals*, ed. U. H. Reichard and C. Boesch (Cambridge: Cam-
bridge University Press, 2003).

7. Moller, "The Evolution of Monogam"; R. Woodroffe and A. Vincent, "Mother's
Little Helpers: Patterns of Male Care in Mammals," *Trends in Ecology and Evolution* 9,
no. 8 (1994), https://doi.org/0169-5347(94)90033-7.

8. H. S. Kaplan, P. L. Hooper, and M. Gurven, "The Evolutionary and Ecological Roots of Human Social Organization," *Philosophical Transactions of the Royal Society B: Biological Sciences* 364, no. 1533 (2009), https://doi.org/10.1098/rstb.2009.0115; F. W. Marlowe, "Paternal Investment and the Human Mating System," *Behavioural Processes* 51, no. 1–3 (2000), https://doi.org/10.1016/S0376–6357(00)00118–2.

9. Hrdy, *Mothers and Others.*

10. Barry S. Hewlett, *Intimate Fathers: The Nature and Context of Aka Pygmy Paternal Infant Care* (Ann Arbor: University of Michigan, 1991).

11. R. L. Coles, "Single-Father Families: A Review of the Literature," *Journal of Family Theory and Review* 7, no. 2 (2015), https://doi.org/10.1111/jftr.1206; S. Golombok et al., "Adoptive Gay Father Families: Parent–Child Relationships and Children's Psychological Adjustment," *Child Development* 85, no. 2 (014), https://doi.org/10.1111/cdev.12155.

12. R. Sear and R. Mace, "Who Keeps Children Alive? A Review of the Effects of Kin on Child Survival," *Evolution and Human Behavior* 29, no. 1 (2008), https://doi.org/10.1016/j.evolhumbehav.2007.10.001.

13. J. A. Gaudino, B. Jenkins, and R. W. Rochat, "No Fathers' Names: A Risk Factor for Infant Mortality in the State of Georgia, USA," *Social Science and Medicine* 48, no. 2 (1999), https://doi.org/10.1016/S0277–9536(98)00342–6.

14. N. J. Cabrera et al., "Fatherhood in the Twenty-First Century," *Child Development* 71, no. 1 (2000), https://doi.org/0.1111/1467–8624.00126; A. Sarkadi et al., "Fathers' Involvement and Children's Developmental Outcomes: A Systematic Review of Longitudinal Studies," *Acta Paediatrica* 97, no. 2 (2008), https://doi.org/10.1111/j.1651–2227.2007.00572.x.

15. Daniel Paquette, "Theorizing the Father–Child Relationship: Mechanisms and Developmental Outcomes," *Human Development* 47 (2004), https://doi.org/10.1159/000078723; G. L. Brown et al., "Associations between Father Involvement and Father–Child Attachment Security: Variations Based on Timing and Type of Involvement," *Journal of Family Psychology* 32, no. 8 (2018), https://doi.org/10.1037/fam0000472.

16. I. Alger et al., "Paternal Provisioning Results from Ecological Change," *Proceedings of the National Academy of Sciences* 117, no. 20 (2020), https://doi.org/10.1073/pnas.1917166117.

17. S. Gavrilets, "Human Origins and the Transition from Promiscuity to Pair-Bonding," *Proceedings of the National Academy of Sciences* 109, no. 25 (2012), https://doi.org/10.1073/pnas.1200717109.

18. C. Helfrecht et al., "Life History and Socioecology of Infancy," *American Journal of Physical Anthropology* 173, no. 4 (2020), https://doi.org/10.1002/ajpa.24145.

19. K. L. Kramer, "Cooperative Breeding and Its Significance to the Demographic Success of Humans," *Annual Review of Anthropology* 39 (2010), https://doi.org/10.1146/annurev.anthro.012809.105054; H. Kaplan et al., "A Theory of Human Life History Evolution: Diet, Intelligence, and Longevity," *Evolutionary Anthropology* 9, no. 4 (2000); K. Hawkes et al., "Grandmothering, Menopause, and the Evolution of Human Life Histories," *Proceedings of the National Academy of Sciences* 95, no. 3 (1998), https://doi.org/10.1073/pnas.95.3.1336, https://www.ncbi.nlm.nih.gov/pubmed/9448332.

20. L. T. Gettler, "Applying Socioendocrinology to Evolutionary Models: Fatherhood and Physiology," *Evolutionary Anthropology* 23, no. 4 (2014), https://doi.org/10.1002/evan.21412; J. K. Rilling and J. S. Mascaro, "The Neurobiology of Fatherhood," *Current Opinion in Psychology* 15 (2017), https://doi.org/10.1016/j.copsyc.2017.02.013.

21. L. T. Gettler et al., "Longitudinal Evidence That Fatherhood Decreases Testosterone in Human Males," *Proceedings of the National Academy of Sciences* 108, no. 39 (2011), https://doi.org/10.1073/pnas.110540310; M. N. Muller et al., "Testosterone and Paternal Care in East African Foragers and Pastoralists," *Philosophical Transactions of the Royal Society B: Biological Sciences* 276, no. 1655, https://doi.org/10.1098/rspb.2008.1028; Peter B. Gray and Kermyt G. Anderson, *Fatherhood: Evolution and Human Paternal Behavior* (Cambridge, MA: Harvard University Press, 2010).

22. J. S. Mascaro, P. D. Hackett, and J. K. Rilling, "Differential Neural Responses to Child and Sexual Stimuli in Human Fathers and Non-Fathers and Their Hormonal Correlates," *Psychoneuroendocrinology* 46 (2014), https://doi.org/10.1016/j.psyneuen.2014.04.014; Rilling and Mascaro, "The Neurobiology of Fatherhood"; O. Weisman, O. Zagoory-Sharon, and R. Feldman, "Oxytocin Administration, Salivary Testosterone, and Father-Infant Social Behavior," *Progress in Neuro-Psychopharmacology and Biological Psychiatry* 49 (2014), https://doi.org/10.1016/j.pnpbp.2013.11.006.

23. Michael Numan, *The Parental Brain: Mechanism, Development and Evolution* (New York: Oxford University Press, 2020); R. Feldman, K. Braun, and F. A. Champagne, "The Neural Mechanisms and Consequences of Paternal Caregiving," *Nature Reviews Neuroscience* 20, no. 4 (2019), https://doi.org/10.1038/s41583-019-0124-6.

24. Rilling and Mascaro, "The Neurobiology of Fatherhood."

25. P. Kim et al., "Neural Plasticity in Fathers of Human Infants," *Society for Neuroscience* 9, no. 5 (2014), https://doi.org/10.1080/17470919.2014.933713.

Chapter 1

1. Jung Chang and Jon Halliday, *Mao: The Unknown Story* (New York: Anchor Books, 2006).

2. Alan Bullock, *Hitler and Stalin: Parallel Lives* (New York: Knopf, 1992).

3. Henry Murray, *Analysis of the Personality of Adolph Hitler* (Washington, DC: US Office of Strategic Services, 1943).

4. Philip Short, *Mao: A Life* (New York: Owl Books, 2001); Bullock, *Hitler and Stalin*.

5. Joseph Bucklin Bishop, *Theodore Roosevelt and His Time Shown in His Own Letters* (New York: Scribner's, 1920).

6. Kim Hill and Magdalena Hurtado, *Ache Life History: The Ecology and Demography of a Foraging People* (London: Routledge, 1996); K. Hill, "Altruistic Cooperation during Foraging by the Ache, and the Evolved Human Predisposition to Cooperate," *Human Nature* 13, no. 1 (2002), https://doi.org/10.1007/s12110-002-1016-3.

7. R. Sear and R. Mace, "Who Keeps Children Alive? A Review of the Effects of Kin on Child Survival," *Evolution and Human Behavior* 29, no. 1 (2008): 1–18. https://doi.org/10.1016/j.evolhumbehav.2007.10.001.

8. J. A. Gaudino Jr., B. Jenkins, and R. W. Rochat, "No Fathers' Names: A Risk Factor for Infant Mortality in the State of Georgia, USA," *Social Science and Medicine* 48, no. 2 (1999), https://doi.org/10.1016/s0277-9536(98)00342-6.

9. A. H. Boyette, S. Lew-Levy, and L. T. Gettler, "Dimensions of Fatherhood in a Congo Basin Village: A Multimethod Analysis of Intracultural Variation in Men's Parenting and Its Relevance for Child Health," *Current Anthropology* 59, no. 6 (2018), https://doi.org/10.1086/700717.

10. J. Winking and J. Koster, "The Fitness Effects of Men's Family Investments: A Test of Three Pathways in a Single Population," *Human Nature* 26, no. 3 (2015), https://doi.org/10.1007/s12110-015-9237-4.

11. K. Schmeer, "Father Absence due to Migration and Child Illness in Rural Mexico," *Social Science and Medicine* 69, no. 8 (2009), https://doi.org/10.1016/j.socscimed.2009.07.030.

12. Michael E. Lamb, ed., *The Role of the Father in Child Development*, 5th ed. (Hoboken, NJ: Wiley, 2010).

13. "Obama Father's Day Speech." 2008, https://www.c-span.org/video/?205980-1/obama-fathers-day-speech.

14. Wendy Sigle-Rushton and Sara McLanahan, "Father Absence and Child Wellbeing: A Critical Review," in *The Future of the Family*, ed. D. P. Moynihan, L. Rainwater, and T. Smeeding (New York: Russell Sage Foundation, 2004).

15. J. Belsky and M. H. van IJzendoorn, "Genetic Differential Susceptibility to the Effects of Parenting," *Current Opinion in Psychology* 15 (2017), https://doi.org/10.1016/j.copsyc.2017.02.021.

16. W. Thomas Boyce, *The Orchid and the Dandelion: Why Some Children Struggle and How All Can Thrive* (New York: Knopf, 2019).

17. Charlotte Patterson, "Parents' Sexual Orientation and Children's Development," *Child Development Perspectives* 11, no. 1 (2016).

18. "My Life as an NBA Superstar Single Dad," 2011, https://www.allprodad.com /my-life-as-an-nba-superstar-single-dad/.

19. Patterson, "Parents' Sexual Orientation."

20. B. MacFarlan Hewlett, "Fathers' Roles in Hunter-Gatherers and Other Small-Scale Cultures," in *The Role of the Father in Child Development*, ed. M. E. Lamb (Hoboken, NJ: Wiley, 2010).

21. S. McLanahan, L. Tach, and D. Schneider, "The Causal Effects of Father Absence," *Annual Review of Sociology* 39 (2013), https://doi.org/10.1146/annurev-soc -071312-145704.

22. R. Jia et al., "Effects of Neonatal Paternal Deprivation or Early Deprivation on Anxiety and Social Behaviors of the Adults in Mandarin Voles," *Behavioural Processes* 82, no. 3 (2009), https://doi.org/10.1016/j.beproc.2009.07.006.

23. R. Jia et al., "Neonatal Paternal Deprivation or Early Deprivation Reduces Adult Parental Behavior and Central Estrogen Receptor Alpha Expression in Mandarin voles (*Microtus mandarinus*)," *Behavioural Brain Research* 224, no. 2 (2011), https://doi .org/10.1016/j.bbr.2011.05.042; P. Yu et al., "Early Social Deprivation Impairs Pair Bonding and Alters Serum Corticosterone and the NAcc Dopamine System in Mandarin Voles," *Psychoneuroendocrinology* 38, no. 12 (2013), https://doi.org/10.1016/j .psyneuen.2013.09.012.

24. Biography.com and Tyler Piccotti, "Oprah Winfrey," last updated February 4, 2024, https://www.biography.com/movies-tv/oprah-winfrey.

25. Richard B. Lee and Richard Daly, eds., *The Cambridge Encyclopedia of Hunters and Gatherers* (Cambridge: Cambridge University Press, 2004).

26. F. W. Marlowe, "A Critical Period for Provisioning by Hadza Men—Implications for Pair Bonding," *Evolution and Human Behavior* 24, no. 3 (2003), https://doi.org/10 .1016/S1090–5138(03)00014-X.

27. P. R. Amato and J. G. Gilbreth, "Nonresident Fathers and Children's Well-Being: A Meta-Analysis," *Journal of Marriage and the Family* 61, no. 3 (1999), https://doi.org /10.2307/353560.

28. M. D. S. Ainsworth, S. M. Bell, and D. J. Stayton, "Infant-Mother Attachment and Social Development," in *The Introduction of the Child into a Social World*, ed. M. P. Richards (Cambridge: Cambridge University Press, 1974); N. Lucassen et al., "The

Association between Paternal Sensitivity and Infant-Father Attachment Security: A Meta-Analysis of Three Decades of Research," *Journal of Family Psychology* 25, no. 6 (2011), https://doi.org/10.1037/a0025855; Kenna E. Ranson and Liana J. Urichuk, "The Effect of Parent–Child Attachment Relationships on Child Biopsychosocial Outcomes: A Review," *Early Child Development and Care* 178, no. 2 (2008), https://doi.org/10.1080/03004430600685282; Geoffrey L. Brown and Hasan Alp Aytuglu, "Father–Child Attachment Relationships," in *Handbook of Fathers and Child Development*, ed. H. E. Fitzgerald, K. von Klitzing, N. Cabrera, J. Scarano de Mendonça, and Th. Skjøthaug (Berlin: Springer, 2020).

29. Amber Paulk and Ryan Zayac, "Attachment Style as a Predictor of Risky Sexual Behavior in Adolesents," *Journal of Social Sciences* 9, no. 2 (2013).

30. Karin Grossman et al., "The Uniqueness of the Child-Father Attachment Relationship: Fathers' Sensitive and Challenging Play as a Pivotal Variable in a 16-Year Longitudinal Study" *Social Development* 11, no. 3 (2002).

31. A. Vakrat, Y. Apter-Levy, and R. Feldman, "Sensitive Fathering Buffers the Effects of Chronic Maternal Depression on Child Psychopathology," *Child Psychiatry and Human Development* (2018), https://doi.org/10.1007/s10578-018-0795-7.

32. D. Baumrind, "Authoritarian vs. Authoritative Parental Control," *Adolescence* 3, no. 11 (1968); L. Steinberg, I. Blatt-Eisengart, and E. Cauffman, "Patterns of Competence and Adjustment among Adolescents from Authoritative, Authoritarian, Indulgent, and Neglectful Homes: A Replication in a Sample of Serious Juvenile Offenders," *Journal of Research on Adolescence* 16, no. 1 (2006), https://doi.org/10.1111/j.1532–7795.2006.00119.x; S. D. Lamborn et al., "Patterns of Competence and Adjustment among Adolescents from Authoritative, Authoritarian, Indulgent, and Neglectful Families," *Child Development* 62, no. 5 (1991), https://doi.org/10.1111/j.1467–8624.1991.tb01588.x; K. Luyckx et al., "Parenting and Trajectories of Children's Maladaptive Behaviors: A 12-Year Prospective Community Study," *Journal of Clinical Child and Adolescent Psychology* 40, no. 3 (2011), https://doi.org/93701620910.1080/15374416.2011.5634701.

33. Paquette, "Theorizing the Father–Child Relationship"; Daniel Paquette, Carole Gagnon, and Julio Macario de Medeiros, "Fathers and the Activation Relationship," in *Handbook of Fathers and Child Development*, ed. H. E. Fitzgerald, K. von Klitzing, N. Cabrera, N., J.Scarano de Mendonça, and Th. Skjøthaug (Berlin: Springer, 2020).

34. E. L. Moller et al., "Associations between Maternal and Paternal Parenting Behaviors, Anxiety and Its Precursors in Early Childhood: A Meta-Analysis," *Clinical Psychology Review* 45 (2016), https://doi.org/10.1016/j.cpr.2016.03.002.

35. M. Majdandzic et al., "Fathers' Challenging Parenting Behavior Prevents Social Anxiety Development in Their 4-Year-Old Children: A Longitudinal Observational

Study," *Journal of Abnormal Child Psychology* 42, no. 2 (2014), https://doi.org/10.1007/s10802-013-9774-4.

36. Paquette et al., "Fathers and the Activation Relationship."

37. Paquette et al., "Fathers and the Activation Relationship."

38. N. Islamiah et al., "The Role of Fathers in Children's Emotion Regulation Development: A Systematic Review," *Infant and Child Development* 32, no. 2 (2023), https://doi.org/10.1002/icd.2397.

39. A. Sarkadi, R. Kristiansson, F. Oberklaid, and S. Bremberg, "Fathers' Involvement and Children's Developmental Outcomes: A Systematic Review of Longitudinal Studies," *Acta Paediatrica* 97, no. (2008): 153–158, https://doi.org/10.1111/j.1651-2227.2007.00572.x.

40. N. Vadenkiernan et al., "Household Family-Structure and Children's Aggressive-Behavior: A Longitudinal-Study of Urban Elementary-School Children," *Journal of Abnormal Child Psychology* 23, no. 5 (1995), https://doi.org/10.1007/Bf01447661.

41. K. M. Harris and J. K. Marmer, "Poverty, Paternal Involvement, and Adolescent Well-Being," *Journal of Family Issues* 17, no. 5 (1996), https://doi.org/10.1177/019251396017005003.

42. R. L. Coley and B. L. Medeiros, "Reciprocal Longitudinal Relations between Nonresident Father Involvement and Adolescent Delinquency," *Child Development* 78, no. 1 (2007), https://doi.org/10.1111/j.1467–8624.2007.00989.x.

43. E. Flouri and A. Buchanan, "Father Involvement in Childhood and Trouble with the Police in Adolescence—Findings from the 1958 British Cohort," *Journal of Interpersonal Violence* 17, no. 6 (2002), https://doi.org 10.1177/0886260502017006006.

44. M. Novak and H. Harlow, "Social Recovery of Monkeys Isolated for the First Year of Life," *Developmental Psychology* 11 (1975).

45. M. E. Lamb, "Father-Infant and Mother-Infant Interaction in 1st Year of Life," *Child Development* 48, no. 1 (1977), https://doi.org/10.2307/1128896.

46. N. S. Gordon et al., "Socially-Induced Brain 'Fertilization': Play Promotes Brain Derived Neurotrophic Factor Transcription in the Amygdala and Dorsolateral Frontal Cortex in Juvenile Rats," *Neuroscience Letter* 341, no. 1 (2003), https://doi.org/10.1016/s0304-3940(03)00158-7.

47. Jaak Panksepp, "Play, ADHD, and the Construction of the Social Brain: Should the First Class Each Day Be Recess?," *American Journal of Play* 1, no. 1 (2008).

48. Paquette et al., "Fathers and the Activation Relationship."

49. Melvin J. Konner, *The Evolution of Childhood: Relationships, Emotion, Mind* (Cambridge, MA: Harvard University Press, 2010).

50. K. Macdonald and R. D. Parke, "Bridging the Gap—Parent–Child Play Interaction and Peer Interactive Competence," *Child Development* 55, no. 4 (1984), https://doi.org/10.2307/1129996.

51. Melvin Konner, "Hunter-Gatherer Infancy and Childhood," in *Hunter-Gatherer Childhoods*, ed. Barry S. Hewlett and Michael E. Lamb (London: Routledge, 2005).

52. E. T. Gershoff and A. Grogan-Kaylor, "Spanking and Child Outcomes: Old Controversies and New Meta-Analyses," *Journal of Family Psychology* 30, no. 4 (2016), https://doi.org/10.1037/fam0000191.

53. G. T. Pace, S. J. Lee, and A. Grogan-Kaylor, "Spanking and Young Children's Socioemotional Development in Low- and Middle-Income Countries," *Child Abuse and Neglect* 88 (2019), https://doi.org/10.1016/j.chiabu.2018.11.003.

54. J. K. Rilling and C. Hadley, "A Mixed Methods Study of the Challenges and Rewards of Fatherhood in a Diverse Sample of U.S. Fathers," *Sage Open* 13, no. 3 (2023). https://doi.org/10.1177/21582440231193939.

55. James M. Herzog, *Father Hunger : Explorations with Adults and Children* (Hillsdale, NJ: Analytic Press, 2001).

56. B. B. Whiting, "Sex Identity Conflict and Physical Violence—a Comparative-Study," *American Anthropologist* 67, no. 6 (1965), https://www.jstor.org/stable/668842

57. C. R. Ember and M. Ember, "Father Absence and Male Aggression: A Re-Examination of the Comparative Evidence," *Ethos* 29, no. 3 (2001), https://doi.org/10.1525/eth.2001.29.3.296\.

58. G. J. Broude, "Protest Masculinity—a Further Look at the Causes and the Concept," *Ethos* 18, no. 1 (1990), https://doi.org/10.1525/eth.1990.18.1.02a00040\.

59. P. A. Graziano et al., "The Role of Emotion Regulation in Children's Early Academic Success," *Journal of School Psychology* 45, no. 1 (2007), https://doi.org/10.1016/j.jsp.2006.09.002.

60. Macdonald and Parke, "Bridging the Gap."

61. B. J. Ellis et al., "Does Father Absence Place Daughters at Special Risk for Early Sexual Activity and Teenage Pregnancy?," *Child Development* 74, no. 3 (2003), https://doi.org/10.1111/1467–8624.00569.

62. W. L. Rostad, P. Silverman, and M. K. McDonald, "Daddy's Little Girl Goes to College: An Investigation of Females' Perceived Closeness with Fathers and Later Risky Behaviors," *Journal of American College Health* 62, no. 4 (2014), http://www.ncbi.nlm.nih.gov/pubmed/24527944.

63. J. Mendle, K. P. Harden, E. Turkheimer, C. A. Van Hulle, B. M. D'Onofrio, J. Brooks-Gunn, J. L. Rodgers, R. E. Emery, and B. B. Lahey, "Associations between Father Absence and Age of First Sexual Intercourse," *Child Development* 80, no. 5 (2009): 1463–1480. https://doi.org/10.1111/j.1467-8624.2009.01345.x.

64. M. L. Rowe, D. Coker, and B. A. Pan, "A Comparison of Fathers' and Mothers' Talk to Toddlers in Low-Income Families," *Social Development* 13, no. 2 (2004), https://doi.org/10.1111/j.1467–9507.2004.000267.x.

65. M. L. Rowe, K. A. Leech, and N. Cabrera, "Going Beyond Input Quantity: Wh-Questions Matter for Toddlers' Language and Cognitive Development," *Cognitive Science* 41 (2017), https://doi.org/10.1111/cogs.12349.

66. N. Pancsofar and L. Vernon-Feagans, "Mother and Father Language Input to Young Children: Contributions to Later Language Development," *Journal of Applied Developmental Psychology* 27, no. 6 (2006), https://doi.org/10.1016/j.appdev.2006.08 .003.

67. D. Nettle, "Why Do Some Dads Get More Involved Than Others? Evidence from a Large British Cohort," *Evolution and Human Behavior* 29, no. 6 (2008): 416–423, https://doi.org/10.1016/j.evolhumbehav.2008.06.002.

68. Peter Gray and Alyssa Crittenden, "Father Darwin: Effects of Children on Men, Viewed from an Evolutionary Perspective," *Fathering* 12, no. 2 (2014), https://psyc-net.apa.org/record/2014-35882-002.

69. Konner, "Hunter-Gatherer Infancy and Childhood."

70. Fisher Stevens, dir. *Beckham*. Docuseries. Los Gatos, CA: Netflix, 2023.

71. B. A. Scelza, "Fathers' Presence Speeds the Social and Reproductive Careers of Sons," *Current Anthropology* 51, no. 2 (2010), https://doi.org/10.1086/651051.

72. Lamb, *The Role of the Father in Child Development.*

73. Leonard Pitts, *Becoming Dad: Black Men and the Journey to Fatherhood* (Marietta, GA: Longstreet, Inc., 2006).

74. Hewlett, "Fathers' Roles."

75. K. Hawkes, J. F. O'Connell, and N. G. B. Jones, "Hadza Women's Time Allocation, Offspring Provisioning, and the Evolution of Long Postmenopausal Life Spans," *Current Anthropology* 38, no. 4 (1997), https://doi.org/10.1086/204646.

76. E. O. Chung et al., "The Contribution of Grandmother Involvement to Child Growth and Development: An Observational Study in Rural Pakistan," *BMJ Global Health* 5, no. 8 (2020), https://doi.org/10.1136/bmjgh-2019–002181.

77. S. Schrijner and J. Smits, "Grandmothers and Children's Schooling in Sub-Saharan Africa," *Human Nature* 29, no. 1 (2018), https://doi.org/10.1007/s12110-017-9306-y.

78. S. K. Pope et al., "Low-Birth-Weight Infants Born to Adolescent Mothers—Effects of Coresidency with Grandmother on Child-Development," *Journal of the American Medical Association* 269, no. 11 (1993), 1396–1400, https://doi.org/10.1001/jama.269.11.1396.

79. J. Jeong et al., "Paternal Stimulation and Early Child Development in Low- and Middle-Income Countries," *Pediatrics* 138, no. 4 (Ot 2016), https://doi.org/10.1542/peds.2016-1357.

80. J. K. Choi and H. S. Pyun, "Nonresident Fathers' Financial Support, Informal Instrumental Support, Mothers' Parenting, and Child Development in Single-Mother Families with Low Income," *Journal of Family Issues* 35, no. 4 (2014), https://doi.org/10.1177/0192513x13478403.

81. Amato and Gilbreth, "Nonresident Fathers and Children's Well-Being."

82. S. Golombok, L. Mellish, S. Jennings, P. Casey, F. Tasker, and M. E. Lamb, "Adoptive Gay Father Families: Parent–Child Relationships and Children's Psychological Adjustment," *Child Development* 85, no. 2 (2014): 456–468, https://doi.org/10.1111/cdev.12155.

83. Benjamin Graham Miller, Stephanie Kors, and Jenny Macfie, "No Differences? Meta-Analytic Comparisons of Psychological Adjustment in Children of Gay Fathers and Heterosexual Parents," *Psychology of Sexual Orientation and Gender Diversity* 4, no. 1 (2017), https://doi.org/10.1037/sgd0000203

84. Lancet Public, "Single Fathers: Neglected, Growing, and Important," *Lancet Public Health* 3, no. 3 (2018): e100, https://doi.org/10.1016/S2468-2667(18)30032-X.

85. R. L. Coles, "Single-Father Families: A Review of the Literature," *Journal of Family Theory & Review* 7, no. 2 (2015): 144–166, https://doi.org/10.1111/jftr.12069.

86. T. J. Wade, S. Veldhuizen, and J. Cairney, "Prevalence of Psychiatric Disorder in Lone Fathers and Mothers: Examining the Intersection of Gender and Family Structure on Mental Health," *Canadian Journal of Psychiatry–Revue Canadienne de Psychiatrie* 56, no. 9 (2011), https://doi.org/10.1177/070674371105600908.

87. K. A. Kong and S. I. Kim, "Mental Health of Single Fathers Living in an Urban Community in South Korea," *Comprehensive Psychiatry* 56 (2015), https://doi.org/10.1016/j.comppsych.2014.09.012.

88. Leonard Pitts, *Becoming Dad: Black Men and the Journey to Fatherhood* (Marietta, GA: Longstreet, 2006); Daniel Patrick Moynihan, *The Negro Family: The Case for National Action* (Washington, DC: US Department of Labor,1965).

89. J. Jones and W. D. Mosher, "Fathers' Involvement with Their Children: United States, 2006–2010," *National Health Statistics Report*, no. 71 (2013), 1–21, https://www.ncbi.nlm.nih.gov/pubmed/24467852.

90. C. Z. Ellerbe, J. B. Jones, and M. J. Carlson, "Race/Ethnic Differences in Nonresident Fathers' Involvement after a Nonmarital Birth," *Social Science Quarterly* 99, no. 3 (2018), 1158–1182, https://doi.org/10.1111/ssqu.12482.

91. Jones and Mosher, "Fathers' Involvement with Their Children."

92. J. F. Paulson and S. D. Bazemore, "Prenatal and Postpartum Depression in Fathers and Its Association with Maternal Depression: A Meta-Analysis," *JAMA* 303, no. 19 (2010), 1961–1969, https://doi.org/10.1001/jama.2010.605.

93. James F. Paulson, Kelsey T. Ellis, and Regina Alexander, "Paternal Prenatal Depression and Postpartum Depression," in *Handbook of Fathers and Child Development*, ed. H. E. Fitzgerald, K. von Klitzing, N. Cabrera, J., Scarano de Mendonça, and Th. Skjøthaug (Berlin: Springer, 2020).

94. A. Viktorin et al., "Heritability of Perinatal Depression and Genetic Overlap with Nonperinatal Depression," *American Journal of Psychiatry* 173, no. 2 (2016), https://doi.org/10.1176/appi.ajp.2015.15010085.

95. L. Gutierrez-Galve et al., "Paternal Depression in the Postnatal Period and Child Development: Mediators and Moderators," *Pediatrics* 135, no. 2 (2015): e339–447, https://doi.org/10.1542/peds.2014-2411.

96. Paulson et al., "Paternal Prenatal Depression."

97. Paulson et al., "Paternal Prenatal Depression."

98. Paulson et al., "Paternal Prenatal Depression."

99. Ellen Galinsky, Kerstin Aumann, and James T. Bond, *Times Are Changing* (Hillsborough, NJ: Families and Work Institute, 2011), www.familiesandwork.org.

100. Kerstin Aumann, Ellen Galinsky, and Kenneth Matos, *The New Male Mystique* (Hillsborough, NJ: Families and Work Institute, 2011), www.familiesandwork.org.

101. Paulson et al., "Paternal Prenatal Depression."

102. Paulson et al., , "Paternal Prenatal Depression."

103. L. T. Gettler, T. W. McDade, A. B. Feranil, and C. W. Kuzawa, "Longitudinal Evidence That Fatherhood Decreases Testosterone in Human Males," *Proceedings of the National Academy of Sciences of the United States of America* 108, no. 39 (2011): 16194–16199, https://doi.org/10.1073/pnas.1105403108.

104. M. Ebinger et al., "Is There a Neuroendocrinological Rationale for Testosterone as a Therapeutic Option in Depression?," *Journal of Psychopharmacology* 23, no. 7 (2009), 841–853, https://doi.org/10.1177/0269881108092337.

105. Paulson et al., "Paternal Prenatal Depression."

106. R. G. Barr, "Preventing Abusive Head Trauma Resulting from a Failure of Normal Interaction between Infants and Their Caregivers," *Proceedings of the National Academy of Sciences* 109, Suppl. 2 (2012), https://doi.org/10.1073/pnas.1121267109.

107. Paulson et al., "Paternal Prenatal Depression."

108. K. C. Thomson et al., "Adolescent Antecedents of Maternal and Paternal Perinatal Depression: A 36-Year Prospective Cohort," *Psychological Medicine* 51, no. 12 (2021), https://doi.org/10.1017/S0033291720000902.

109. Gray and Anderson, *Fatherhood*

110. William Marsiglio and Ramon Hinojosa, "Stepfathers' Lives," in *The Role of the Father in Child Development* (Hoboken, NJ: Wiley, 2010).

111. Martin Daly and Margo Wilson, *Homicide* (New York: Aldine de Gruyter, 1988).

112. Gray and Anderson, *Fatherhood*; Marsiglio and Hinojosa, "Stepfathers' Lives."

113. M. V. Flinn and B. G. England, "Social Economics of Childhood Glucocorticoid Stress Response and Health," *American Journal of Physical Anthropology* 102, no. 1 (1997): 33–53, https://doi.org/10.1002/(SICI)1096-8644(199701)102:1<33::AID-AJPA4>3.0.CO;2-E.

114. Gray and Anderson, *Fatherhood*.

115. W. Marsiglio, "Paternal Engagement Activities with Minor Children," *Journal of Marriage and the Family* 53, no. 4 (1991), https://doi.org/10.2307/353001.

Chapter 2

1. A. P. Moller, "The Evolution of Monogamy: Mating Relationships, Parental Care, and Sexual Selection," in *Monogamy: Mating Strategies and Partnerships in Birds, Humans, and Other Mammals,* edited by U. H. Reichard and C. Boesch, 29–41 (Cambridge: Cambridge University Press, 2003).

2. Trivers, Robert, "Parental Investment and Sexual Selection," in *Sexual Selection and the Descent of Man, 1871–1971,* edited by B. Campbell, 136–179 (Chicago: Aldine, 1972).

3. Timothy H. Clutton-Brock, *The Evolution of Parental Care* (Princeton, NJ: Princeton University Press, 1991).

4. T. K. Wood, "Aggregation Behavior of *Umbonia crassicornis* (Homoptera: Membracidae)," *Canadian Entomologist* 106, no. 2 (1974), https://doi.org/10.4039/Ent106169-2; T. K. Wood, "Biology and Presocial Behavior of *Platycotis vittata*

(Homoptera: Membracidae)," *Annals of the Entomological Society of America* 69, no. 5 (1976), https://doi.org/10.1093/aesa/69.5.807.

5. M. Salomon, J. Schneider, and Y. Lubin, "Maternal Investment in a Spider with Suicidal Maternal Care, *Stegodyphus lineatus* (Araneae, Eresidae)," *Oikos* 109, no. 3 (2005), https://doi.org/10.1111/j.0030–1299.2005.13004.x.

6. Z. Q. Chen et al., "Prolonged Milk Provisioning in a Jumping Spider," *Science* 362, no. 6418 (2018), https://doi.org/10.1126/science.aat3692.

7. H. Sato, "Male Participation in Nest Building in the Dung Beetle *Scarabaeus catenatus* (Coleoptera: Scarabaeidae): Mating Effort versus Paternal Effort," *Journal of Insect Behavior* 11, no. 6 (1998), https://doi.org/10.1023/A:1020860010165.

8. R. L. Smith, "Brooding Behavior of a Male Water Bug *Belostoma flumineum* (Hemiptera: Belostomatidae)," *Journal of the Kansas Entomological Society* 49 (1976), https://www.jstor.org/stable/25082829.

9. S. Y. Ohba, N. Okuda, and S. Kudo, "Sexual Selection of Male Parental Care in Giant Water Bugs," *Royal Society Open Science* 3, no. 5 (2016), https://doi.org/10.1098/rsos.150720.

10. J. D. Reynolds, N. B. Goodwin, and R. P. Freckleton, "Evolutionary Transitions in Parental Care and Live Bearing in Vertebrates," *Philosophical Transactions of the Royal Society B: Biological Sciences* 357, no. 1419 (2002), https://doi.org/10.1098/rstb.2001.0930.

11. Clutton-Brock, *Evolution of Parental Care.*

12. Clutton-Brock, *Evolution of Parental Care.*

13. J. J. A. Van Iersel, "An Analysis of the Parental Behaviour of the Male Three-Spined Stickleback (*Gasterosteus aculeatus L.*)," *Behaviour Suppl.* 3 (1953).

14. A. B. Wilson et al., "The Dynamics of Male Brooding, Mating Patterns, and Sex Roles in Pipefishes and Seahorses (family Syngnathidae)," *Evolution* 57, no. 6 (2003), https://doi.org/10.1111/j.0014-3820.2003.tb00345.x; Sigal Balshine, "Patterns of Parental Care in Vertebrates," in *The Evolution of Parental Care*, ed. Nick J. Royle, Per T. Smiseth, and Mathias Kölliker (New York: Oxford University Press, 2012).

15. Clutton-Brock, *Evolution of Parental Care.*

16. P. Weygoldt, "Evolution of Parental Care in Dart Poison Frogs (Amphibia, Anura, Dendrobatidae)," *Zeitschrift Fur Zoologische Systematik Und Evolutionsforschung* 25, no. 1 (1987); S. Limerick, "Courtship Behavior and Oviposition of the Poison-Arrow Frog Dendrobates-Pumilio," *Herpetologica* 36, no. 1 (1980).

17. "Rhinoderma Darwinii." Animal Diveristy Web, 2000, accessed January 30, 2024, https://animaldiversity.org/accounts/Rhinoderma_darwinii/.

18. "Pouched Frog." Australian Museum, https://australian.museum/learn/animals/frogs/pouched-frog/.

19. Richard Shine, "Parental Care in Reptiles," in *Biology of Reptilia*, ed. Carl Gans and Raymond B. Huey (New York: Alan R. Liss, 1987).

20. Clutton-Brock, *The Evolution of Parental Care*.

21. Clutton-Brock, *The Evolution of Parental Care*.

22. A. P. Moller and J. J. Cuervo, "The Evolution of Paternity and Paternal Care in Birds," *Behavioral Ecology* 11, no. 5 (2000), https://doi.org/10.1093/beheco/11.5.472.

23. M. R. Servedio, T. D. Price, and R. Lande, "Evolution of Displays within the Pair Bond," *Philosophical Transactions of the Royal Society B: Biological Sciences* 280, no. 1757 (2013), https://doi.org/10.1098/rspb.2012.3020.

24. Clutton-Brock, *Evolution of Parental Care*.

25. A. M. Muldal, J. D. Moffatt, and R. J. Robertson, "Parental Care of Nestlings by Male Red-Winged Blackbirds," *Behavioral Ecology and Sociobiology* 19, no. 2 (1986), https://doi.org/0.1007/Bf00299945.

26. D. A. Christie, A Folch, F. Jutglar, and E. F. J. Garcia, "Southern Cassowary (*Casuarius casuarius*)," in *Handbook of the Birds of the World Alive*, ed. Elliott del Hoyo J, A. J. Sargatal, D. A. Christie, and E. de Juan (Barcelona: Lynx Edicions, 2017).

27. L. A. Moore, "Population Ecology of the Southern Cassowary *Casuarius casuarius johnsonii*, Mission Beach North Queensland," *Journal of Ornithology* 148, no. 3 (2007), https://doi.org/10.1007/s10336-007-0137-1.

28. I. P. F. Owens, "Male-Only Care and Classical Polyandry in Birds: Phylogeny, Ecology and Sex Differences in Remating Opportunities," *Philosophical Transactions of the Royal Society B: Biological Sciences* 357, no. 1419 (2002a), https://doi.org/10.1098/rstb.2001.0929.

29. Sarah Blaffer Hrdy, *Mothers and Others* (Cambridge, MA: Harvard University Press, 2009).

30. Walter Koenig and Janis Dickinson, *Ecology and Evolution of Cooperatively Breeding Birds* (Cambridge: Cambridge University Press, 2004).

31. R. L. Mumme, "Do Helpers Increase Reproductive Success: An Experimental-Analysis in the Florida Scrub Jay," *Behavioral Ecology and Sociobiology* 31, no. 5 (1992); Glen Everett Woolfenden and John W. Fitzpatrick, *The Florida Scrub Jay: Demography of a Cooperatively Breeding Bird* (Princeton, NJ: Princeton University Press, 1985).

32. John Pickrell, "How the Earliest Mammals Thrived Alongside Dinosaurs," *Nature* 574 (2019), https://www.nature.com/articles/d41586-019-03170-7

33. D. Lukas and T. Clutton-Brock, "Evolution of Social Monogamy in Primates Is Not Consistently Associated with Male Infanticide," *Proceedings of the National Academy of Sciences.* 111, no. 17 (2014), https://doi.org/10.1073/pnas.1401012111; C. Opie et al., "Male Infanticide Leads to Social Monogamy in Primates," *Proceedings of the National Academy of Sciences* 110, no. 33 (2013), https://doi.org/10.1073/pnas .1307903110.

34. R. Woodroffe and A. Vincent, "Mother's Little Helpers: Patterns of Male Care in Mammals," *Trends Ecol Evol* 9, no. 8 (1994): 294–297, https://doi.org/10.1016/0169 -5347(94)90033-7.

35. D. J. Gubernick and T. Teferi, "Adaptive Significance of Male Parental Care in a Monogamous Mammal," *Philosophical Transactions of the Royal Society B: Biological Sciences* 267, no. 1439 (2000), https://doi.org/10.1098/rspb.2000.0979.

36. D. Dudley, "Paternal Behavior in the California Mouse, *Peromyscus californicus,*" *Behavioral Biology* 11, no. 2 (1974), http://www.ncbi.nlm.nih.gov/pubmed/4847526.

37. Wenda Trevathan and James J. McKenna, "Evolutionary Environments of Human Birth and Infancy: Insights to Apply to Contemporary Life," *Children's Environments* 11, no. 2 (1994). https://www.jstor.org/stable/41514918

38. J. S. Jones and K. E. Wynne-Edwards, "Paternal Hamsters Mechanically Assist the Delivery, Consume Amniotic Fluid and Placenta, Remove Fetal Membranes, and Provide Parental Care during the Birth Process," *Hormones and Behavior* 37, no. 2 (2000a), https://doi.org/10.1006/hbeh.1999.1563, http://www.ncbi.nlm.nih.gov /pubmed/10753581.

39. C. J. Burgin et al., "How Many Species of Mammals Are There?," *Journal of Mammalogy* 99, no. 1 (2018), https://doi.org/10.1093/jmammal/gyx147.

40. D. W. Macdonald et al., "Monogamy: Cause, Consequence, or Corollary of Success in Wild Canids?," *Frontiers in Ecology and Evolution* 7 (2019), https://doi.org/10 .3389/fevo.2019.00341.

41. H. W. Y. Wright, "Paternal Den Attendance Is the Best Predictor of Offspring Survival in the Socially Monogamous Bat-Eared Fox," *Animal Behaviour* 71 (2006), https://doi.org/10.1016/j.anbehav.2005.03.043.

42. S. Dolotovskaya, C. Roos, and E. W. Heymann, "Genetic Monogamy and Mate Choice in a Pair-Living Primate," *Scientific Reports* 10, no. 1 (2020), https://doi.org /10.1038/s41598-020-77132-9.

43. C. T. Lambert, A. C. Sabol, and N. G. Solomon, "Genetic Monogamy in Socially Monogamous Mammals Is Primarily Predicted by Multiple Life History Factors: A Meta-Analysis," *Frontiers in Ecology and Evolution* 6 (2018), https://doi.org/10.3389 /fevo.2018.00139.

44. P.C. Wright, "Biparental Care in *Aotus trivirgatus* and *Callicebus moloch*," in *Female Primates: Studies by Women Primatologists*, ed. M. F. Small (New York: Alan R. Liss, 1984); A. Spence-Aizenberg, A. Di Fiore, and E. Fernandez-Duque, "Social Monogamy, Male-Female Relationships, and Biparental Care in Wild Titi Monkeys (*Callicebus discolor*)," *Primates* 57, no. 1 (2016), https://doi.org/10.1007/s10329-015 -0489-8.

45. K. A. Hoffman et al., "Responses of Infant Titi Monkeys, Callicebus Moloch, to Removal of One or Both Parents—Evidence for Paternal Attachment," *Developmental Psychobiology* 28, no. 7 (1995), https://doi.org/10.1002/dev.420280705.

46. Hrdy, *Mothers and Others*.

47. W. Saltzman, L. J. Digby, and D. H. Abbott, "Reproductive Skew in Female Common Marmosets: What Can Proximate Mechanisms Tell Us about Ultimate Causes?," *Philosophical Transactions of the Royal Society B: Biological Sciences* 276, no. 1656 (2009), https://doi.org/10.1098/rspb.2008.1374.

48. J. Terborgh and A. W. Goldizen, "On the Mating System of the Cooperatively Breeding Saddle-Backed Tamarin (*Saguinus fuscicollis*)," *Behavioral Ecology and Sociobiology* 16, no. 4 (1985), https://doi.org/10.1007/Bf00295541; R. W. Sussman and P. A. Garber, "A New Interpretation of the Social-Organization and Mating System of the Callitrichidae," *International Journal of Primatology* 8, no. 1 (1987), https://doi.org/10 .1007/Bf02737114.

49. T. E. Ziegler et al., "Pregnancy Weight Gain: Marmoset and Tamarin Dads Show It Too," *Biology Letters* 2, no. 2 (2006), https://doi.org/10.1098/rsbl.2005.0426.

50. K. Bales et al., "Effects of Allocare-Givers on Fitness of Infants and Parents in Callitrichid Primates," *Folia Primatologica* 71, no. 1–2 (2000), https://doi.org/10.1159 /000021728; P. A. Garber, "One for All and Breeding for One: Cooperation and Competition as a Tamarin Reproductive Strategy," *Evolutionary Anthropology* 5, no. 6 (1997). https://doi.org/10.1002/(SICI)1520-6505(1997)5:6<187::AID-EVAN1>3.0.CO;2-A

51. S. F. Lunn and A. S. McNeilly, "Failure of Lactation to Have a Consistent Effect on Interbirth Interval in the Common Marmoset, *Callithrix jacchus jacchus*," *Folia Primatologica* 37 (1982), https://doi.org/10.1159/000156023

52. Garber, "One for All and Breeding for One."

53. R. I. M. Dunbar, "The Mating System of Callitrichid Primates 2; The Impact of Helpers," *Animal Behaviour* 50 (1995), https://doi.org/10.1016/0003–3472(95) 80107-3.

54. J. A. French et al., "Social Monogamy in Nonhuman Primates: Phylogeny, Phenotype, and Physiology," *Journal of Sex Research* 55, no. 4–5 (2018), https://doi.org /10.1080/00224499.2017.1339774.

55. S. B. Hrdy, "Male-Male Competition and Infanticide among the Langurs (*Presbytis entellus*) of Abu, Rajasthan," *Folia Primatologica* 22, no. 1 (1974), https://doi.org/10.1159/000155616.

56. L. Swedell and T. Tesfaye, "Infant Mortality after Takeovers in Wild Ethiopian Hamadryas Baboons," *American Journal of Primatology* 60, no. 3 (2003), https://doi.org/10.1002/ajp.10096; C. Borries et al., "Males as Infant Protectors in Hanuman langurs (*Presbytis entellus*) Living in Multimale Groups—Defence Pattern, Paternity and Sexual Behaviour," *Behavioral Ecology and Sociobiology* 46, no. 5 (1999), https://doi.org/10.1007/s002650050629.

57. Sarah Blaffer Hrdy, "Infanticide as a Primate Reproductive Strategy," *American Scientist* 65, no. 1 (1977). https://www.jstor.org/stable/27847641

58. Sarah Hrdy, "Infanticide among Animals: A Review, Classification and Examination of the Implications for the Reproductive Strategies of Females," *Ethology and Sociobiology* 1 (1979), https://doi.org/10.1016/0162-3095(79)90004-9; Sarah Blaffer Hrdy, *Mother Nature* (New York: Ballantine Books, 1999).

59. J. C. Buchan et al., "True Paternal Care in a Multi-Male Primate Society," *Nature* 425, no. 6954 (2003); C. Minge et al., "Patterns and Consequences of Male-Infant Relationships in Wild Assamese Macaques (*Macaca assamensis*)," *International Journal of Primatology* 37, no. 3 (2016), https://doi.org/10.1007/s10764-016-9904-2; D. Langos et al., "Does Male Care, Provided to Immature Individuals, Influence Immature Fitness in Rhesus Macaques?," *PloS One* 10, no. 9 (2015), https://doi.org/10.1371/journal.pone.0137841.

60. L. A. Parr et al., "Visual Kin Recognition in Nonhuman Primates (*Pan troglodytes* and *Macaca mulatta*) Inbreeding Avoidance or Male Distinctiveness?," *Journal of Comparative Psychology* 124, no. 4 (2010), https://doi.org/10.1037/a0020545.

61. Barbara B. Smuts, *Sex and Friendship in Baboons: With a New Preface* (Cambridge, MA: Harvard University Press, 1999); N. Menard et al., "Is Male-Infant Caretaking Related to Paternity and/or Mating Activities in Wild Barbary Macaques (*Macaca sylvanus*)?," *Comptes Rendu de l'Académie des Sciences III* 324, no. 7 (2001), https://doi.org/10.1016/s0764-4469(01)01339-7.

62. L. Pozzi et al., "Primate Phylogenetic Relationships and Divergence Dates Inferred from Complete Mitochondrial Genomes," *Molecular Phylogenetics Evolution* 75 (2014), https://doi.org/10.1016/j.ympev.2014.02.023.

63. S. Lappan, "Male Care of Infants in a Siamang (*Symphalangus syndactylus*) Population including Socially Monogamous and Polyandrous Groups," *Behavioral Ecology and Sociobiology* 62, no. 8 (2008), https://doi.org/10.1007/s00265-008-0559-7.

64. U. Reichard, "Extra-Pair Copulations in Monogamous Wild White-Handed Gibbons (*Hylobates lar*)," *Zeitschrift für Saugetierkunde/International Journal of Mammalian Biology* 60, no. 3 (1995).

65. Pozzi et al., "Primate Phylogenetic Relationships."

66. S. S. Utami et al., "Male Bimaturism and Reproductive Success in Sumatran Orangutans," *Behavioral Ecology* 13, no. 5 (2002), https://doi.org/10.1093/beheco/13.5.643.

67. I. Singleton and C. P. van Schaik, "The Social Organisation of a Population of Sumatran Orangutans," *Folia Primatologica* 73, no. 1 (2002), https://doi.org/10.1159/000060415.

68. T. M. Smith et al., "Cyclical Nursing Patterns in Wild Orangutans," *Science Advances* 3, no. 5 (2017), https://doi.org/10.1126/sciadv.1601517; Serge A. Wich, Han de Vries, Marc Ancrenaz, Lori Perkins, Robert W. Shumaker, Akira Suzuki, and Carel P. van Schaik, "Orangutan Life History Variation," in *Orangutans: Geographic Variation in Behavioral Ecology and Conservation*, ed. Serge A. Wich, S. Suci Utami Atmoko, Tatang Mitra Setia, and Carel P. van Schaik (New York: Oxford University Press, 2009).

69. Pozzi et al., "Primate Phylogenetic Relationships."

70. D. P. Watts, "Infanticide in Mountain Gorillas—New Cases and a Reconsideration of the Evidence," *Ethology* 81, no. 1 (1989).

71. S. Rosenbaum, J. B. Silk, and T. S. Stoinski, "Male-Immature Relationships in Multi-Male Groups of Mountain Gorillas (*Gorilla beringei beringei*)," *American Journal of Primatology* 73, no. 4 (2011), https://doi.org/10.1002/ajp.20905.

72. M. M. Robbins and A. M. Robbins, "Variation in the Social Organization of Gorillas: Life History and Socioecological Perspectives," *Evolutionary Anthropology* 27, no. 5 (2018), https://doi.org/10.1002/evan.21721.

73. S. Rosenbaum et al., "Caring for Infants Is Associated with Increased Reproductive Success for Male Mountain Gorillas," *Scientific Reports* 8 (2018), https://doi.org/10.1038/s41598-018-33380-4.

74. Martin N. Muller, Richard W. Wrangham, and David R. Pilbeam, eds., *Chimpanzees and Human Evolution* (Cambridge, MA: Belknap Press of Harvard University Press, 2017).

75. E. E. Wroblewski et al., "Male Dominance Rank," *Animal Behavior* 77, no. 4 (2009), https://doi.org/10.1016/j.anbehav.2008.12.014.

76. F. B. de Waal, *Chimpanzee Politics* (London: Jonathan Cape, 1982).

77. J. T. Feldblum et al., "Sexually Coercive Male Chimpanzees Sire More Offspring," *Current Biology* 24, no. 23 (2014), https://doi.org/10.1016/j.cub.2014.10.039.

78. Jane Goodall, *The Chimpanzees of Gombe: Patterns of Behavior* (Cambridge, MA: Harvard University Press, 1986); J. H. Manson and R. W. Wrangham, "Intergroup

Aggression in Chimpanzees and Humans," *Current Anthropology* 32, no. 4 (1991), https://doi.org/10.1086/203974; C. Boesch et al., "Intergroup Conflicts among Chimpanzees in Tai National Park: Lethal Violence and the Female Perspective," *American Journal of Primatology* 70, no. 6 (2008), https://doi.org/10.1002/ajp.20524.

79. A. E. Lowe et al., "Intra-Community Infanticide in Wild, Eastern Chimpanzees: A 24-Year Review," *Primates* 61, no. 1 (2020), https://doi.org/10.1007/s10329-019 -00730-3.

80. C. M. Murray et al., "Chimpanzee Fathers Bias Their Behaviour towards Their Offspring," *Royal Society Open Science* 3, no. 11 (2016), https://doi.org/10.1098 /rsos.160441; J. Lehmann, G. Fickenscher, and C. Boesch, "Kin Biased Invest- ment in Wild Chimpanzees," *Behaviour* 143 (2006), https://doi.org/10.1163 /156853906778623635.

81. B. Hare, V. Wobber, and R. Wrangham, "The Self-Domestication Hypothesis: Evolution of Bonobo Psychology Is Due to Selection against Aggression," *Animal Behaviour* 83, no. 3 (2012), https://doi.org/10.1016/j.anbehav.2011.12.007.

82. B. G. Blount, "Issues in Bonobo (*Panpaniscus*) Sexual-Behavior," *American Anthropologist* 92, no. 3 (1990), https://doi.org/10.1525/aa.1990.92.3.02a00100.

83. Hare et al., "The Self-Domestication Hypothesis."

84. Hare et al., "The Self-Domestication Hypothesis."

85. F. B. de Waal, *Bonobo: The Forgotten Ape* (Berkeley: University of California Press, 1997).

86. Nicholas B. Davies, John R. Krebs, and Stuart A. West, *An Introduction to Behavioural Ecology*, 4th ed. (West Sussex, UK: Wiley-Blackwell, 2012).

87. H. Gottfried et al., "Aggression by Male Bonobos against Immature Individuals Does Not Fit with Predictions of Infanticide," *Aggressive Behavior* 45, no. 3 (2019), https://doi.org/10.1002/ab.21819.

88. T. Gruber and Z. Clay, "A Comparison between Bonobos and Chimpanzees: A Review and Update," *Evolutionary Anthropology* 25, no. 5 (016), https://doi.org/10 .1002/evan.21501.

89. R. Schacht and K. L. Kramer, "Are We Monogamous? A Review of the Evolution of Pair-Bonding in Humans and Its Contemporary Variation Cross-Culturally," *Fron- tiers in Ecology and Evolution* 7 (2019), https://doi.org/10.3389/fevo.2019.00230.

90. H. Kaplan et al., "A Theory of Human Life History Evolution: Diet, Intelligence, and Longevity," *Evolutionary Anthropology* 9, no. 4 (2000).

91. H. S. Kaplan, P. L. Hooper, and M. Gurven, "The Evolutionary and Ecological Roots of Human Social Organization," *Philosophical Transactions of the Royal Society B: Biological Sciences* 364, no. 1533 (2009), https://doi.org/10.1098/rstb.2009.0115.

92. K. Hawkes, "Showing Off—Tests of an Hypothesis about Men's Foraging Goals," *Ethology and Sociobiology* 12, no. 1 (1991), https://doi.org/10.1016/0162-3095(91) 90011-E.

93. E. A. Smith, "Why Do Good Hunters Have Higher Reproductive Success?," *Human Nature* 15, no. 4 (2004), https://doi.org/10.1007/s12110-004-1013-9.

94. M. Gurven and K. Hill, "Why Do Men Hunt? A Reevaluation of 'Man the Hunter' and the Sexual Division of Labor," *Current Anthropology* 50, no. 1 (2009), https://doi.org/10.1086/595620.

95. K. Hawkes and J. E. Coxworth, "Grandmothers and the Evolution of Human Longevity: A Review of Findings and Future Directions," *Evolutionary Anthropology* 22, no. 6 (2013), https://doi.org/10.1002/evan.21382; K. Hawkes, J. F. O'Connell, and N. G. B. Jones, "Hadza Women's Time Allocation, Offspring Provisioning, and the Evolution of Long Postmenopausal Life Spans," *Current Anthropology* 38, no. 4 (1997), https://doi.org/10.1086/204646.

96. Martin N. Muller, Richard W. Wrangham, and David R. Pilbeam, *Chimpanzees and Human Evolution* (Cambridge, MA: Belknap Press of Harvard University Press, 2017a).

97. A. Fischer et al., "Bonobos Fall within the Genomic Variation of Chimpanzees," *PloS One* 6, no. 6 (011), https://doi.org/10.1371/journal.pone.0021605; Brian Hare and Richard Wrangham, "Equal, Similar, But Different: Convergent Bonobos and Conserved Chimpanzees," in *Chimpanzees and Human Evolution*, ed. Martin Muller, Richard Wrangham, and David Pilbeam (Cambridge, MA: Harvard University Press, 2017a).

98. This sexual division of labor is not absolute, however. Among modern-day hunting and gathering, or foraging, populations, it is not the case that men *exclusively* hunt and women *exclusively* gather. There are many societies in which women have been reported to hunt, most often focusing on smaller game (Anderson, Chilczuk, Nelson, Ruther, & Wall-Scheffler, 2023; Reyes-García, Díaz-Reviriego, Duda, Fernández-Llamazares, & Gallois, 2020). Men also gather plant foods. However, men consistently hunt more than women do, and women consistently gather more than men do across foraging societies (Murdock & Provost, 1973; Reyes-García et al., 2020). A. Anderson, S. Chilczuk, K. Nelson, R. Ruther, and C. Wall-Scheffler, "The Myth of Man the Hunter: Women's Contribution to the Hunt across Ethnographic Contexts, *PLoS One* 18, no. 6 (2023), https://doi.org/10.1371/journal.pone.0287101; G. P. Murdock and C. Provost, "Factors in the Division of Labor by Sex: A Cross-Cultural Analysis," *Ethnology* 12 no. 2: (1973): 203–225; V. Reyes-García, I. Díaz-Reviriego, R. Duda, A. Fernández-Llamazares, and S. Gallois. "Hunting Otherwise: Women's Hunting in Two Contemporary Forager-Horticulturalist Societies," *Human Nature: An Interdisciplinary Biosocial Perspective* 31 no. 3 (2020): 203–221, https://doi .org/10.1007/s12110-020-09375-4.

99. I. Alger et al., "Paternal Provisioning Results from Ecological Change," *Proceedings of the National Academy of Sciences* 117, no. 20 (2020), https://doi.org/10.1073 /pnas.1917166117.

100. C. O. Lovejoy, "The Origin of Man," *Science* 211, no. 4480 (1981), https://doi .org/10.1126/science.211.4480.341.

101. S. Gavrilets, "Human Origins and the Transition from Promiscuity to Pair-Bonding," *Proceedings of the National Academy of Sciences* 109, no. 25 (2012), https:// doi.org/10.1073/pnas.1200717109; Beverly Strassman, "Sexual Selection, Paternal Care, and Concealed Ovultion in Humans," *Ethology and Sociobiology* 2 (1981), https://doi.org/10.1016/0162-3095(81)90020-0.

102. C. M. Gomes and C. Boesch, "Wild Chimpanzees Exchange Meat for Sex on a Long-Term Basis," *PloS One* 4, no. 4 (2009), https://doi.org/10.1371/journal.pone .0005116.

103. Gurven and Hill, "Why Do Men Hunt?"

104. John Marshall, "The Hunters," Documentary Educational Resources, 1958.

105. Marjorie Shostak, *Nisa, the Life and Words of a !Kung Woman* (New York: Vintage Books, 1983).

106. Gavrilets, "Human Origins and the Transition from Promiscuity to Pair-Bonding."

107. R. Nielsen et al., "A Scan for Positively Selected Genes in the Genomes of Humans and Chimpanzees," *PLoS Biology* 3, no. 6 (2005), https://doi.org/10.1371 /journal.pbio.0030170.

108. Lovejoy, "The Origin of Man"; R. D. Alexander and KM. Nonnan, "Concealment of Ovulation, Parental Care and Human Social Evolution," in *Evolutionary Biology and Human Social Behavior: An Anthropological Perspective*, ed. N.A. Chagnon and W.G. Irons (North Scituate, MA: Duxbury Press, 1979).

109. Melissa Emery Thompson, Martin N. Muller, Sonya M. Kahlenbergand, and Richard W. Wrangham "Sexual Coercion by Male Chimps Shows That Female Choice May Be More Apparent Than Real," *Behavioral Ecology and Sociobiology* 65 (2011).

110. G. Hohmann and B. Fruth, "Intra- and Inter-Sexual Aggression by Bonobos in the Context of Mating," *Behaviour* 140 (2003), https://doi.org 10.1163 /156853903771980648; M. Surbeck et al., "Male Reproductive Skew Is Higher in Bonobos Than Chimpanzees," *Current Biology* 27, no. 13 (2017), https://doi.org/10 .1016/j.cub.2017.05.039.

111. R. W. Wrangham, "Hypotheses for the Evolution of Reduced Reactive Aggression in the Context of Human Self-Domestication," *Frontiers in Psychology* 10 (2019), https://doi.org/ARTN191410.3389/fpsyg.2019.01914.

112. R. L. Cieri et al., "Craniofacial Feminization, Social Tolerance, and the Origins of Behavioral Modernity," *Current Anthropology* 55, no. 4 (2014), https://doi.org/10 .1086/677209.

113. Wrangham, "Hypotheses."

114. Richard W. Wrangham, *The Goodness Paradox: The Strange Relationship between Virtue and Violence in Human Evolution* (New York: Pantheon Books, 2019).

115. H. Kaplan et al., "A Theory of Human Life History Evolution: Diet, Intelligence, and Longevity," *Evolutionary Anthropology* 9, no. 4 (2000); Smith et al., "Cyclical Nursing Patterns"; Serge A. Wich, Han de Vries, Marc Ancrenaz, Lori Perkins, Robert W. Shumaker, Akira Suzuki, and Carel P. van Schaik, "Orangutan Life History Variation," in *Orangutans: Geographic Variation in Behavioral Ecology and Conservation*, ed. Serge A. Wich, S. Suci Utami Atmoko, Tatang Mitra Setia, and Carel P. van Schaik (New York: Oxford University Press, 2009); M. Emery Thompson et al., "Aging and Fertility Patterns in Wild Chimpanzees Provide Insights into the Evolution of Menopause," *Current Biology* 17, no. 24 (2007), https://doi.org/10.1016/j.cub.2007.11.033; M. D. Gurven and R. J. Davison, "Periodic Catastrophes over Human Evolutionary History Are Necessary to Explain the Forager Population Paradox," *Proceedings of the National Academy of Sciences* 116, no. 26 (2019), https://doi.org/10.1073/pnas .1902406116.

116. Kaplan et al., "A Theory of Human Life History Evolution"; Gurven and Davison, "Periodic Catastrophes over Human Evolutionary History Are Necessary to Explain the Forager Population Paradox."

117. J. C. Mitani and D. Watts, "The Evolution of Non-maternal Caretaking among Anthropoid Primates: Do Helpers Help?" *Behavioral Ecology and Sociobiology* 40, no. 4 (1997): 213–220, https://doi.org/10.1007/s002650050335.

118. K. Isler and C. P. van Schaik, "Allomaternal Care, Life History and Brain Size Evolution in Mammals," *Journal of Human Evolution* 63, no. 1 (2012), https://doi.org /10.1016/j.jhevol.2012.03.009.

119. Y. M. Bar-On, R. Phillips, and R. Milo, "The Biomass Distribution on Earth," *Proceedings of the National Academy of Sciences* 115, no. 25 (2018), https://doi.org/10 .1073/pnas.1711842115.

120. K. L. Kramer, "Cooperative Breeding and Its Significance to the Demographic Success of Humans," *Annual Review of Anthropology* 39 (2010), https://doi.org/10 .1146/annurev.anthro.012809.105054; R. Boyd, P. J. Richerson, and J. Henrich, "The Cultural Niche: Why Social Learning Is Essential for Human Adaptation," *Proceedings of the National Academy of Sciences* 108, suppl. 2 (2011), https://doi.org/10.1073 /pnas.1100290108.

121. L. C. Aiello and P. Wheeler, "The Expensive-Tissue Hypothesis: The Brain and the Digestive-System in Human and Primate Evolution," *Current Anthropology* 36, no. 2 (1995), https://doi.org/0.1086/204350.

122. J. Michael Plavcan, "Sexual Dimorphism in Hominin Ancestors," in *The International Encyclopedia of Anthropology*, ed. Hilary Callan (Hoboken, NJ: Wiley, 2018); M. Grabowski et al., "Body Mass Estimates of Hominin Fossils and the Evolution of Human Body Size," *Journal of Human Evolution* 85 (2015), https://doi.org/10.1016/j.jhevol.2015.05.005; H. M. Mchenry, "Behavioral Ecological Implications of Early Hominid Body-Size," *Journal of Human Evolution* 27, no. 1–3 (1994), https://doi.org/0.1006/jhev.1994.1036.

123. Alger et al., "Paternal Provisioning Results from Ecological Change."

124. D. J. Kruger, "Male Facial Masculinity Influences Attributions of Personality and Reproductive Strategy," *Personal Relationships* 13, no. 4 (2006), https://doi.org/10.1111/j.1475–6811.2006.00129.x.

125. Kruger, "Male Facial Masculinity Influences Attributions."

126. Marlowe, "Paternal Investment and the Human Mating System."

127. A. N. Crittendena and F. W. Marlowe, "Allomaternal Care among the Hadza of Tanzania.," *American Journal of Physical Anthropology* (2006).

128. Barry S. Hewlett, *Intimate Fathers: The Nature and Context of Aka Pygmy Paternal Infant Care* (Ann Arbor: University of Michigan, 1991).

129. C. Helfrecht et al., "Life History and Socioecology of Infancy," *American Journal of Physical Anthropology* 173, no. 4 (2020), https://doi.org/10.1002/ajpa.24145.

130. C. L. Meehan, "The Effects of Residential Locality on Parental and Alloparental Investment among the Aka Foragers of the Central African Republic," *Human Nature* 16, no. 1 (2005), https://doi.org/10.1007/s12110-005-1007-2.

131. H. N. Fouts, "Father Involvement with Young Children among the Aka and Bofi Foragers," *Cross-Cultural Research* 42, no. 3 (2008), https://doi.org/10.1177/1069397108317484.

132. J. Winking et al., "The Goals of Direct Paternal Care among a South Amerindian Population," *American Journal of Physical Anthropology* 139, no. 3 (2009), https://doi.org/10.1002/ajpa.20981.

Chapter 3

1. D. J. Handelsman, A. L. Hirschberg, and S. Bermon, "Circulating Testosterone as the Hormonal Basis of Sex Differences in Athletic Performance," *Endocrine Reviews* 39, no. 5 (2018), https://doi.org/10.1210/er.2018-00020.

2. Carole Hooven, *T: The Story of Testosterone, the Hormone That Dominates and Divides Us* (New York: Holt, 2021); Rebecca M. Jordan-Young and Katrina Alicia Karkazis, *Testosterone: An Unauthorized Biography* (Cambridge, MA: Harvard University Press, 2019); Cordelia Fine, *Testosterone Rex: Myths of Sex, Science, and Society* (New York: Norton, 2017).

3. J. Archer, "Testosterone and Human Aggression: An Evaluation of the Challenge Hypothesis," *Neuroscience and Biobehavioral Review* 30, no. 3 (2006), https://doi.org /10.1016/j.neubiorev.2004.12.007; Jordan-Young and Karkazis, *Testosterone.*

4. This example is from Dan Gilbert at Harvard University. "Testosterone: The Hormone that Dominates and Divides Us," YouTube video, 1:02:58, October 25, 2021, https://youtu.be/cbMjHLpc0Sc.

5. Hooven, *T: The Story of Testosterone.*

6. Handelsman et al., "Circulating Testosterone."

7. Archer, "Testosterone and Human Aggression."

8. S. N. Geniole and J. M. Carre, "Human Social Neuroendocrinology: Review of the Rapid Effects of Testosterone," *Hormones and Behavior* (2018), https://doi.org/10 .1016/j.yhbeh.2018.06.001.

9. Handelsman et al., "Circulating Testosterone."

10. M. Becker and V. Hesse, "Minipuberty: Why Does it Happen?," *Hormone Research in Paediatrics* 93, no. 2 (2020), https://doi.org/10.1159/000508329.

11. J. M. Tanner, *Fetus into Man* (Cambridge, MA: Harvard University Press, 1990).

12. K. J. Ottenbacher et al., "Androgen Treatment and Muscle Strength in Elderly Men: A Meta-Analysis," *Journal of the American Geriatrics Society* 54, no. 11 (2006), https://doi.org/10.1111/j.1532-5415.2006.00938.x; M. M. Fernandez-Balsells et al., "Clinical Review 1: Adverse Effects of Testosterone Therapy in Adult Men: A Systematic Review and Meta-Analysis," *Journal of Clinical Endocrinology and Metabolism* 95, no. 6 (2010), https://doi.org/10.1210/jc.2009-25756; J. L. Vingren, W. J. Kraemer, N. A. Ratamess, J. M. Anderson, J. S. Volek, and C. M. Maresh, "Testosterone Physiology in Resistance Exercise and Training: The Up-stream Regulatory Elements" *Sports Med* 40, no. 12 (2010), 1037–1053. https://doi.org/10.2165/11536910-000000000-00000.

13. Tanner, *Fetus into Man.*

14. B. J. Dixson and P. L. Vasey, "Beards Augment Perceptions of Men's Age, Social Status, and Aggressiveness, But Not Attractiveness," *Behavioral Ecology* 23, no. 3 (2012), https://doi.org/10.1093/beheco/arr214.

15. P. J. Rizk et al., "Testosterone Therapy Improves Erectile Function and Libido in Hypogonadal Men," *Current Opinion in Urology* 27, no. 6 (2017), https://doi.org

/10.1097/MOU.0000000000000442; M. N. Muller, "Testosterone and Reproductive Effort in Male Primates," *Hormones and Behavior* 91 (2017), https://doi.org/10.1016/j .yhbeh.2016.09.001; C. Eisenegger, J. Haushofer, and E. Fehr, "The Role of Testosterone in Social Interaction," *Trends in Cognitive Sciences* 15, no. 6 (2011), https://doi .org/10.1016/j.tics.2011.04.008; E. Domonkos et al., "On the Role of Testosterone in Anxiety-Like Behavior across Life in Experimental Rodents," *Frontiers in Endocrinology* 9 (2018), https://doi.org/10.3389/fendo.2018.004417.

16. Eisenegger et al., "The Role of Testosterone in Social Interaction"; Allan Mazur, *Biosociology of Dominance and Deference* (New York: Rowman and Littlefield, 2005); Muller, "Testosterone and Reproductive Effort in Male Primates."

17. J. D. Ligon et al., "Male Male Competition, Ornamentation and the Role of Testosterone in Sexual Selection in Red Jungle Fowl," *Animal Behaviour* 40 (1990), https://doi.org/10.1016/S0003–3472(05)80932–7.

18. P. B. Gray et al., "Human Reproductive Behavior, Life History, and the Challenge Hypothesis: A 30-Year Review, Retrospective and Future Directions," *Hormones and Behavior* 123 (2020), https://doi.org/10.1016/j.yhbeh.2019.04.017.

19. Timothy H. Clutton-Brock, *The Evolution of Parental Care* (Princeton, NJ: Princeton University Press, 1991).

20. John C. Wingfield et al., "The "Challenge Hypothesis": Theoretical Implications for Patterns of Testosterone Secretion, Mating Systems and Breeding Strategies," *American Naturalist* 136, no. 6 (1990), https://www.jstor.org/stable/2462170.

21. B. Apfelbeck, H. Flinks, and W. Goymann, "Variation in Circulating Testosterone during Mating Predicts Reproductive Success in a Wild Songbird," *Frontiers in Ecology and Evolution* 4 (2016), https://doi.org/10.3389/fevo.2016.00107.

22. J. C. Wingfield, "Androgens and Mating Systems—Testosterone-Induced Polygyny in Normally Monogamous Birds," *Auk* 101, no. 4 (1984), https://doi.org/10 .2307/4086893.

23. P. Marler et al., "The Role of Sex Steroids in the Acquisition and Production of Birdsong," *Nature* 336, no. 6201 (1988), https://doi.org/10.1038/336770a0.

24. G. F. Ball, L. V. Riters, and J. Balthazart, "Neuroendocrinology of Song Behavior and Avian Brain Plasticity: Multiple Sites of Action of Sex Steroid Hormones," *Frontiers in Neuroendocrinology* 23, no. 2 (2002), https://doi.org/10.1006/frne.2002.0230.

25. F. Mougeot et al., "Testosterone, Immunocompetence, and Honest Sexual Signaling in Male Red Grouse," *Behavioral Ecology* 15, no. 6 (2004), https://doi.org/10 .1093/beheco/arh087.

26. W. L. Reed et al., "Physiological Effects on Demography: A Long-Term Experimental Study of Testosterone's Effects on Fitness," *American Naturalist* 167, no. 5 (2006), https://doi.org/10.1086/503054.

27. A. M. Dufty, "Testosterone and Survival—a Cost of Aggressiveness," *Hormones and Behavior* 23, no. 2 (1989), https://doi.org/10.1016/0018–506x(89)90059–7.

28. Mougeot et al., "Testosterone, Immunocompetence, and Honest Sexual Signaling."

29. E. D. Ketterson et al., "Testosterone and Avian Life Histories: The Effect of Experimentally Elevated Testosterone on Corticosterone and Body Mass in Dark-Eyed Juncos," *Hormones and Behavior* 25, no. 4 (1991), https://doi.org/10.1016/0018-506x (91)90016-b.

30. M. B. Mathur et al., "Perceived Stress and Telomere Length: A Systematic Review, Meta-Analysis, and Methodologic Considerations for Advancing the Field," *Brain, Behavior, and Immunity* 54 (2016), https://doi.org/10.1016/j.bbi.2016.02.002.

31. B. J. Heidinger et al., "Experimentally Elevated Testosterone Shortens Telomeres across Years in a Free-Living Songbird," *Molecular Ecology* (2021), https://doi.org/10 .1111/mec.15819.

32. C. Alonso-Alvarez et al., "Testosterone and Oxidative Stress: The Oxidation Handicap Hypothesis," *Philosophical Transactions of the Royal Society B: Biological Sciences* 274, no. 1611 (2007), https://doi.org/10.1098/rspb.2006.3764.

33. E. D. Ketterson et al., "Testosterone and Avian Life Histories—Effects of Experimentally Elevated Testosterone on Behavior and Correlates of Fitness in the Dark-Eyed Junco (*Junco hyemalis*)," *American Naturalist* 140, no. 6 (1992), https://doi.org /10.1086/285451.

34. B. Silverin, "Effects of Long-Acting Testosterone Treatment on Free-Living Pied Flycatchers, *Ficedula hypoleuca*, during the Breeding Period," *Animal Behaviour* 28, no. (1980), https://doi.org/10.1016/S0003–3472(80)80152–7; R. E. Hegner and J. C. Wingfield, "Effects of Experimental Manipulation of Testosterone Levels on Parental Investment and Breeding Success in Male House Sparrows," *Auk* 104, no. 3 (1987), https://doi.org/10.2307/4087545.

35. W. Goymann and P. F. Davila, "Acute Peaks of Testosterone Suppress Paternal Care: Evidence from Individual Hormonal Reaction Norms," *Philosophical Transactions of the Royal Society B: Biological Sciences* 284, no. 1857 (2017), https://doi.org/10 .1098/rspb.2017.0632

36. Hegner and Wingfield, "Effects of Experimental Manipulation."

37. E. M. Tuttle, "Alternative Reproductive Strategies in the White-Throated Sparrow: Behavioral and Genetic Evidence," *Behavioral Ecology* 14, no. 3 (2003), https:// doi.org/10.1093/beheco/14.3.425.

38. B. M. Horton, I. T. Moore, and D. L. Maney, "New Insights into the Hormonal and Behavioural Correlates of Polymorphism in White-Throated Sparrows,

Zonotrichia albicollis," *Animal Behaviour* 93 (2014), https://doi.org/10.1016/j.anbehav .2014.04.015.

39. F. Macrides, A. Bartke, and S. Dalterio, "Strange Females Increase Plasma Testosterone Levels in Male Mice," *Science* 189, no. 4208 (1975), https://doi.org/10.1126 /science.1162363.

40. K. Purvis and N. B. Haynes, "Effect of Odor of Female Rat Urine on Plasma Testosterone Concentrations in Male Rats," *Journal of Reproduction and Fertility* 53, no. 1 (1978), https://doi.org/10.1530/jrf.0.0530063.

41. P. J. James and J. G. Nyby, "Testosterone Rapidly Affects the Expression of Copulatory Behavior in House Mice (*Mus musculus*)," *Physiology and Behavior* 75, no. 3 (2002), https://doi.org/10.1016/S0031-9384(01)00666-7.

42. W. J. Zielinski and J. G. Vandenbergh, "Testosterone and Competitive Ability in Male House Mice, *Mus musculus*—Laboratory and Field Studies," *Animal Behaviour* 45, no. 5 (1993), https://doi.org/10.1006/anbe.1993.1108.

43. Sarah Hrdy, "Infanticide among Animals: A Review, Classification and Examination of the Implications for the Reproductive Strategies of Females," *Ethology and Sociobiology* 1 (1979): 13–40; Sarah Blaffer Hrdy, *Mothers and Others* (Cambridge, MA: Harvard University Press, 2009).

44. G. Perrigo, W. C. Bryant, and F. S. Vomsaal, "Fetal, Hormonal and Experiential Factors Influencing the Mating-Induced Regulation of Infanticide in Male House Mice," *Physiology and Behavior* 46, no. 2 (1989), https://doi.org/10.1016/0031-9384 (89)90244-8.

45. R. E. Brown, "Social and Hormonal Factors Influencing Infanticide and Its Suppression in Adult Male Long-Evans Rats (*Rattus*)," *Journal of Comparative Psychology* 100, no. 2 (1986), https://doi.org/10.1037/0735-7036.100.2.155.

46. J. S. Schneider et al., "Progesterone Receptors Mediate Male Aggression toward Infants," *Proceedings of the National Academy of Sciences* 100, no. 5 (2003), https:// doi.org/10.1073/pnas.0130100100; J. S. Schneider et al., "Effects of Progesterone on Male-Mediated Infant-Directed Aggression," *Behavioural Brain Research* 199, no. 2 (2009), https://doi.org/10.1016/j.bbr.2008.12.019.

47. M. Garratt, H. Try, and R. C. Brooks, "Access to Females and Early Life Castration Individually Extend Maximal But Not Median Lifespan in Male Mice," *Geroscience* 43, no. 3 (2021), https://doi.org/10.1007/s11357-020-00308-8.

48. V. L. Hughes and S. E. Randolph, "Testosterone Depresses Innate and Acquired Resistance to Ticks in Natural Rodent Hosts: A Force for Aggregated Distributions of Parasites," *Journal of Parasitology* 87, no. 1 (2001), https://doi.org/10.1645/0022-3395 (2001)087[0049:TDIAAR]2.0.CO;2.

49. A. J. Bradley, I. R. Mcdonald, and A. K. Lee, "Stress and Mortality in a Small Marsupial (*Antechinus stuartii, Macleay*)," *General and Comparative Endocrinology* 40, no. 2 (1980), https://doi.org/0.1016/0016–6480(80)90122–7; R. Naylor, S. J. Richardson, and B. M. McAllan, "Boom and Bust: A Review of the Physiology of the Marsupial Genus Antechinus," *Journal of Comparative Physiology B: Biochemical Systems and Environmental Physiology* 178, no. 5 (2008), https://doi.org/10.1007/s00360-007-0250-8.

50. R. E. Brown et al., "Hormonal Responses of Male Gerbils to Stimuli from Their Mate and Pups," *Hormones and Behavior* 29, no. 4 (1995), https://doi.org/10.1006/hbeh.1995.1275.

51. M. M. Clark and B. G. Galef, "Why Some Male Mongolian Gerbils May Help at the Nest: Testosterone, Asexuality and Alloparenting," *Animal Behaviour* 59 (2000), https://doi.org/10.1006/anbe.1999.1365.

52. M. M. Clark and B. G. Galef, "A Testosterone-Mediated Trade-Off between Parental and Sexual Effort in Male Mongolian Gerbils (*Meriones unguiculatus*)," *Journal of Comparative Psychology* 113, no. 4 (1999), https://doi.org/10.1037/0735–7036.113.4.388.

53. B. C. Trainor et al., "Variation in Aromatase Activity in the Medial Preoptic Area and Plasma Progesterone Is Associated with the Onset of Paternal Behavior," *Neuroendocrinology* 78, no. 1 (2003), https://doi.org/10.1159/000071704.

54. B. C. Trainor and C. A. Marler, "Testosterone Promotes Paternal Behaviour in a Monogamous Mammal via Conversion to Estrogen," *American Zoologist* 41, no. 6 (2001), https://doi.org/10.1098/rspb.2001.1954

55. Estrogen is really a class of hormones. Estradiol is the specific hormone we are referring to here.

56. Trainor et al., "Variation in Aromatase Activity."

57. B. C. Trainor and C. A. Marler, "Testosterone Promotes Paternal Behaviour in a Monogamous Mammal via Conversion to Oestrogen," *Philosophical Transactions of the Royal Society B: Biological Sciences* 269, no. 1493 (2002), https://doi.org/10.1098/rspb.2001.1954.

58. G. A. Lincoln, F. Guinness, and R. V. Short, "Way in Which Testosterone Controls the Social and Sexual Behavior of the Red Deer Stag (*Cervus elaphus*)," *Hormones and Behavior* 3, no. 4 (1972), https://doi.org/10.1016/0018–506x(72)90027-X.

59. A. F. Malo et al., "What Does Testosterone Do for Red Deer Males?," *Philosophical Transactions of the Royal Society B: Biological Sciences* 276, no. 1658 (2009), https://doi.org/10.1098/rspb.2008.13672.

60. Malo et al., "What Does Testosterone Do for Red Deer Males?"

61. C. M. Murray, M. A. Stanton, E. V. Lonsdorf, E. E. Wroblewski, and A. E. Pusey, "Chimpanzee Fathers Bias Their Behaviour towards Their Offspring," *R Soc Open Sci* 3, no. (2016): 160441, https://doi.org/10.1098/rsos.160441; J. Lehmann, G. Fickenscher, and C. Boesch, "Kin Biased Investment in Wild Chimpanzees," *Behaviour* 143 (2006): 931–955. https://doi.org/10.1163/156853906778623635.

62. Martin N. Muller and Richard W. Wrangham, "Dominance, Aggression and Testosterone in wild Chimpanzees: A Test of the 'Challenge Hypothesis,'" *Animal Behavior* 67 (2004), https://doi.org/10.1016/j.anbehav.2003.03.013; M. E. Sobolewski, J. L. Brown, and J. C. Mitani, "Female Parity, Male Aggression, and the Challenge Hypothesis in Wild Chimpanzees," *Primates* 54, no. 1 (2013), https://doi.org/10.1007/s10329-012-0332-4.

63. Wroblewski et al., "Male Dominance Rank and Reproductive Success in Chimpanzees, *Pan troglodytes schweinfurthii*."

64. M. P. Muehlenbein, D. P. Watts, and P. L. Whitten, "Dominance Rank and Fecal Testosterone Levels in Adult Male Chimpanzees (*Pan troglodytes schweinfurthii*) at Ngogo, Kibale National Park, Uganda," *American Journal of Primatology* 64, no. 1 (2004); Muller and Wrangham, "Dominance, Aggression and Testosterone"; Sobolewski et al., "Female Parity, Male Aggression."

65. R. M. Sapolsky, "Endocrine Aspects of Social Instability in the Olive Baboon (*Papio anubis*)," *American Journal of Primatology* 5, no. 4 (1983), https://doi.org/10.1002/ajp.1350050406; R. M. Sapolsky, "Stress-Induced Elevation of Testosterone Concentration in High Ranking Baboons: Role of Catecholamines," *Endocrinology* 118, no. 4 (1986), https://doi.org/10.1210/endo-118-4-1630; Robert M. Sapolsky, "Testicular Function, Social Rank and Personality among Wild Baboons," *Psychoneuroendocrinology* 16, no. 4 (1991), https://doi.org/10.1016/0306-4530(91)90015-L.

66. Muller, "Testosterone and Reproductive Effort in Male Primates."

67. Muller, "Testosterone and Reproductive Effort in Male Primates."

68. P. Fedurek et al., "The Relationship between Testosterone and Long-Distance Calling in Wild Male Chimpanzees," *Behavioral Ecology and Sociobiology* 70, no. 5 (2016), https://doi.org/10.1007/s00265-016-2087-1.

69. M. E. Sobolewski, J. L. Brown, and J. C. Mitani, "Territoriality, Tolerance and Testosterone in Wild Chimpanzees," *Animal Behaviour* 84, no. 6 (2012), https://doi.org/10.1016/j.anbehav.2012.09.018.

70. J. M. Williams et al., "Causes of Death in the Kasekela Chimpanzees of Gombe National Park, Tanzania," *American Journal of Primatology* 70, no. 8 (2008), https://doi.org/10.1002/ajp.20573.

71. M. P. Muehlenbein and D. P. Watts, "The Costs of Dominance: Testosterone, Cortisol and Intestinal Parasites in Wild Male Chimpanzees," *Biopsychosocial Medicine* 4 (2010), https://doi.org/10.1186/1751-0759-4-21.

72. M. N. Muller and R. W. Wrangham, "Mortality Rates among Kanyawara Chimpanzees," *Journal of Human Evolution* 66 (2014), https://doi.org/10.1016/j.jhevol .2013.10.004.

73. T. E. Ziegler, S. L. Prudom, and S. R. Zahed, "Variations in Male Parenting Behavior and Physiology in the Common Marmoset," *American Journal of Human Biology* 21, no. 6 2009), https://doi.org/10.1002/ajhb.20920.

74. T. E. Ziegler et al., "Neuroendocrine Response to Female Ovulatory Odors Depends upon Social Condition in Male Common Marmosets, *Callithrix jacchus*," *Hormones and Behavior* 47, no. 1 (2005), https://doi.org/10.1016/j.yhbeh.2004.08 .009.

75. S. L. Prudom et al., "Exposure to Infant Scent Lowers Serum Testosterone in Father Common Marmosets *(Callithrix jacchus),*" *Biology Letters* 4, no. 6 (2008), https://doi.org/10.1098/rsbl.2008.0358.

76. T. E. Ziegler et al., "Differential Endocrine Responses to Infant Odors in Common Marmoset *(Callithrix jacchus)* Fathers," *Hormones and Behavior* 59, no. 2 (2011), https://doi.org/10.1016/j.yhbeh.2010.12.0010.

77. S. Nunes, J. E. Fite, and J. A. French, "Variation in Steroid Hormones Associated with Infant Care Behaviour and Experience in Male Marmosets *(Callithrix kuhlii),*" *Animal Behaviour* 60 (2000), https://doi.org/10.1006/anbe.2000.1524; S. Nunes et al., "Interactions among Paternal Behavior, Steroid Hormones, and Parental Experience in Male Marmosets *(Callithrix kuhlii),*" *Hormones and Behavior* 39, no. 1 (2001), https://doi.org/10.1006/hbeh.2000.1631.

78. C. N. Ross, J. A. French, and K. J. Patera, "Intensity of Aggressive Interactions Modulates Testosterone in Male Marmosets," *Physiology and Behavior* 83, no. 3 (2004), https://doi.org/10.1016/j.physbeh.2004.08.036.

79. J. R. Roney, A. W. Lukaszewski, and Z. L. Simmons, "Rapid Endocrine Responses of Young Men to Social Interactions with Young Women," *Hormones and Behavior* 52, no. 3 (2007), https://doi.org/10.1016/j.yhbeh.2007.05.008.

80. J. R. Roney, S. V. Mahler, and D. Maestripieri, "Behavioral and Hormonal Responses of Men to Brief Interactions with Women," *Evolution and Human Behavior* 24, no. 6 (2003), https://doi.org/10.1016/S1090-5138(03)00053-9.

81. R. Ronay and W. von Hippel, "The Presence of an Attractive Woman Elevates Testosterone and Physical Risk Taking in Young Men," *Social Psychological and Personality Science* 1, no. 1 (2010), https://doi.org/10.1177/1948550609352807.

82. S. L. Miller, J. K. Maner, and J. K. McNulty, "Adaptive Attunement to the Sex of Individuals at a Competition: The Ratio of Opposite- to Same-Sex Individuals Correlates with Changes in Competitors' Testosterone Levels," *Evolution and Human Behavior* 33, no. 1 (2012), https://doi.org/10.1016/j.evolhumbehav.2011.05.006.

83. S. L. Miller and J. K. Maner, "Scent of a Woman: Men's Testosterone Responses to Olfactory Ovulation Cues," *Psychological Science* 21, no. 2 (2010), https://doi.org/10.1177/0956797609357733.

84. M. V. Flinn, D. Ponzi, and M. P. Muehlenbein, "Hormonal Mechanisms for Regulation of Aggression in Human Coalitions," *Human Nature* 23, no. 1 (2012), https://doi.org/10.1007/s12110-012-9135-y.

85. S. Zilioli et al., "Interest in Babies Negatively Predicts Testosterone Responses to Sexual Visual Stimuli among Heterosexual Young Men," *Psychological Science* 27, no. 1 (2016), https://doi.org/10.1177/0956797615615868.

86. James R. Roney and Lee T. Gettler, "The Role of Testosterone in Human Romantic Relationships," *Current Opinion in Psychology* 1 (2015), https://doi.org/10.1016/j.copsyc.2014.11.003.

87. Gray et al., "Human Reproductive Behavior."

88. M. Peters, L. W. Simmons, and G. Rhodes, "Testosterone Is Associated with Mating Success But Not Attractiveness or Masculinity in Human Males," *Animal Behaviour* 76 (2008), https://doi.org/10.1016/j.anbehav.2008.02.008; T. V. Pollet et al., "Testosterone Levels and Their Associations with Lifetime Number of Opposite Sex Partners and Remarriage in a Large Sample of American Elderly Men and Women," *Hormones and Behavior* 60, no. 1 (2011), https://doi.org/10.1016/j.yhbeh.2011.03.005.

89. C. Klimas et al., "Higher Testosterone Levels Are Associated with Unfaithful Behavior in Men," *Biological Psychology* 146 (2019), https://doi.org/10.1016/j.biopsycho.2019.107730.

90. A. Booth and J. M. Dabbs, "Testosterone and Men's Marriages," *Social Forces* 72, no. 2 (1993), https://doi.org/10.2307/2579857.

91. Pollet et al., "Testosterone Levels."

92. Roney and Gettler, "The Role of Testosterone."

93. L. T. Gettler, T. W. McDade, A. B. Feranil, and C. W. Kuzawa, "Longitudinal Evidence That Fatherhood Decreases Testosterone in Human Males," *Proceedings of the National Academy of Sciences of the United States of America* 108, no. 39 (2011): 16194–16199. https://doi.org/10.1073/pnas.1105403108.

94. M. R. Fales, K. A. Gildersleeve, and M. G. Haselton, "Exposure to Perceived Male Rivals Raises Men's Testosterone on Fertile Relative to Nonfertile Days of their

Partner's Ovulatory Cycle," *Hormones and Behavior* 65, no. 5 (2014), https://doi.org /10.1016/j.yhbeh.2014.04.002.

95. J. R. Roney et al., "Reading Men's Faces: Women's Mate Attractiveness Judgments Track Men's Testosterone and Interest in Infants," *Philosophical Transactions of the Royal Society B: Biological Sciences* 273, no. 1598 (2006), https://doi.org/10.1098 /rspb.2006.3569.

96. J. J. M. O'Connor et al., "Perceptions of Infidelity Risk Predict Women's preferences for Low Male Voice Pitch in Short-Term over Long-Term Relationship Contexts," *Personality and Individual Differences* 56 (2014), https://doi.org/10.1016/j.paid .2013.08.029.

97. E. C. Shattuck and M. P. Muehlenbein, "Human Sickness Behavior: Ultimate and Proximate Explanations," *American Journal of Physical Anthropology* 157, no. 1 (2015a), https://doi.org/10.1002/ajpa.22698; E. C. Shattuck and M. P. Muehlenbein, "Mood, Behavior, Testosterone, Cortisol, and Interleukin-6 in Adults during Immune Activation: A Pilot Study to Assess Sickness Behaviors in Humans," *American Journal of Human Biology* 27, no. 1 (2015b), https://doi.org/10.1002/ajhb.22608; Z. L. Simmons and J. R. Roney, "Androgens and Energy Allocation: Quasi-Experimental Evidence for Effects of Influenza Vaccination on Men's Testosterone," *American Journal of Human Biology* 21, no. 1 (2009), https://doi.org/10.1002/ajhb.20837.

98. G. Rhodes et al., "Does Sexual Dimorphism in Human Faces Signal Health?," *Philosophical Transactions of the Royal Society B: Biological Sciences* 270 (2003), https:// doi.org/10.1098/rsbl.2003.0023; R. Thornhill and S. W. Gangestad, "Facial Sexual Dimorphism, Developmental Stability, and Susceptibility to Disease in Men and Women," *Evolution and Human Behavior* 27, no. 2 (2006), https://doi.org/10.1016/j .evolhumbehav.2005.06.001.

99. K. G. Phalane et al., "Facial Appearance Reveals Immunity in African Men," *Scientific Reports* 7 (2017), https://doi.org/10.1038/s41598-017-08015-9.

100. H. Greiling and D. M. Buss, "Women's Sexual Strategies: The Hidden Dimension of Extra-Pair Mating," *Personality and Individual Differences* 28, no. 5 (2000), https://doi.org/10.1016/S0191–8869(99)00151–8.

101. D. Cohen et al., "Insult, Aggression, and the Southern Culture of Honor: An 'Experimental Ethnography,'" *Journal of Personality and Social Psychology* 70, no. 5 (1996), https://doi.org/10.1037//0022-3514.70.5.945.

102. K. V. Casto and D. A. Edwards, "Testosterone, Cortisol, and Human Competition," *Hormones and Behavior* 82 (2016), https://doi.org/10.1016/j.yhbeh.2016.04 .004.

103. P. H. Mehta and R. A. Josephs, "Testosterone Change after Losing Predicts the Decision to Compete Again," *Hormones and Behavior* 50, no. 5 (2006), https://doi.org /10.1016/j.yhbeh.2006.07.001.

104. G. D. Sherman et al., "The Interaction of Testosterone and Cortisol Is Associated with Attained Status in Male Executives," *Journal of Personality and Social Psychology* 110, no. 6 (2016), https://doi.org/10.1037/pspp00000639.

105. P. H. Mehta and R. A. Josephs, "Testosterone and Cortisol Jointly Regulate Dominance: Evidence for a Dual-Hormone Hypothesis," *Hormones and Behavior* 58, no. 5 (2010), https://doi.org/10.1016/j.yhbeh.2010.08.020.

106. T. J. Scaramella and W. A. Brown, "Serum Testosterone and Aggressiveness in Hockey Players," *Psychosomatic Medicine* 40, no. 3 (1978), http://www.ncbi.nlm.nih.gov/pubmed/663054.

107. K. Christiansen and E. M. Winkler, "Hormonal, Anthropometrical, and Behavioral-Correlates of Physical Aggression in Kung-San Men of Namibia," *Aggressive Behavior* 18, no. 4 (1992), https://doi.org/10.1002/1098-2337(1992)18:4<271::AID-AB2480180403>3.0.CO;2-6.

108. J. C. Dreher et al., "Testosterone Causes Both Prosocial and Antisocial Status-Enhancing Behaviors in Human Males," *Proceedings of the National Academy of Sciences* 113, no. 41 (2016), https://doi.org/10.1073/pnas.1608085113.

109. Gray et al., "Human Reproductive Behavior."

110. Gettler et al., "Longitudinal Evidence."

111. A. Alvergne, C. Faurie, and M. Raymond, "Variation in Testosterone Levels and Male Reproductive Effort: Insight from a Polygynous Human Population," *Hormones and Behavior* 56, no. 5 (2009), https://doi.org/10.1016/j.yhbeh.2009.07.013; P. B. Gray, "Marriage, Parenting, and Testosterone Variation among Kenyan Swahili Men," *American Journal of Physical Anthropology* 122, no. 3 (2003), https://doi.org/10.1002/ajpa.10293.

112. E. J. Hermans, P. Putman, and J. van Honk, "Testosterone Administration Reduces Empathetic Behavior: A Facial Mimicry Study," *Psychoneuroendocrinology* 31, no. 7 (2006), https://doi.org/10.1016/j.psyneuen.2006.04.002.

113. D. Olweus et al., "Testosterone, Aggression, Physical, and Personality Dimensions in Normal Adolescent Males," *Psychosomatic Medicine* 42, no. 2 (1980), http://www.ncbi.nlm.nih.gov/pubmed/7454920; J. Agrawal et al., "Chronic Testosterone Increases Impulsivity and Influences the Transcriptional Activity of the Alpha-2A Adrenergic Receptor Signaling Pathway in Rat Brain," *Molecular Neurobiology* 56, no. 6 (2019), https://doi.org/10.1007/s12035-018-1350-z; A. Aluja et al., "Interactions among Impulsiveness, Testosterone, Sex Hormone Binding Globulin and Androgen Receptor Gene CAG Repeat Length," *Physiology and Behavior* 147 (2015), https://doi.org/10.1016/j.physbeh.2015.04.022.

114. Alison S. Fleming et al., "Testosterone and Prolactin Are Associated with Emotional Responses to Infant Cries in New Fathers," *Hormones and Behavior* 42, no. 4 (2002), https://doi.org/10.1006/hbeh.2002.1840.

115. Muller et al., "Testosterone and Paternal Cares."

116. L. T. Gettler et al., "Does Cosleeping Contribute to Lower Testosterone Levels in Fathers? Evidence from the Philippines," *PloS One* 7, no. 9 (2012), https://doi.org/10.1371/journal.pone.0041559.

117. S. Gelstein et al., "Human Tears Contain a Chemosignal," *Science* 331, no. 6014 (2011), https://doi.org/10.1126/science.1198331.

118. J. S. Mascaro, P. D. Hackett, and J. K. Rilling, "Testicular Volume Is Inversely Correlated with nurturing-related brain activity in human fathers," *Proceedings of the National Academy of Sciences* 110, no. 39 (2013), https://doi.org/10.1073/pnas.1305579110; Weisman, Zagoory-Sharon, and Feldman, "Oxytocin Administration, Salivary Testosterone, and Father-Infant Social Behavior"; O. Weisman, O. Zagoory-Sharon, and R. Feldman, "Oxytocin Administration to Parent Enhances Infant Physiological and Behavioral Readiness for Social Engagement," *Biological Psychiatry* 72, no. 12 (2012): 982–989, https://doi.org/10.1016/j.biopsych.2012.06.011.

119. D. E. Saxbe et al., "Fathers' Decline In Testosterone and Synchrony with Partner Testosterone during Pregnancy Predicts Greater Postpartum Relationship Investment," *Hormones and Behavior* 90 (2017), https://doi.org/10.1016/j.yhbeh.2016.07.005, https://www.ncbi.nlm.nih.gov/pubmed/27469070.

120. R. S. Edelstein et al., "Prospective and Dyadic Associations between Expectant Parents' Prenatal Hormone Changes and Postpartum Parenting Outcomes," *Developmental Psychobiology* 59, no. 1 (2017), https://doi.org/10.1002/dev.21469.

121. D. E. Saxbe et al., "High Paternal Testosterone May Protect against Postpartum Depressive Symptoms in Fathers, But Confer Risk to Mothers and Children," *Hormones and Behavior* 95 (2017), https://doi.org/10.1016/j.yhbeh.2017.07.014.

122. S. J. Berg and K. E. Wynne-Edwards, "Changes in Testosterone, Cortisol, and Estradiol Levels in Men Becoming Fathers," *Mayo Clinic Proceedings* 76, no. 6 (2001), https://doi.org/10.4065/76.6.582.

123. S. M. van Anders, R. M. Tolman, and B. L. Volling, "Baby Cries and Nurturance Affect Testosterone in Men," *Hormones and Behavior* 61, no. 1 (2012), https://doi.org/10.1016/j.yhbeh.2011.09.012.

124. Fleming et al., "Testosterone and Prolactin"; van Anders et al., "Baby Cries and Nurturance."

125. J. Thornton, J. L. Zehr, and M. D. Loose, "Effects of Prenatal Androgens on Rhesus Monkeys: A Model System to Explore the Organizational Hypothesis in

Primates," *Hormones and Behavior* 55, no. 5 (2009), https://doi.org/10.1016/j.yhbeh.2009.03.015.

126. L. Ahnert et al., "Fathering Behavior, Attachment, and Engagement in Childcare Predict Testosterone and Cortisol," *Developmental Psychobiology* 63, no. 6 (2021), https://doi.org/10.1002/dev.22149.

127. C. M. Worthman and M. J. Konner, "Testosterone Levels Change with Subsistence Hunting Effort in Kung San Men," *Psychoneuroendocrinology* 12, no. 6 (1987), https://doi.org/10.1016/0306-4530(87)90079-5.

128. E. T. Schroeder et al., "Are Acute Post-Resistance Exercise Increases in Testosterone, Growth Hormone, and IGF-1 Necessary to Stimulate Skeletal Muscle Anabolism and Hypertrophy?," *Medicine and Science in Sports and Exercise* 45, no. 11 (2013), https://doi.org/10.1249/Mss.0000000000000147.

129. B. C. Trumble et al., "Successful Hunting Increases Testosterone and Cortisol in a Subsistence Population," *Philosophical Transactions of the Royal Society B: Biological Sciences* 281, no. 1776 (2014), https://doi.org/10.1098/rspb.2013.2876.

130. A. H. Boyette et al., "Testosterone, Fathers as Providers and Caregivers, and Child Health: Evidence from Fisher-Farmers in the Republic of the Congo," *Hormones and Behavior* 107 (2019), https://doi.org/10.1016/j.yhbeh.2018.09.006.

131. B. C. Trumble et al., "Age-Independent Increases in Male Salivary Testosterone during Horticultural Activity among Tsimane Forager-Farmers," *Evolution and Human Behavior* 34, no. 5 (2013), https://doi.org/10.1016/j.evolhumbehav.2013.06.002.

132. L. C. Alvarado et al., "The Paternal Provisioning Hypothesis: Effects of Workload and Testosterone Production on Men's Musculature," *American Journal of Physical Anthropology* 158, no. 1 2015), https://doi.org/10.1002/ajpa.22771.

133. P. X. Kuo et al., "Individual Variation in Fathers' Testosterone Reactivity to Infant Distress Predicts Parenting Behaviors with Their 1-Year-Old Infants," *Developmental Psychobiology* 58, no. 3 (2016); van Anders et al., "Baby Cries and Nurturance."

134. E. Roellke et al., "Infant Crying Levels Elicit Divergent Testosterone Response in Men," *Parenting: Science and Practice* 19, no. 1–2 (2019), https://doi.org/10.1080/15295192.2019.1555425.

135. Richard G. Bribiescas, *Men: Evolutionary and Life History* (Cambridge, MA: Harvard University Press, 2006).

136. B. C. Campbell et al., "Androgen Receptor CAG Repeats and Body Composition among Ariaal Men," *International Journal of Andrology* 32, no. 2 (2009), https://doi.org/10.1111/j.1365-2605.2007.00825.x.

137. L. T. Gettler et al., "Adiposity, CVD Risk Factors and Testosterone Variation by Partnering Status and Residence with Children in US Men," *Evolution Medicine and Public Health*, no. 1 (2017), https://doi.org/10.1093/emph/eox005.

138. M. Z. Hossin, "The Male Disadvantage in Life Expectancy: Can We Close the Gender Gap?," *International Health* 13, no. 5 (2021), https://doi.org/10.1093/inthealth/ihaa106.

139. J. F. Lemaitre et al., "Sex Differences in Adult Lifespan and Aging Rates of Mortality across Wild Mammals," *Proceedings of the National Academy of Sciences* 117, no. 15 (2020), https://doi.org/10.1073/pnas.1911999117.

140. Worldometer, "Life Expectancy of the World Population," https://www.worldometers.info/demographics/life-expectancy/.

141. Lemaitre et al., "Sex Differences in Adult Lifespan."

142. I. P. F. Owens, "Sex Differences in Mortality Rate," *Science* 297, no. 5589 (2002b), https://doi.org/10.1126/science.1076813.

143. Martin Daly and Margo Wilson, *Homicide* (New York: Aldine de Gruyter, 1988).

144. Owens, "Sex Differences in Mortality Rate"

145. Hossin, "The Male Disadvantage"

146. K. J. Min, C. K. Lee, and H. N. Park, "The Lifespan of Korean Eunuchs," *Current Biology* 22, no. 18 (2012), https://doi.org/10.1016/j.cub.2012.06.036.

147. H. Horwitz, J. T. Andersen, and K. P. Dalhoff, "Health Consequences of Androgenic Anabolic Steroid Use," *Journal of Internal Medicine* 285, no. 3 (2019), https://doi.org/10.1111/joim.12850.

148. M. Parssinen et al., "Increased Premature Mortality of Competitive Powerlifters Suspected to Have Used Anabolic Agents," *Internal Journal of Sports Medicine* 21, no. 3 (2000), https://doi.org/10.1055/s-2000-3048.

149. J. A. Kettunen et al., "All-Cause and Disease-Specific Mortality among Male, Former Elite Athletes: An Average 50-Year Follow-Up," *British Journal of Sports Medicine* 49, no. 13 (2015), https://doi.org/10.1136/bjsports-2013-093347.

150. A. Bjornebekk et al., "Long-Term Anabolic-Androgenic Steroid Use Is Associated with Deviant Brain Aging," *Biological Psychiatry: Cognitive Neuroscience and Neuroimaging* 6, no. 5 (2021), https://doi.org/10.1016/j.bpsc.2021.01.001.

151. M. Estrada, A. Varshney, and B. E. Ehrlich, "Elevated Testosterone Induces Apoptosis in Neuronal Cells," *Journal of Biological Chemistry* 281, no. 35 (2006), https://doi.org/10.1074/jbc.M603193200 9.

152. F. Saad et al., "Differential Effects of 11 Years of Long-Term Injectable Testosterone Undecanoate Therapy on Anthropometric and Metabolic Parameters in Hypogonadal Men with Normal Weight, Overweight and Obesity in Comparison with Untreated Controls: Real-World Data from a Controlled Registry Study," *International Journal of Obesity* 44, no. 6 (2020), https://doi.org/10.1038/s41366-019-0517 -7; F. Comhaire, "Hormone Replacement Therapy and Longevity," *Andrologia* 48, no. 1 (2016), https://doi.org/10.1111/and.12419.

153. Z. L. Simmons and J. R. Roney, "Variation in CAG Repeat Length of the Androgen Receptor Gene Predicts Variables Associated with Intrasexual Competitiveness in Human Males," *Hormones and Behavior* 60, no. 3 (2011), https://doi.org/10.1016/j .yhbeh.2011.06.006.

154. L. T. Gettler et al., "The Role of Testosterone in Coordinating Male Life History Strategies: The Moderating Effects of the Androgen Receptor CAG Repeat Polymorphism," *Hormones and Behavior* 87 (2017), https://doi.org/10.1016/j.yhbeh.2016.10 .012.

Chapter 4

1. H. J. Lee et al., "Oxytocin: The Great Facilitator of Life," *Progress in Neurobiology* 88, no. 2 (2009), https://doi.org/10.1016/j.pneurobio.2009.04.001.

2. A. Meyer-Lindenberg et al., "Oxytocin and Vasopressin in the Human Brain: Social Neuropeptides for Translational Medicine," *Nature Reviews Neuroscience* 12, no. 9 (2011), https://doi.org/10.1038/nrn3044; H. E. Ross and L. J. Young, "Oxytocin and the Neural Mechanisms Regulating Social Cognition and Affiliative Behavior," *Frontiers in Neuroendocrinology* 30, no. 4 (2009), https://doi.org/10.1016/j.yfrne .2009.05.004.

3. J. K. Rilling and L. J. Young, "The Biology of Mammalian Parenting and Its Effect on Offspring Social Development," *Science* 345, no. 6198 (2014), https://doi.org/10 .1126/science.1252723.

4. M. Numan, "Motivational Systems and the Neural Circuitry of Maternal Behavior in the Rat," *Developmental Psychobiology* 49, no. 1 (2007), https://doi.org/10.100 2/dev.20198.

5. D. E. Olazabal, "Role of Oxytocin in Parental Behaviour," *Journal of Neuroendocrinology* 30, no. 7 (2018), https://doi.org/10.1111/jne.12594.

6. F. Champagne et al., "Naturally Occurring Variations in Maternal Behavior in the Rat Are Associated with Differences in Estrogen-Inducible Central Oxytocin Receptors," *Proceedings of the National Academy of Sciences* 98, no. 22 (2001), https://doi.org /10.1073/pnas.221224598.

7. R. Feldman, I. Gordon, and O. Zagoory-Sharon, "Maternal and Paternal Plasma, Salivary, and Urinary Oxytocin and Parent-Infant Synchrony: Considering Stress and Affiliation Components of Human Bonding," *Developmental Science* 14, no. 4 (2011), https://doi.org/10.1111/j.1467-7687.2010.01021.x.

8. T. A. Thul et al., "Oxytocin and Postpartum Depression: A Systematic Review," *Psychoneuroendocrinology* 120 (2020), https://doi.org/10.1016/j.psyneuen.2020 .104793.

9. K. Bernard et al., "Association between Maternal Depression and Maternal Sensitivity from Birth to 12 Months: A Meta-Analysis," *Attachment and Human Development* 20, no. 6 (2018), https://doi.org/10.1080/14616734.2018.1430839; S. Brummelte and L. A. Galea, "Postpartum Depression: Etiology, Treatment and Consequences for Maternal Care," *Hormones and Behavior* 77 (2016), https://doi.org/10 .1016/j.yhbeh.2015.08.008; M. C. Lovejoy et al., "Maternal Depression and Parenting Behavior: A Meta-Analytic Review," *Clinical Psychology Review* 20, no. 5 (2000), https://doi.org/10.1016/S0272–7358(98)00100–7; J. L. Pawluski, J. S. Lonstein, and A. S. Fleming, "The Neurobiology of Postpartum Anxiety and Depression," *Trends in Neuroscience* 40, no. 2 (2017), https://doi.org/10.1016/j.tins.2016.11.009.

10. R. Feldman et al., "Natural Variations in Maternal and Paternal Care Are Associated with Systematic Changes in Oxytocin following Parent-Infant Contact," *Psychoneuroendocrinology* 35, no. 8 (2010), https://doi.org/10.1016/j.psyneuen.2010.01 .013; S. Kim et al., "Maternal Oxytocin Response Predicts Mother-to-Infant Gaze," *Brain Research* 1580 (2014), https://doi.org/10.1016/j.brainres.2013.10.050.

11. L. Strathearn et al., "Adult Attachment Predicts Maternal Brain and Oxytocin Response to Infant Cues," *Neuropsychopharmacology* 34, no. 13 (2009), https://doi .org/10.1038/npp.2009.103.

12. J. K. Rilling, "A Potential Role for Oxytocin in the Intergenerational Transmission of Secure Attachment," *Neuropsychopharmacology* 34, no. 13 (2009), https://doi .org/10.1038/npp.2009.136.

13. A. S. McNeilly et al., "Release of Oxytocin and Prolactin in Response to Suckling," *British Medical Journal (Clinical Research Edition)* 286, no. 6361 (1983), http:// www.ncbi.nlm.nih.gov/pubmed/6402061; A. M. Stuebe, K. Grewen, and S. Meltzer-Brody, "Association between Maternal Mood and Oxytocin Response to Breastfeeding," *Journal of Women's Health* 22, no. 4 (2013), https://doi.org/10.1089/jwh.2012 .3768.

14. L. Strathearn et al., "Does Breastfeeding Protect against Substantiated Child Abuse and Neglect? A 15-Year Cohort Study," *Pediatrics* 123, no. 2 (2009), https://doi .org/10.1542/peds.2007–3546.

15. K. J. Michalska et al., "Genetic Imaging of the Association of Oxytocin Receptor Gene (OXTR) Polymorphisms with Positive Maternal Parenting," *Frontiers in*

Behavioral Neuroscience 8 (2014), https://doi.org/10.3389/fnbeh.2014.00021; M. J. Bakermans-Kranenburg and M. H. van Ijzendoorn, "Oxytocin Receptor (OXTR) and Serotonin Transporter (5-HTT) Genes Associated with Observed Parenting," *Social Cognitive and Affective Neuroscience* 3, no. 2 (2008), https://doi.org/10.1093/scan /nsn004; A. M. Klahr, K. Klump, and S. A. Burt, "A Constructive Replication of the Association between the Oxytocin Receptor Genotype and Parenting," *Journal of Family Psychology* 29, no. 1 (2015), https://doi.org/10.1037/fam0000034.

16. W. Yuan et al., "Role of Oxytocin in the Medial Preoptic area (MPOA) in the Modulation of Paternal Behavior in Mandarin Voles," *Hormones and Behavior* 110 (2019), https://doi.org/10.1016/j.yhbeh.2019.02.014.

17. W. M. Kenkel et al., "Neuroendocrine and Behavioural Responses to Exposure to an Infant in Male Prairie Voles," *Journal of Neuroendocrinology* 24, no. 6 (2012), https://doi.org/10.1111/j.1365-2826.2012.02301.x.

18. J. M. Borland et al., "Sex-Dependent Regulation of Social Reward by Oxytocin: An Inverted U Hypothesis," *Neuropsychopharmacology* 44, no. 1 (2019), https://doi .org/10.1038/s41386-018-0129-2.

19. K. Macdonald and D. Feifel, "Oxytocin's Role in Anxiety: A Critical Appraisal," *Brain Research* (2014), https://doi.org/10.1016/j.brainres.2014.01.025

20. W. M. Kenkel, G. Suboc, and C. S. Carter, "Autonomic, Behavioral and Neuroen-docrine Correlates of Paternal Behavior in Male Prairie Voles," *Physiological Behavior* 128 (2014), https://doi.org/10.1016/j.physbeh.2014.02.006.

21. M. J. Woller et al., "Differential Hypothalamic Secretion of Neurocrines in Male Common Marmosets: Parental Experience Effects?," *Journal of Neuroendocrinology* 24, no. 3 (2012), https://doi.org/10.1111/j.1365-2826.2011.02252.x.

22. J. H. Taylor and J. A. French, "Oxytocin and Vasopressin Enhance Respon-siveness to Infant Stimuli in Adult Marmosets," *Hormones and Behavior* 75 (2015), https://doi.org/10.1016/j.yhbeh.2015.10.002.

23. A. Saito and K. Nakamura, "Oxytocin Changes Primate Paternal Tolerance to Offspring in Food Transfer," *Journal of Comparative Physiology. A Neuroethology, Sensory, Neural, and Behavior Physiology* 197, no. 4 (2011), https://doi.org/10.1007 /s00359-010-0617-2.

24. Sarah Blaffer Hrdy, *Mothers and Others* (Cambridge, MA: Harvard University Press, 2009).

25. J. R. Madden and T. H. Clutton-Brock, "Experimental Peripheral Administration of Oxytocin Elevates a Suite of Cooperative Behaviours in a Wild Social Mammal," *Philosophical Transactions of the Royal Society B: Biological Sciences* 278, no. 1709 (2011), https://doi.org/10.1098/rspb.2010.1675.

26. C. Finkenwirth et al., "Oxytocin Is Associated with Infant-Care Behavior and Motivation in Cooperatively Breeding Marmoset Monkeys," *Hormones and Behavior* 80 (2016), https://doi.org/10.1016/j.yhbeh.2016.01.008.

27. I. Gordon et al., "Oxytocin and the Development of Parenting in Humans," *Biological Psychiatry* 68, no. 4 (2010), https://doi.org/10.1016/j.biopsych.2010.02.005.

28. J. S. Mascaro, P. D. Hackett, and J. K. Rilling, "Differential Neural Responses to Child and Sexual Stimuli in Human Fathers and Non-Fathers and Their Hormonal Correlates." *Psychoneuroendocrinology* 46 (2014): 153–163, https://doi.org/10.1016/j.psyneuen.2014.04.014; Gordon et al., "Oxytocin and the Development of Parenting in Hhumans."

29. Gordon et al., "Oxytocin and the Development of Parenting in Humans."

30. Gordon et al., "Oxytocin and the Development of Parenting in Humans."

31. A. R. Morris et al., "Physical Touch during Father-Infant Interactions Is Associated with Paternal Oxytocin Levels," *Infant Behavior and Development* 64 (2021), https://doi.org/10.1016/j.infbeh.2021.101613. In studies involving humans, oxytocin is usually measured in blood or saliva. These measures tell us about oxytocin levels in the body (i.e., the periphery), but not necessarily about levels in the brain. This is because oxytocin has a very limited ability to pass from the body to the brain, crossing the blood-brain barrier. Generally, when trying to understand oxytocin effects on human behaviors, such as parental behavior, we expect that brain oxytocin levels are most relevant. However, it is very difficult to measure brain oxytocin levels in human participants. It requires obtaining a sample of cerebral spinal fluid, a painful and highly invasive procedure. Although oxytocin has trouble passing between brain and body, there are some neurons that seem able to release oxytocin into both the body and the brain, which could lead to coordinated release into these two compartments. There are also studies showing positive correlations between oxytocin levels in the brain and the blood. So oxytocin levels in the body may serve as a kind of proxy for levels in the brain. It may also be the case that peripheral oxytocin itself has behavioral effects. For example, oxytocin could work through the vagus nerve in the periphery to influence brain function, potentially even stimulating brain oxytocin release.

32. X. Cong et al., "Parental Oxytocin Responses during Skin-to-Skin Contact in Pre-Term Infants," *Early Human Development* 91, no. 7 (2015), https://doi.org/10.1016/j.earlhumdev.2015.04.012; D. Vittner et al., "Increase in Oxytocin from Skin-to-Skin Contact Enhances Development of Parent-Infant Relationship," *Biological Research for Nursing* 20, no. 1 (2018), https://doi.org/10.1177/1099800417735633.

33. L. T. Gettler et al., "Fathers' Oxytocin Responses to First Holding Their Newborns: Interactions with Testosterone Reactivity to Predict Later Parenting Behavior

and Father-Infant Bonds," *Developmental Psychobiology* 63, no. 5 (2021), https://doi.org/10.1002/dev.22121.

34. Feldman et al., "Natural Variations."

35. Y. Rassovsky et al., "Martial Arts Increase Oxytocin Production," *Science Reports* 9, no. 1 (2019), https://doi.org/10.1038/s41598-019-49620-0.

36. Researchers who study oxytocin effects in humans usually administer oxytocin intranasally, based on the assumption that nasal administration allows oxytocin to get past the blood-brain barrier and enter the brain. Intranasal oxytocin seems capable of increasing oxytocin levels in both the brain and in the body, at least in some cases. So it is not clear if the behavioral effects of intranasal oxytocin are mediated by oxytocin actions in the body or the brain, or both. One way to evaluate this is to compare intravenous versus intranasal administration, since only the latter should achieve entry into the brain.

37. Results of intranasal oxytocin studies should be regarded with caution since it is now clear that many of these studies are statistically underpowered, meaning they have too few participants to detect the expected effect size. This means that some, or even many, positive findings may actually be false positives. The remedy for this problem is to conduct studies with larger sample sizes that afford greater statistical power. When results from intranasal oxytocin studies are consistent with expectations based on animal studies, we can have increased confidence in their validity.

38. O. Weisman, O. Zagoory-Sharon, and R. Feldman, "Oxytocin Administration to Parent Enhances Infant Physiological and Behavioral Readiness for Social Engagement," *Biological Psychiatry* 72, no. 12 (2012), https://doi.org/10.1016/j.biopsych.2012.06.011.

39. O. Weisman et al., "Oxytocin Shapes Parental Motion during Father-Infant Interaction," *Biological Letters* 9, no. 6 (2013), https://doi.org/10.1098/rsbl.2013.0828.

40. F. Naber et al., "Intranasal Oxytocin Increases Fathers' Observed Responsiveness during Play with Their Children: A Double-Blind within-Subject Experiment," *Psychoneuroendocrinology* 35, no. 10 (2010), https://doi.org/10.1016/j.psyneuen.2010.04.007.

41. Weisman et al., "Oxytocin Administration to Parent."

42. S. G. Shamay-Tsoory and A. Abu-Akel, "The Social Salience Hypothesis of Oxytocin," *Biological Psychiatry* 79, no. 3 (2016), https://doi.org/10.1016/j.biopsych.2015.07.020.

43. A. J. Guastella, P. B. Mitchell, and M. R. Dadds, "Oxytocin Increases Gaze to the Eye Region of Human Faces," *Biological Psychiatry* 63, no. 1 (2008), https://doi.org/10.1016/j.biopsych.2007.06.026.

44. G. Domes et al., "Oxytocin Improves `Mind-Reading" in Humans,' *Biological Psychiatry* 61, no. 6 (2007), https://doi.org/10.1016/j.biopsych.2006.07.015.

45. S. Korb et al., "Sniff and Mimic—Intranasal Oxytocin Increases Facial Mimicry in a Sample of Men," *Hormones and Behavior* 84 (2016), https://doi.org/10.1016/j.yhbeh.2016.06.003.

46. A. Abu-Akel et al., "Oxytocin Increases Empathy to Pain When Adopting the Other—But Not the Self-Perspective," *Society for Neuroscience* 10, no. 1 (2015), https://doi.org/10.1080/17470919.2014.948637

47. Numan, "Motivational Systems."

48. J. S. Mascaro, P. D. Hackett, and J. K. Rilling, "Testicular Volume is Inversely Correlated with Nurturing-Related Brain Activity in Human Fathers," *Proc Natl Acad Sci U S A* 110, no. 39 (2013): 15746–15751, https://doi.org/10.1073/pnas.1305579110; T. Li et al., "Intranasal Oxytocin, But Not Vasopressin, Augments Neural Responses to Toddlers in Human Fathers," *Hormones and Behavior* 93 (2017), https://doi.org/10.1016/j.yhbeh.2017.01.006.

49. Macdonald and Feifel, "Oxytocin's Role in Anxiety."

50. M. Heinrichs et al., "Social Support and Oxytocin Interact to Suppress Cortisol and Subjective Responses to Psychosocial Stress," *Biological Psychiatry* 54, no. 12 (2003), https://doi.org/10.1016/S0006-3223(03)00465-7.

51. Bosch, "Maternal Aggression in Rodents: Brain Oxytocin and Vasopressin Mediate Pup Defence," *Philosophical Transactions of the Royal Society of London B: Biological Sciences* 368, no. 1631 (2013), https://doi.org/10.1098/rstb.2013.0085.

52. B. L. Mah et al., "Oxytocin Promotes Protective Behavior in Depressed Mothers: A Pilot Study with the Enthusiastic Stranger Paradigm," *Depression and Anxiety* 32, no. 2 (2014), https://doi.org/10.1002/da.22245.

53. J. N. Ferguson et al., "Social Amnesia in Mice Lacking the Oxytocin Gene," *Nature Genetics* 25, no. 3 (2000), https://doi.org/10.1038/77040.

54. O. J. Bosch and L. J. Young, "Oxytocin and Social Relationships: From Attachment to Bond Disruption," *Current Topics in Behavioral Neuroscience* 35 (2018), https://doi.org/10.1007/7854_2017_10.

55. R. Hurlemann et al., "Oxytocin Enhances Amygdala-Dependent, Socially Reinforced Learning and Emotional Empathy in Humans," *Journal of Neuroscience* 30, no. 14 (2010), https://doi.org/10.1523/JNEUROSCI.5538–09.2010.

56. Ann Pusey, "Magnitude and Sources of Variation in Female Reproductive Performance," in *The Evolution of Primate Societies*, ed. John C. Mitani et al. (Chicago: University of Chicago Press, 2012).

57. K. A. Lynch and J. B. Greenhouse, "Risk-Factors for Infant-Mortality in 19th-Century Sweden," *Population Studies* 48, no. 1 (1994):117–133, https://doi.org/10 .1080/0032472031000147506; S. K. Mishra et al., "Birth Order, Stage of Infancy and Infant Mortality in India," *Journal of Biosocial Science* 50, no. 5 (2018), https://doi.org /10.1017/S0021932017000487.

58. A. P. Moller, "The Evolution of Monogamy: Mating Relationships, Parental Care, and Sexual Selection," in *Monogamy: Mating Strategies and Partnerships in Birds, Humans, and Other Mammals,* edited by U. H. Reichard and C. Boesch, 29–41. Cambridge: Cambridge University Press, 2003.

59. J. R. Williams et al., "Oxytocin Administered Centrally Facilitates Formation of a Partner Preference in Female Prairie Votes (*Microtus ochrogaster*)," *Journal of Neuroendocrinology* 6, no. 3 (1994), https://doi.org/10.1111/j.1365–2826.1994.tb00579.x.

60. Bosch and Young, "Oxytocin and Social Relationships."

61. Bosch and Young, "Oxytocin and Social Relationships."

62. I. Schneiderman et al., "Oxytocin during the Initial Stages of Romantic Attachment: Relations to Couples' Interactive Reciprocity," *Psychoneuroendocrinology* 37, no. 8 (2012), https://doi.org/10.1016/j.psyneuen.2011.12.021.

63. J. Holt-Lunstad, W. C. Birmingham, and K. C. Light, "Relationship Quality and Oxytocin: Influence of Stable and Modifiable Aspects of Relationships," *Journal of Social and Personal Relationships* 32, no. 4 (2015), https://doi.org/10.1177 /0265407514536294; K. M. Grewen et al., "Effects of Partner Support on Resting Oxytocin, Cortisol, Norepinephrine, and Blood Pressure before and after Warm Partner Contact," *Psychosomatic Medicine* 67, no. 4 (2005), https://doi.org/10.1097 /01.psy.0000170341.88395.47.

64. D. Scheele et al., "Oxytocin Enhances Brain Reward System Responses in Men Viewing the Face of their Female Partner," *Proceedings of the National Academy of Sciences* 110, no. 50 (2013), https://doi.org/10.1073/pnas.1314190110.

65. M. Kosfeld et al., "Oxytocin Increases Trust in Humans," *Nature* 435, no. 7042 (2005), https://doi.org/10.1038/nature03701.

66. G. Nave, C. Camerer, and M. McCullough, "Does Oxytocin Increase Trust in Humans? A Critical Review of Research," *Perspectives on Psychological Science* 10, no. 6 (2015), https://doi.org/10.1177/1745691615600138.

67. H. Kurokawa et al., "Oxytocin-Trust Link in Oxytocin-Sensitive Participants and Those without Autistic Traits," *Frontiers in Neuroscience* 15 (2021), https://doi.org/10 .3389/fnins.2021.659737; J. A. Bartz et al., "Social effects of oxytocin in Humans: Context and Person Matter," *Trends in Cognitive Sciences* 15, no. 7 (2011), https://doi .org/10.1016/j.tics.2011.05.002.

68. X. Zheng et al., "Intranasal Oxytocin May Help Maintain Romantic Bonds by Decreasing Jealousy Evoked by Either Imagined or Real Partner Infidelity," *Journal of Psychopharmacology* 35, no. 6 (2021), https://doi.org/10.1177/0269881121991576.

69. B. Ditzen et al., "Intranasal Oxytocin Increases Positive Communication and Reduces Cortisol Levels during Couple Conflict," *Biological Psychiatry* 65, no. 9 (2009), https://doi.org/10.1016/j.biopsych.2008.10.011.

70. D. Scheele et al., "Oxytocin Modulates Social Distance between Males and Females," *Journal of Neuroscience* 32, no. 46 (14 2012), https://doi.org/10.1523/JNEU-ROSCI.2755-12.2012.

71. M. S. Carmichael et al., "Plasma Oxytocin Increases in the Human Sexual Response," *Journal of Clinical Endocrinology and Metabolism* 64, no. 1 (1987), https://doi.org/10.1210/jcem-64-1-27.

72. R. M. Costa and S. Brody, "Women's Relationship Quality Is Associated with Specifically Penile-Vaginal Intercourse Orgasm and Frequency," *Journal of Sex and Marital Therapy* 33, no. 4 (2007), https://doi.org/10.1080/00926230701385548; J. K. McNulty, C. A. Wenner, and T. D. Fisher, "Longitudinal Associations among Relationship Satisfaction, Sexual Satisfaction, and Frequency of Sex in Early Marriage," *Archives of Sexual Behavior* 45, no. 1 (2016), https://doi.org/10.1007/s10508-014 -0444-6.

73. C. Heim et al., "Lower CSF Oxytocin Concentrations in Women with a History of Childhood Abuse," *Molecular Psychiatry* 14, no. 10 (2009), https://doi.org/10.1038 /mp.2008.1120.

74. L. B. King et al., "Variation in the Oxytocin Receptor Gene Predicts Brain Region-Specific Expression and Social Attachment," *Biological Psychiatry* 80, no. 2 (2016), https://doi.org/10.1016/j.biopsych.2015.12.008.

75. A. M. Perkeybile et al., "Early Nurture Epigenetically Tunes the Oxytocin Receptor," *Psychoneuroendocrinology* 99 (2019), https://doi.org/10.1016/j.psyneuen.2018 .08.037.

76. GTEx Consortium, "The GTEx Consortium Atlas of Genetic Regulatory Effects across Human Tissues," *Science* 369, no. 6509 (2020): 1318–1330, https://doi.org/10 .1126/science.aaz1776.

77. K. M. Krol et al., "Epigenetic Dynamics in Infancy and the Impact of Maternal Engagement," *Science Advances* 5, no. 10 (2019), https://doi.org/10.1126/sciadv .aay0680.

78. Z. Wang, C. F. Ferris, and G. J. De Vries, "Role of Septal Vasopressin Innervation in Paternal Behavior in Prairie Voles (*Microtus ochrogaster*)," *Proceedings of the National Academy of Sciences* 91, no. 1 (1994), https://doi.org/10.1073/pnas.91.1.400.

79. E. A. Hammock and L. J. Young, "Oxytocin, Vasopressin and Pair Bonding: Implications for Autism," *Philosophical Transactions of the Royal Society of London Series B: Biological Sciences* 361, no. 1476 (2006), https://doi.org/10.1098/rstb.2006 .1939.

80. Y. Kozorovitskiy et al., "Fatherhood Affects Dendritic Spines and Vasopressin V1a Receptors in the Primate Prefrontal Cortex," *Nature Neuroscience* 9, no. 9 (2006), https://doi.org/10.1038/nn1753.

81. Y. Apter-Levi, O. Zagoory-Sharon, and R. Feldman, "Oxytocin and Vasopressin Support Distinct Configurations of Social Synchrony," *Brain Research* 1580 (2014), https://doi.org/10.1016/j.brainres.2013.10.052.

82. C. C. Cohen-Bendahan et al., "Explicit and Implicit Caregiving Interests in Expectant Fathers: Do Endogenous and Exogenous Oxytocin and Vasopressin Matter?," *Infant Behavior and Development* 41 (2015), https://doi.org/10.1016/j .infbeh.2015.06.007.

83. M. Nagasawa et al., "Social Evolution. Oxytocin-Gaze Positive Loop and the Coevolution of Human-Dog Bonds," *Science* 348, no. 6232 (17 2015), https://doi.org /10.1126/science.1261022.

84. Tom McGrath, dir. *The Boss Baby*. Century City, CA: 20th Century Fox, 2017.

85. C. K. De Dreu et al., "The Neuropeptide Oxytocin Regulates Parochial Altruism in Intergroup Conflict among Humans," *Science* 328, no. 5984 (2010), https://doi .org/328/5984/10.1126/science.1189047.

86. Sebastian Junger, *War* (New York: Twelve, 2010).

87. C. Feng et al., "Oxytocin and Vasopressin Effects on the Neural Response to Social Cooperation Are Modulated by Sex in Humans," *Brain Imaging and Behavior* 9, no. 4 (2015), https://doi.org/10.1007/s11682-014-9333-9.

88. L. Samuni et al., "Oxytocin Reactivity during Intergroup Conflict in Wild Chimpanzees," *Proceedings of the National Academy of Sciences* 114, no. 2 (2016), https://doi .org/10.1073/pnas.1616812114.

Chapter 5

1. Z. Wu et al., "Galanin Neurons in the Medial Preoptic Area Govern Parental Behaviour," *Nature* 509, no. 7500 (2014), https://doi.org/10.1038/nature13307.

2. Paul D. MacLean, *The Triune Brain in Evolution: Role in Paleocerebral Functions* (New York: Plenum Press, 1990).

3. M. R. Murphy, P. D. MacLean, and S. C. Hamilton, "Species-Typical Behavior of Hamsters Deprived from Birth of the Neocortex," *Science* 213, no. 4506 (1981), http://www.ncbi.nlm.nih.gov/pubmed/7244642.

4. Michael Numan, *The Parental Brain: Mechanisms, Development and Evolution* (New York: Oxford University Press, 2020).

5. Numan, *The Parental Brain*.

6. Numan, *The Parental Brain*.

7. C. A. Pedersen et al., "Oxytocin Activates the Postpartum Onset of Rat Maternal Behavior in the Ventral Tegmental and Medial Preoptic Areas," *Behavioral and Brain Sciences* 108, no. 6 (1994), https://doi.org/10.1037//0735-7044.108.6.1163.

8. Numan, *The Parental Brain*; J. Kohl, A. E. Autry, and C. Dulac, "The Neurobiology of Parenting: A Neural Circuit Perspective," *Bioessays* 39, no. 1 (2017), https://doi .org/10.1002/bies.201600159.

9. Numan, *The Parental Brain*.

10. T. W. Robbins and B. J. Everitt, "Neurobehavioural Mechanisms of Reward and Motivation," *Current Opinion in Neurobiology* 6, no. 2 (1996), https://doi.org/10.1016 /s0959-4388(96)80077-8.

11. J. Kohl and C. Dulac, "Neural Control of Parental Behaviors," *Current Opinion in Neurobiology* 49 (2018), https://doi.org/10.1016/j.conb.2018.02.002.

12. Numan, *The Parental Brain*.

13. N. D. Volkow, G. F. Koob, and A. T. McLellan, "Neurobiologic Advances from the Brain Disease Model of Addiction," *New England Journal of Medicine* 374, no. 4 (2016), https://doi.org/10.1056/NEJMra1511480.

14. N. D. Volkow, G. J. Wang, and R. D. Baler, "Reward, Dopamine and the Control of Food Intake: Implications for Obesity," *Trends in Cognitive Sciences*15, no. 1 (2011), https://doi.org/10.1016/j.tics.2010.11.001.

15. Kohl and Dulac, "Neural Control of Parental Behaviors."

16. W. Romero-Fernandez et al., "Evidence for the Existence of Dopamine D2-Oxytocin Receptor Heteromers in the Ventral and Dorsal Striatum with Facilitatory Receptor-Receptor Interactions," *Molecular Psychiatry* 18, no. 8 (2013), https://doi .org/10.1038/mp.2012.103.

17. O. J. Bosch, "Maternal Aggression in Rodents: Brain Oxytocin and Vasopressin Mediate Pup Defence," *Philosophical Transactions of the Royal Society of London B: Biological Sciences* 368, no. 1631 (2013), https://doi.org/10.1098/rstb.2013.0085.

18. Numan, *The Parental Brain*; Bosch, "Maternal Aggression in Rodents."

19. Numan, *The Parental Brain*.

20. M. L. Glocker et al., "Baby Schema Modulates the Brain Reward System in Nulliparous Women," *Proceedings of the National Academy of Sciences* 106, no. 22 (2009), https://doi.org/10.1073/pnas.0811620106.

21. N. K. Logothetis, "The Neural Basis of the Blood-Oxygen-Level-Dependent Functional Magnetic Resonance Imaging Signal.," *Philosophical Transactions of the Royal Society B: Biological Sciences* 357, no. 1424 (2002), https://doi.org/10.1098/rstb.2002.1114.

22. Hrdy, Sarah Blaffer, *Mothers and Others* (Cambridge, MA: Harvard University Press, 2009).

23. L. Strathearn, P. Fonagy, J. Amico, and P. R. Montague, "Adult Attachment Predicts Maternal Brain and Oxytocin Response to Infant Cues," *Neuropsychopharmacology* 34, no. 13 (2009): 2655–2666, https://doi.org/10.1038/npp.2009.103.

24. S. Atzil, T. Hendler, and R. Feldman, "Specifying the Neurobiological Basis of human Attachment: Brain, Hormones, an d Behavior in Synchronous and Intrusive Mothers," *Neuropsychopharmacology* 36, no. 13 (2011), https://doi.org/10.1038/npp.2011.172.

25. S. Atzil et al., "Dopamine in the Medial Amygdala Network Mediates Human Bonding," *Proceedings of the National Academy of Sciences* 114, no. 9 (2017), https://doi.org/10.1073/pnas.1612233114.

26. Atzil et al., "Dopamine in the Medial Amygdala Network."

27. P. Rigo et al., "Specific Maternal Brain Responses to Their Own Child's Face: An fMRI Meta-Analysis," *Developmental Review* 51 (2019), https://doi.org/10.1016/j.dr.2018.12.001.

28. M. H. Bornstein et al., "Neurobiology of Culturally Common Maternal Responses to Infant Cry," *Proceedings of the National Academy of Sciences* 114, no. 45 (2017), https://doi.org/10.1073/pnas.1712022114; J. P. Lorberbaum et al., "A Potential Role for Thalamocingulate Circuitry in Human Maternal Behavior," *Biological Psychiatry* 51, no. 6 (2002), https://doi.org/10.1016/s0006-3223(01)01284-7; H. K. Laurent and J. C. Ablow, "A Cry in the Dark: Depressed Mothers Show Reduced Neural Activation to Their Own Infant's Cry," *Social Cognitive and Affective Neuroscience* 7, no. 2 (2012), https://doi.org/10.1093/scan/nsq091.

29. C. Post and B. Leuner, "The Maternal Reward System in Postpartum Depression," *Archives of Women's Mental Health* 22, no. 3 (2019), https://doi.org/10.1007/s00737-018-0926-y; H. K. Laurent and J. C. Ablow, "A Cry in the Dark: Depressed Mothers Show Reduced Neural Activation to Their Own Infant's Cry," *Social Cognitive and Affective Neuroscience* 7, no. 2 (2012), https://doi.org/10.1093/scan/nsq091; H. K. Laurent and J. C. Ablow, "A Face a Mother Could Love: Depression-Related

Maternal Neural Responses to Infant Emotion Faces," *Social Neuroscience* 8, no. 3 (2013), https://doi.org/10.1080/17470919.2012.762039.

30. R. Gregory et al., "Oxytocin Increases VTA Activation to Infant and Sexual Stimuli in Nulliparous and Postpartum Women," *Hormones and Behavior* 69 (2015), https://doi.org/10.1016/j.yhbeh.2014.12.009.

31. M. M. Riem et al., "Oxytocin Modulates Amygdala, Insula, and Inferior Frontal Gyrus Responses to Infant Crying: A Randomized Controlled Trial," *Biological Psychiatry* 70, no. 3 (2011), https://doi.org/10.1016/j.biopsych.2011.02.006.

32. G. Rizzolatti and M. Fabbri-Destro, "Mirror Neurons: From Discovery to Autism," *Experimental Brain Research* 200, no. 3–4 (), https://doi.org/10.1007/s00221-009-2002-3.

33. D. Lenzi et al., "Neural Basis of Maternal Communication and Emotional Expression Processing during Infant Preverbal Stage," *Cerebral Cortex* 19, no. 5 (2009), https://doi.org/10.1093/Cercor/Bhn153.

34. H. K. Laurent, A. Stevens, and J. C. Ablow, "Neural Correlates of Hypothalamic-Pituitary-Adrenal Regulation of Mothers with Their Infants," *Biological Psychiatry* 70, no. 9 (2011), https://doi.org/10.1016/j.biopsych.2011.06.011.

35. J. S. Rosenblatt, S. Hazelwood, and J. Poole, "Maternal Behavior in Male Rats: Effects of Medial Preoptic Area Lesions and Presence of Maternal Aggression," *Hormones and Behavior* 30, no. 3 (1996), https://doi.org/10.1006/hbeh.1996.0025; J. D. Sturgis and R. S. Bridges, "N-Methyl-DL-Aspartic Acid Lesions of the Medial Preoptic Area Disrupt Ongoing Parental Behavior in Male Rats," *Physiological Behavior* 62, no. 2 (1997), https://doi.org/10.1016/s0031-9384(97)88985-8.

36. J. S. Rosenblatt and K. Ceus, "Estrogen Implants in the Medial Preoptic Area Stimulate Maternal Behavior in Male Rats," *Hormones and Behavior* 33, no. 1 (1998), https://doi.org/10.1006/hbeh.1997.1430.

37. A. W. Lee and R. E. Brown, "Medial Preoptic Lesions Disrupt Parental Behavior in Both Male and Female California Mice (*Peromyscus californicus*)," *Behavioral Neuroscience* 116, no. 6 (2002), https://doi.org/10.1037/0735-7044.116.6.968; T. R. de Jong et al., "From Here to Paternity: Neural Correlates of the Onset of Paternal Behavior in California Mice (*Peromyscus californicus*)," *Hormones and Behavior* 56, no. 2 (2009), https://doi.org/10.1016/j.yhbeh.2009.05.001.

38. E. R. Glasper et al., "More Than Just Mothers: The Neurobiological and Neuroendocrine Underpinnings of Allomaternal Caregiving," *Frontiers in Neuroendocrinology* 53 (2019), https://doi.org/10.1016/j.yfrne.2019.02.005.

39. Wu et al., "Galanin Neurons."

40. B. Wang et al., "Behavioral Responses to Pups in Males with Different Reproductive Experiences Are Associated with Changes in Central OT, TH and OTR, D1R, D2R mRNA Expression in Mandarin Voles," *Hormones and Behavior* 67 (2015), https://doi.org/10.1016/j.yhbeh.2014.11.013; Yuan et al., "Role of Oxytocin in the Medial Preoptic Area (MPOA) in the Modulation of Paternal Behavior in Mandarin Voles."

41. Wang et al., "Behavioral Responses to Pups."

42. B. C. Trainor, I. M. Bird, N. A. Alday, B. A. Schlinger, and C. A. Marler, "Variation in Aromatase Activity in the Medial Preoptic Area and Plasma Progesterone Is Associated with the Onset of Paternal Behavior," *Neuroendocrinology* 78, no. 1 (2003): 36–44. https://doi.org/10.1159/000071704.

43. M. M. Hyer et al., "Neurogenesis and Anxiety-Like Behavior in Male California Mice during the Mate's Postpartum Period," *European Journal of Neuroscience* 43, no. 5 (2016), https://doi.org/10.1111/ejn.13168.

44. C. L. Franssen et al., "Fatherhood Alters Behavioural and Neural Responsiveness in a Spatial Task," *Journal of Neuroendocrinology* 23, no. 11 (2011), https://doi.org/10.1111/j.1365-2826.2011.02225.x.

45. G. K. Mak and S. Weiss, "Paternal Recognition of Adult Offspring Mediated by Newly Generated CNS Neurons," *Nature Neuroscience* 13, no. 6 (2010), https://doi.org/10.1038/nn.2550.

46. Y. Kozorovitskiy, M. Hughes, K. Lee, and E. Gould, "Fatherhood Affects Dendritic Spines and Vasopressin V1a Receptors in the Primate Prefrontal Cortex," *Nature Neuroscience* 9, no. 9 (2006): 1094–1095, https://doi.org/10.1038/nn1753.

47. J. S. Mascaro, P. D. Hackett, and J. K. Rilling, "Testicular Volume Is Inversely Correlated with Nurturing-Related Brain Activity in Human Fathers," *Proc Natl Acad Sci U S A* 110, no. 39 (2013): 15746–15751, https://doi.org/10.1073/pnas.1305579110.

48. T. Li, X. Chen, J. Mascaro, E. Haroon, and J. K. Rilling, "Intranasal Oxytocin, but Not Vasopressin, Augments Neural Responses to Toddlers in Human Fathers," *Hormones and Behavior* 93 (2017): 193–202, https://doi.org/10.1016/j.yhbeh.2017.01.006.

49. P. X. Kuo et al., "Neural Responses to Infants Linked with Behavioral Interactions and Testosterone in Fathers," *Biological Psychiatry* 91, no. 2 (Aug 10 2012), https://doi.org/10.1016/j.biopsycho.2012.08.002.

50. P. Kim et al., "A Prospective Longitudinal Study of Perceived Infant Outcomes at 18–24 Months: Neural and Psychological Correlates of Parental Thoughts and Actions Assessed during the First Month Postpartum," *Frontiers in Psychology* 6 (2015), https://doi.org/10.3389/fpsyg.2015.01772.

51. M. L. Kringelbach and E. T. Rolls, "The Functional Neuroanatomy of the Human Orbitofrontal Cortex: Evidence from Neuroimaging and Neuropsychology," *Progress in Neurobiology* 72, no. 5 (2004), https://doi.org/10.1016/j.pneurobio.2004.03.006.

52. Mascaro, Hackett, and Rilling, "Differential Neural Responses to Child and Sexual Stimuli in Human Fathers and Non-Fathers and Their Hormonal Correlates."

53. P. X. Kuo, J. Carp, K. C. Light, and K. M. Grewen, "Neural Responses to Infants Linked with Behavioral Interactions and Testosterone in Fathers," *Biological Psychiatry* 91, no. 2 (2012): 302–306, https://doi.org/10.1016/j.biopsycho.2012.08.002.

54. J. S. Mascaro et al., "Child Gender Influences Paternal Behavior, Language, and Brain Function," *Behavioral Neuroscience* 131, no. 3 (2017), https://doi.org/10.1037/bne0000199.

55. L. Cossette et al., "Emotional Expressions of Female and Male Infants in a Social and a Nonsocial Context," *Sex Roles* 35, no. 11–12 (1996), https://doi.org/10.1007/Bf01544087; M. LaFrance, M. A. Hecht, and E. L. Paluck, "The Contingent Smile: A Meta-Analysis of Sex Differences in Smiling," *Psychological Bulletin* 129, no. 2 (2003), https://doi.org/10.1037/0033-2909.129.2.305.

56. R. G. Barr, "Preventing Abusive Head Trauma Resulting from a Failure of Normal Interaction between Infants and Their Caregivers," *Proc Natl Acad Sci U S A* 109 Suppl 2 (2012): 17294–17301, https://doi.org/10.1073/pnas.1121267109.

57. J. S. Mascaro et al., "Behavioral and Genetic Correlates of the Neural Response to Infant Crying among Human Fathers," *Social Cognitive and Affective Neuroscience* 9, no. 11 (2014), https://doi.org/10.1093/scan/nst1669.

58. T. Li et al., "Explaining Individual Variation in Paternal Brain Responses to Infant Cries," *Physiology and Behavior* 193, pt. A (2018), https://doi.org/10.1016/j.physbeh.2017.12.033.

59. A. M. Witte et al., "The Effects of Oxytocin and Vasopressin Administration on Fathers' Neural Responses to Infant Crying: A Randomized Controlled Within-Subject Study," *Psychoneuroendocrinology* 140 (2022), https://doi.org/10.1016/j.psyneuen.2022.105731.

60. J. K. Rilling et al., "The Neural Correlates of Paternal Consoling Behavior and Frustration in Response to Infant Crying," *Developmental Psychobiology* (2021), https://doi.org/10.1002/dev.22092.

61. E. Abraham et al., "Father's Brain Is Sensitive to Childcare Experiences," *Proceedings of the National Academy of Sciences* 111, no. 27 (2014), https://doi.org/10.1073/pnas.1402569111.

62. Numan, *The Parental Brain*.

63. Abraham et al., "Father's Brain Is Sensitive."

64. Abraham et al., "Father's Brain Is Sensitive."

65. M. M. E. Riem et al., "A Soft Baby Carrier Intervention Enhances Amygdala Responses to Infant Crying in Fathers: A Randomized Controlled trial," *Psychoneuroendocrinology* 132 (2021), https://doi.org/10.1016/j.psyneuen.2021.105380.

66. E. A. Amadei et al., "Dynamic Corticostriatal Activity Biases Social Bonding in Monogamous Female Prairie Voles," *Nature* 546, no. 7657 (2017), https://doi.org/10.1038/nature22381.

67. J. K. Rilling et al., "Intranasal Oxytocin Modulates Neural Functional Connectivity during Human Social Interaction," *American Journal of Primatology* 80, no. 10 (2018), https://doi.org/10.1002/ajp.22740.

68. E. Abraham et al., "Empathy Networks in the Parental Brain and Their Long-Term Effects on Children's Stress Reactivity and Behavior Adaptation," *Neuropsychologia* 116 (2018), https://doi.org/0.1016/j.neuropsychologia.2017.04.015.

69. E. Hoekzema et al., "Pregnancy Leads to Long-Lasting Changes in Human Brain Structure," *Nat Neuroscience* 20, no. 2 (2017), https://doi.org/10.1038/nn.4458.

70. Tomas Paus et al., "Structural Maturation of Neural Pathways in Children and Adolescents: In Vivo Study," *Science* 283 (1999): 1908–1911, https://doi.org/10.1126/science.283.5409.1908.

71. A. S. LaMantia and P. Rakic, "Axon Overproduction and Elimination in the Cerebral Cortex of the Developing Rhesus Monkey," *Journal of Neuroscience* 10 (1990), https://doi.org/10.1523/JNEUROSCI.10-07-02156.

72. J. L. Pawluski et al., "Less Can Be More: Fine Tuning the Maternal Brain," *Neuroscience and Biobehavioral Reviews* 133 (2022), https://doi.org/10.1016/j.neubiorev.2021.11.045.

73. E. Hoekzema et al., "Becoming a Mother Entails Anatomical Changes in the Ventral Striatum of the Human Brain That Facilitate Its Responsiveness to Offspring Cues," *Psychoneuroendocrinology* 112 (2020), https://doi.org/10.1016/j.psyneuen.2019.104507.

74. M. Paternina-Die et al., "The Paternal Transition Entails Neuroanatomic Adaptations That Are Associated with the Father's Brain Response to His Infant Cues," *Cerebral Cortex Communications* 1, no. 1 (2020), https://doi.org/10.1093/texcom/tgaa082.

75. M. Martinez-Garcia et al., "Characterizing the Brain Structural Adaptations across the Motherhood Transition," *Frontiers in Global Women's Health* 2 (2021), https://doi.org/10.3389/fgwh.2021.742775.

76. P. Kim et al., "The Plasticity of Human Maternal Brain: Longitudinal Changes in Brain Anatomy during the Early Postpartum Period," *Behavioral Neuroscience* 124, no. 5 (2010), https://doi.org/10.1037/a0020884.

77. Kim et al., "Neural Plasticity in Fathers of Human Infants."

78. Y. Cao et al., "Neonatal Paternal Deprivation Impairs Social Recognition and Alters Levels of Oxytocin and Estrogen Receptor Alpha mRNA Expression in the MeA and NAcc, and Serum Oxytocin in Mandarin Voles," *Hormones and Behavior* 65, no. 1 (2014), https://doi.org/10.1016/j.yhbeh.2013.11.005.

79. R. Jia et al., "Neonatal Paternal Deprivation or Early Deprivation Reduces Adult Parental Behavior and Central Estrogen Receptor Alpha Expression in Mandarin Voles (*Microtus mandarinus*)," *Behavioural Brain Research* 224, no. 2 (31 2011), https://doi.org/10.1016/j.bbr.2011.05.042.

80. Cao et al., "Neonatal Paternal Deprivation."

81. C. Helmeke et al., "Paternal Deprivation during Infancy Results in Dendrite- and Time-Specific Changes of Dendritic Development and Spine Formation in the Orbitofrontal Cortex of the Biparental Rodent *Octodon degus*," *Neuroscience* 163, no. 3 (2009), https://doi.org/10.1016/j.neuroscience.2009.07.008.

82. J. Hiser and M. Koenigs, "The Multifaceted Role of the Ventromedial Prefrontal Cortex in Emotion, Decision Making, Social Cognition, and Psychopathology," *Biological Psychiatry* 83, no. 8 (2018), https://doi.org/10.1016/j.biopsych.2017.10.030.

83. Sergio M. Pellis, Vivien C. Pellis, and Heather C. Bell, "The Function of Play in the Development of the Social Brain," *American Journal of Play* 2, no. 3 (2010), https://eric.ed.gov/?id=EJ1069225.

84. F. D. Rogers and K. L. Bales, "Revisiting Paternal Absence: Female Alloparental Replacement of Fathers Recovers Partner Preference Formation in Female, but Not Male Prairie Voles," *Developmental Psychobiology* 62, no. 5 (2020): 573–590, https://doi.org/10.1002/dev.21943; F. D. Rogers, S. M. Freeman, M. Anderson, M. C. Palumbo, and K. L. Bales, "Compositional Variation in Early-Life Parenting Structures Alters Oxytocin and Vasopressin 1a Receptor Development in Prairie Voles," *Journal of Neuroendocrinology* 33, no. 8 (2021), https://doi.org/10.1111/jne.13001.

Chapter 6

1. G. L. Brown, S. M. Kogan, and J. Kim, "From Fathers to Sons: The Intergenerational Transmission of Parenting Behavior among African American Young Men," *Family Process* 57, no. 1 (2018), https://doi.org/10.1111/famp.12273.

2. E. Pougnet et al., "The Intergenerational Continuity of Fathers' Absence in a Socioeconomically Disadvantaged Sample," *Journal of Marriage and Family* 74, no. 3 (2012), https://doi.org/10.1111/j.1741-3737.2012.00962.x.

3. Brown et al., "From Fathers to Sons."

4. S. L. Hofferth, J. H. Pleck, and C. K. Vesely, "The Transmission of Parenting from Fathers to Sons," *Parenting: Science and Practice* 12, no. 4 (2012), https://doi.org/10.1080/15295192.2012.709153.

5. Joshua Kendall, "Obama's Most Unusual Legacy? Being a Good Dad," *Washington Post*, June 19, 2016.

6. G. V. Pepper and D. Nettle, "The Behavioural Constellation of Deprivation: Causes and Consequences," *Behavioral and Brain Sciences* 40 (2017), https://doi.org/10.1017/S0140525X1600234X; D. S. Roubinov and W. T. Boyce, "Parenting and SES: Relative Values or Enduring Principles?," *Current Opinion in Psychology* 15 (2017), https://doi.org/10.1016/j.copsyc.2017.03.001.

7. J. C. Koops, A. C. Liefbroer, and A. H. Gauthier, "Having a Child within a Cohabiting Union in Europe and North America: What Is the Role of Parents' Socio-Economic Status?," *Population Space and Place* 27, no. 6 (2021), https://doi.org/10.1002/psp.2434.

8. W. E. Johnson, "Paternal Involvement among Unwed Fathers," *Children and Youth Services Review* 23, no. 6–7 (2001), https://doi.org/10.1016/S0190–7409(01)00146–3.

9. D. Nettle, "Dying Young and Living Fast: Variation in Life History across English Neighborhoods," *Behavioral Ecology* 21, no. 2 (2010), https://doi.org/10.1093/beheco/arp202.

10. M. Minkov and K. Beaver, "A Test of Life History Strategy Theory as a Predictor of Criminal Violence across 51 Nations," *Personality and Individual Differences* 97 (2016), https://doi.org/10.1016/j.paid.2016.03.063.

11. N. Barber, "Single Parenthood as a Predictor of Cross-National Variation in Violent Crime," *Cross-Cultural Research* 38, no. 4 (2004), https://doi.org/10.1177/1069397104267479.

12. Pepper and Nettle, "The Behavioural Constellation of Deprivation."

13. S. E. Johns, "Perceived Environmental Risk as a Predictor of Teenage Motherhood in a British Population," *Health and Place* 17, no. 1 (2011), https://doi.org/10.1016/j.healthplace.2010.09.006; M. Imamura et al., "Factors Associated with Teenage Pregnancy in the European Union Countries: A Systematic review," *European Journal of Public Health* 17, no. 6 (2007), https://doi.org/10.1093/eurpub/ckm014; Nettle, "Dying Young and Living Fast."

14. R. Sear, "Do Human 'Life History Strategies' Exist?," *Evolution and Human Behavior* 41, no. 6 (2020), https://doi.org/10.1016/j.evolhumbehav.2020.09.004.

15. B. J. Ellis et al., "Developmental Programming of Oxytocin through Variation in Early-Life Stress: Four Meta-Analyses and a Theoretical Reinterpretation," *Clinical Psychology Review* 86 (2021), https://doi.org/10.1016/j.cpr.2021.101985.

16. Roubinov and Boyce, "Parenting and SES?"

17. B. S. Hewlett, "Demography and Child-Care in Preindustrial Societies," *Journal of Anthropological Research* 47, no. 1 (1991), https://www.jstor.org/stable/3630579

18. Martin Daly and Margo Wilson, *Homicide* (New York: Aldine de Gruyter, 1988).

19. Gray and Anderson, *Fatherhood: Evolution and Human Paternal Behavior*; William Marsiglio and Ramon Hinojosa, "Stepfathers' Lives," in *The Role of the Father in Child Development* (Hoboken, NJ: John Wiley and Sons, 2010); K. G. Anderson, H. Kaplan, and J. Lancaster, "Paternal Care by Genetic Fathers and Stepfathers I: Reports from Albuquerque Men," *Evolution and Human Behavior* 20, no. 6 (1999), https://doi.org/10.1016/S1090–5138(99)00023–9.

20. K. G. Anderson, H. Kaplan, and J. B. Lancaster, "Confidence of Paternity, Divorce, and Investment in Children by Albuquerque Men," *Evolution and Human Behavior* 28, no. 1 (2007), https://doi.org/10.1016/j.evolhumbehav.2006.06.004.

21. C. L. Apicella and F. W. Marlowe, "Men's Reproductive Investment Decisions: Mating, Parenting, and Self-Perceived Mate Value," *Human Nature* 18, no. 1 (2007), https://doi.org/10.1007/BF02820844.

22. J. Hartung, "Matrilineal Inheritance—New Theory and Analysis," *Behavioral and Brain Sciences* 8, no. 4 (1985), https://doi.org/10.1017/S0140525x00045520.

23. B. Hohmann-Marriott, "Coparenting and Father Involvement in Married and Unmarried Coresident Couples," *Journal of Marriage and Family* 73, no. 1 (2011), https://doi.org/10.1111/j.1741-3737.2010.00805.x; G. L. Brown et al., "Observed and Reported Supportive Coparenting as Predictors of Infant-Mother and Infant-Father Attachment Security," *Early Child Development and Care* 180, no. 1–2 (2010), https://doi.org/10.1080/03004430903415015.

24. J. Fagan and M. Barnett, "The Relationship between Maternal Gatekeeping, Paternal Competence, Mothers' Attitudes about the Father Role, and Father Involvement," *Journal of Family Issues* 24, no. 8 (2003), https://doi.org/10.1177/0192513x03256397.

25. S. J. Schoppe-Sullivan et al., "Who Are the Gatekeepers? Predictors of Maternal Gatekeeping," *Parenting: Science and Practice* 15, no. 3 (2015), https://doi.org/10.1080/15295192.2015.1053321.

26. S. J. Schoppe-Sullivan et al., "Maternal Gatekeeping, Coparenting Quality, and Fathering Behavior in Families with Infants," *Journal of Family Psychology* 22, no. 3 (2008), https://doi.org/10.1037/0893-3200.22.3.389.

27. Schoppe-Sullivan et al., "Who Are the Gatekeepers?"

28. John W. M. Whiting and Beatrice Whiting, "Aloofness and Intimacy of Husbands and Wives: A Cross-Cultural Study," *Ethos* 3, no. 2 (1975), https://www.jstor.org/stable/640228.

29. Shawna J. Lee, Joyce Y. Lee, and Olivia D. Chang, "The Characteristics and Lived Experience of Modern Stay-at-Home Fathers," in *Handbook of Fathers and Child Development*, ed. Hiram Fitzgerald et al. (Berlin: Springer, 2020).

30. Melvin Konner, *Women After All: Sex, Evolution, and the End of Male Supremacy* (New York: Norton, 2015).

31. N. Blom and B. Hewitt, "Becoming a Female-Breadwinner Household in Australia: Changes in Relationship Satisfaction," *Journal of Marriage and Family* 82, no. 4 (2020), https://doi.org/10.1111/jomf.12653.

32. Lee et al., "Characteristics and Lived Experience."

33. R. L. Coles, "Single-Father Families: A Review of the Literature," *Journal of Family Theory & Review* 7, no. 2 (2015): 144–166, https://doi.org/10.1111/jftr.12069.

34. T. J. Wade, S. Veldhuizen, and J. Cairney, "Prevalence of Psychiatric Disorder in Lone Fathers and Mothers: Examining the Intersection of Gender and Family Structure on Mental Health," *Canadian Journal of Psychiatry* 56, no. 9 (2011): 567–573, https://doi.org/10.1177/070674371105600908.

35. Kerstin Aumann, Ellen Galinsky, and Kenneth Matos, *The New Male Mystique* (Palisades, NY: Families and Work Institute, 2011), https://www.familiesandwork.org/research.

36. Ellen Galinsky, Kerstin Aumann, and James T. Bond, *Times Are Changing: Gender and Generation at Work and at Home* (Palisades, NY: Families and Work Institute, 2009), https://www.familiesandwork.org/research.

37. H. Dinh et al., "Parents' Transitions into and out of Work–family Conflict and Children's Mental Health: Longitudinal Influence via Family Functioning," *Social Science and Medicine* 194 (2017), https://doi.org/10.1016/j.socscimed.2017.10.017.

38. B. E. Robinson and L. Kelley, "Adult Children of Workaholics: Self-Concept, Anxiety, Depression, and Locus of Control," *American Journal of Family Therapy* 26, no. 3 (1998), https://doi.org/10.1080/01926189808251102.

39. K. Nishiyama and J. V. Johnson, "Karoshi—Death from Overwork: Occupational Health Consequences of Japanese Production Management," *International Journal of Health Services* 27, no. 4 (1997), https://doi.org/10.2190/1jpc-679v-Dynt-Hj6g.

40. Linda Haas and C. Philip Hwang, "The Impact of Taking Parental Leave on Fathers' Participation in Childcare and Relationships with Children: Lessons from

Sweden," *Community, Work and Family* 11, no. 1 (2008), https://doi.org/10.1080/13668800701785346.

41. M. C. Huerta et al., "Fathers' Leave and Fathers' Involvement: Evidence from Four OECD Countries," *European Journal of Social Security* 16, no. 4 (2014), https://doi.org/10.1177/138826271401600403.

42. R. J. Petts and C. Knoester, "Are Parental Relationships Improved if Fathers Take Time Off of Work after the Birth of a Child?," *Social Forces* 98, no. 3 (2020), https://doi.org/10.1093/sf/soz014.

43. C. Knoester, R. J. Petts, and B. Pragg, "Paternity Leave-Taking and Father Involvement among Socioeconomically Disadvantaged U.S. Fathers," *Sex Roles* 81, no. 5–6 (2019), https://doi.org/10.1007/s11199-018-0994-5.

44. Knoester et al., "Paternity Leave-Taking."

45. Claire Cain Miller, "The World 'Has Found a Way to Do This': The U.S. Lags on Paid Leave," *New York Times*, October 25, 2021, https://www.nytimes.com/2021/10/25/upshot/paid-leave-democrats.html.

46. J. Ekberg, R. Eriksson, and G. Friebel, "Parental Leave: A Policy Evaluation of the Swedish 'Daddy-Month' Reform," *Journal of Public Economics* 97 (2013), https://doi.org/10.1016/j.jpubeco.2012.09.001; S. Cools, J. H. Fiva, and L. J. Kirkeboen, "Causal Effects of Paternity Leave on Children and Parents," *Scandinavian Journal of Economics* 117, no. 3 (2015), https://doi.org/10.1111/sjoe.12113.

47. Alexa L. Secrest, "Daddy Issues: Why Do Swedish Fathers Claim Paternity Leave at Higher Rates Than French Fathers?," *Student Publications*, 2020, https://cupola.gettysburg.edu/student_scholarship/784.

48. Joe Pinsker, "Why Icelandic Dads Take Parental Leave and Japanese Dads Don't," *The Atlantic*, January 23, 2020, https://www.theatlantic.com/family/archive/2020/01/japan-paternity-leave-koizumi/605344/.

49. Secrest, "Daddy Issues."

50. Cindy Brooks Dollar, "Sex Ratios, Single Motherhood, and Gendered Structural Relations: Examining Female-Headed Families across Racial-Ethnic Populations," *Sociological Focus* 50, no. 4 (2017), https://doi.org/10.1080/00380237.2017.1313100.

51. F. W. Marlowe, "Male Care and Mating Effort among Hadza Foragers," *Behav Ecol Sociobiol* 46, no. 1 (1999), 57–64, https://doi.org/10.1007/s002650050592.

52. James K. Rilling, Paige Gallagher, and Minwoo Lee, "Mating-Related Stimuli Induce Rapid Shifts in Fathers' Assessments of Infants," *Evolution and Human Behavior* 45, no. 1 (2024): 13–19.

53. R. Schacht and M. B. Mulder, "Sex Ratio Effects on Reproductive Strategies in Humans," *Royal Society Open Science* 2, no. 1 (2015), https://doi.org/10.1098/rsos .140402.

54. Statistics Times, "List of Countries by Sex Ratio," 2023, https://statisticstimes .com/demographics/countries-by-sex-ratio.php.

55. Leila Morsey and Richard Rothstein, *Mass Incarceration and Children's Outcomes: Criminal Justice Policy Is Education Policy* (Washington, DC: Economic Policy Institute, 2016).

56. Morsey and Rothstein, *Mass Incarceration.*

57. Patricia Draper and Henry Harpending, "Father Absence and Reproductive Strategy: An Evolutionary Perspective," *Journal of Anthropological Research* 38 (1982), https://www.jstor.org/stable/3629848.

58. M. Konner, "Is History the Same as Evolution? No. Is it Independent of Evolution? Certainly Not," *Evolutionary Psychology* 20, no. 1 (2022), https://doi.org/10 .1177/1474704921106913.

59. M. Chudek and J. Henrich, "Culture-Gene Coevolution, Norm-Psychology and the Emergence of Human Prosociality," *Trends in Cognitive Sciences* 15, no. 5 (2011), https://doi.org/10.1016/j.tics.2011.03.003.

60. Bradd Shore, *Culture in Mind* (New York: Oxford University Press, 1996).

61. S. P. Prall and B. A. Scelza, "Why Men Invest in Non-Biological Offspring: Paternal Care and Paternity Confidence among Himba Pastoralists," *Philosophical Transactions of the Royal Society B: Biological Sciences* 287, no. 1922 (2020), https://doi.org /10.1098/rspb.2019.2890.

62. Richard Wike, "French More Accepting of Infidelity Than People in Other Countries," Pew Research Center, January 14, 2014, https://www.pewresearch.org /short-reads/2014/01/14/french-more-accepting-of-infidelity-than-people-in-other -countries/.

63. J. Henrich, "Cultural Group Selection, Coevolutionary Processes and Large-Scale Cooperation," *Journal of Economic Behavior and Organization* 53, no. 1 (2004), https://doi.org/10.1016/S0167-2681(03)00094-5; Samuel Bowles and Herbert Gintis, *A Cooperative Species: Human Reciprocity and Its Evolution* (Princeton, NJ: Princeton University Press, 2011).

64. E. Fehr and U. Fischbacher, "The Nature of Human Altruism," *Nature* 425, no. 6960 (2003), https://doi.org/10.1038/nature02043.

65. J. Moll et al., "Opinion: The Neural Basis of Human Moral Cognition," *Nature Reviews Neuroscience* 6, no. 10 (2005), https://doi.org/10.1038/nrn1768; U. Wagner et al., "Guilt-Specific Processing in the Prefrontal Cortex," *Cerebral Cortex* 21, no. 11

(2011), https://doi.org/10.1093/cercor/bhr016; R. Zhu et al., "Differentiating Guilt and Shame in an Interpersonal Context with Univariate Activation and Multivariate Pattern Analyses," *NeuroImage* 186 (2019), https://doi.org/10.1016/j.neuroimage .2018.11.012; F. A. Jonker et al., "The Role of the Orbitofrontal Cortex in Cognition and Behavior," *Reviews in the Neurosciences* 26, no. 1 (2015), https://doi.org/10.1515 /revneuro-2014-0043.

66. Antonio R. Damasio, *Descartes' Error: Emotion, Reason, and the Human Brain* (New York: Putnam, 1994).

67. Jonker et al., "The Role of the Orbitofrontal Cortex."

68. C. J. Donahue et al., "Quantitative Assessment of Prefrontal Cortex in Humans Relative to Nonhuman Primates," *Proceedings of the National Academy of Sciences* 115, no. 22 (2018), https://doi.org/10.1073/pnas.1721653115.

69. J. K. Rilling and T. R. Insel, "The Primate Neocortex in Comparative Perspective Using Magnetic Resonance Imaging," *Journal of Human Evolution* 37, no. 2 (1999), https://doi.org/10.1006/jhev.1999.0313; K. Zilles et al., "The Human Pattern of Gyrification in the Cerebral Cortex," *Anatomy and Embryology* 179, no. 2 (1988), https:// doi.org/10.1007/BF00304699.

70. Richard E. Passingham, *What Is Special about the Human Brain* (Oxford: Oxford University Press, 2008).

71. R. Boyd, P. J. Richerson, and J. Henrich, "The Cultural Niche: Why Social Learning Is Essential for Human Adaptation," *Proc Natl Acad Sci U S A* 108 Suppl 2 (2011): 10918–10925, https://doi.org/10.1073/pnas.1100290108; C. Tennie, J. Call, and M. Tomasello, "Ratcheting Up the Ratchet: On the Evolution of Cumulative Culture," Comparative Study, *Philosophical Transactions of the Royal Society B: Biological Sciences* 364, no. 1528 (2009), https://doi.org/10.1098/rstb.2009.0052.

72. K. Abernathy, L. J. Chandler, and J. J. Woodward, "Alcohol and the Prefrontal Cortex," *International Review of Neurobiology* 91 (2010), https://doi.org/10.1016 /S0074-7742(10)91009-X.

73. M. H. Teicher et al., "The Effects of Childhood Maltreatment on Brain Structure, Function and Connectivity," *Nature Reviews Neuroscience* 17, no. 10 (2016), https:// doi.org/10.1038/nrn.2016.111.

74. J. Henrich, S. J. Heine, and A. Norenzayan, "The Weirdest People in the World?," *Behavioral and Brain Sciences* 33, no. 2–3 (2010), https://doi.org/10.1017 /S0140525X0999152X; J. Henrich, S. J. Heine, and A. Norenzayan, "Most People Are Not WEIRD," *Nature* 466, no. 7302 (2010), https://doi.org/10.1038/466029a.

75. J. P. Henrich, *The WEIRDest People in the World: How the West Became Psychologically Peculiar and Particularly Prosperous* (New York: Farrar, Straus and Giroux, 2020).

76. Mwenda Ntarangwi, David Mills, and Mustafa H. M. Babiker, *African Anthropologies: History, Critique, And Practice* (London: Zed Books, 2006).

77. H. Miner, "Body Ritual among the Nacirema," *American Anthropologist* 58, no. 3 (1956), https://doi.org/10.1525/aa.1956.58.3.02a00080.

78. F. W. Marlowe, "Paternal Investment and the Human Mating System," *Behavioural Processes* 51, nos. 1–3 (2000): 45–61, https://doi.org/10.1016/S0376-6357 (00)00118-2.

79. A. N. Crittenden and F. W. Marlowe, "Allomaternal Care among the Hadza of Tanzania," *Human Nature* 19, no. 3 (2008), https://doi.org/10.1007/s12110-008 -9043-3.

80. Hewlett, *Intimate Fathers*.

81. Hewlett, "Demography and Child-Care in Preindustrial Societies."

82. Crittenden and Marlowe, "Allomaternal Care among the Hadza."

83. Barry S. Hewlett et al., "Intimate Living: Sharing Space among Aka and Other Hunter-Gatherers," in *Towards a Broader View of Hunter-Gatherer Sharing*, ed. Noa Lavi and David E. Friesem (Cambridge: McDonald Institute for Archaeological Research, 2019).

84. Gray and Anderson, *Fatherhood*.

85. Whiting and Whiting, "Aloofness and Intimacy."

86. M. M. Katz and M. J. Konner, "The Role of the Father: An Anthropological Perspective," in *The Role of the Father in Child Development*, ed. M. E. Lamb (New York: Wiley, 1981).

87. B. S. Hewlett and S. J. MacFarlan, "Fathers' Roles in Hunter-Gatherer and Other Small-Scale Cultures," in *The Role of the Father in Child Development*, edited by M. E. Lamb, 413–434 (Hoboken, NJ: John Wiley and Sons, 2010); S. Harkness and C. M. Super, "The Cultural Foundations of Fathers' Roles: Evidence from Kenya and the United States," in *Father–Child Relations: Cultural and Biosocial Contexts* ed. B. S. Hewlett (New York: Aldine de Gruyter, 1992).

88. Katz and Konner, "The Role of the Father."

89. Whiting and Whiting, "Aloofness and Intimacy."

90. B. S. Hewlett, "Culture, History, and Sex: Anthropological Contributions to Conceptualizing Father Involvement," *Marriage and Family Review* 29, no. 2–3 (2000), https://doi.org/10.1300/J002v29n02_05.

91. C. L. Meehan, "The Effects of Residential Locality on Parental and Alloparental Investment among the Aka Foragers of the Central African Republic," *Human Nature* 16, no. 1 (2005): 58–80, https://doi.org/10.1007/s12110-005-1007-2.

92. H. N. Fouts, "Father Involvement with Young Children among the Aka and Bofi Foragers," *Cross-Cultural Research* 42, no. 3 (2008): 290–312, https://doi.org/10.1177/1069397108317484.

93. J. Winking, M. Gurven, H. Kaplan, and J. Stieglitz, "The Goals of Direct Paternal Care among a South Amerindian Population," *Am J Phys Anthropol* 139, no. 3 (2009): 295–304, https://doi.org/10.1002/ajpa.20981.

94. Stephen Beckerman and Paul Valentine, eds., *Cultures of Multiple Fathers: The Theory and Practice of Partible Paternity in Lowland South America* (Gainesville: University of Florida Press, 2002); Sarah B. Hrdy, "The Optimal Number of Fathers: Evolution, Demography and History in the Shaping of Female Mate Preferences," in *Evolutionary Psychology: Alternative Approaches*, ed. Steven J. Scher and Frederick Rauscher (Norwell, MA: Kluwer, 2003).

95. Ryan Ellsworth, "Book Review: The Human That Never Evolved," *Evolutionary Psychology* 9, no. 3 (2011): 325–335, https://doi.org/10.1177/147470491100900305.

96. Hewlett, "Fathers' Roles in Hunter-Gatherer and Other Small-Scale Cultures."

97. X. Li, "Fathers' Involvement in Chinese Societies: Increasing Presence, Uneven Progress," *Child Development Perspectives* 14, no. 3 (2020), https://doi.org/10.1111/cdep.12375.

98. Nandita Chaundhary, "The Father's Role in the Indian Family: A Story That Must Be Told," in *Fathers in Cultural Context*, ed. David W. Shwalb, Barbara J. Shwalb, and Michael E. Lamb (New York: Routledge, 2013).

99. Ana Cecilia de Sousa Bastos et al., "Fathering in Brazil: A Diverse and Unknown Reality," in *Fathers in Cultural Context*, ed. David W. Shwalb, Barbara J. Shwalb, and Michael E. Lamb (New York: Routledge, 2013).

100. Gretchen Livingston and Kim Parker, "8 Facts about American Dads," Pew Research Center, June 12, 2019, https://www.pewresearch.org/short-reads/2019/06/12/fathers-day-facts/.

101. K. E. Cherry and E. D. Gerstein, "Fathering and Masculine Norms: Implications for the Socialization of Children's Emotion Regulation," *Journal of Family Theory and Review* 13, no. 2 (2021), https://doi.org/10.1111/jftr.1241; K. Elliott, "Caring Masculinities: Theorizing an Emerging Concept," *Men and Masculinities* 19, no. 3 (2016), https://doi.org/10.1177/1097184x15576203; J. H. Pleck, "Fatherhood and Masculinity," in *The Role of the Father in Child Development*, ed. Michael Lamb (Hoboken, NJ: Wiley, 2010).

102. Li, "Fathers' Involvement in Chinese Societies"; Chaundhary, "The Father's Role in the Indian Family"; Ramadan A. Ahmed, "The Father's Role in the Arab World: Cultural Perspectives," in *Fathers in Cultural Context*, ed. David W. Shwalb, Barbara J. Shwalb, and Michael E. Lamb (New York: Routledge, 2013); Marcia C. Inhorn, Wendy Chavkin, and Jose-Alberto Navarro, eds., *Globalized Fatherhood* (New York: Berghahn, 2015); David W. Shwalb, Barbara J. Shwalb, and Michael E. Lamb, eds., *Fathers in Cultural Context* (New York: Routledge, 2013).

103. Xuan Li and Michael Lamb, "Fathers in Chinese Culture," in *Fathers in Cultural Context*, ed. David W. Shwalb, Barbara J. Shwalb, and Michael E. Lamb (New York: Routledge, 2013).

104. Ahmed, "The Father's Role in the Arab World."

105. S. Madhavan, N. W. Townsend, and A. I. Garey, "'Absent Breadwinners': Father–Child Connections and Paternal Support in Rural South Africa," *Journal of Southern African Studies* 34 (2008): 647–663.

106. Jennifer Ultrata, Jean M. Ispa, and Simone Ispa-Landa, "Men on The Margins of Family Life: Fathers in Russia," in *Fathers in Cultural Context*, ed. David W. Shwalb, Barbara J. Shwalb, and Michael E. Lamb (New York: Routledge, 2013).

107. E. Brainerd, "Mortality in Russia since the Fall of the Soviet Union," *Comparative Economic Studies* 63, no. 4 (2021), https://doi.org/10.1057/s41294-021-00169-w.

108. Lisa Gulya, "Adolescent Pregnancy in Russia," in *International Handbook of Adolescent Pregnancy*, ed. A. Cherry and M. Dillon (Berlin: Springer, 2014); Brainerd, "Mortality in Russia"; United Nations Population Division, "Adolescent Fertility Rate (Births per 1,000 Women Ages 15–19)—Russian Federation," World Bank, 2024, https://data.worldbank.org/indicator/SP.ADO.TFRT?locations=RU.

109. Juliana Menasce Horowitz, "Despite Challenges at Home and Work, Most Working Moms and Dads Say Being Employed Is What's Best for Them," Pew Research Center, September 12, 2019, https://www.pewresearch.org/short-reads/2019/09/12/despite-challenges-at-home-and-work-most-working-moms-and-dads-say-being-employed-is-whats-best-for-them/.

110. Karen E. McFadden and Catherine S. Tamis-LeMonda, "Fathers in the U.S.," in *Fathers in Cultural Context*, ed. David W. Shwalb, Barbara J. Shwalb, and Michael E. Lamb (New York: Routledge, 2013).

111. Gretchen Livingston and Kim Parker, "8 Facts about American Dads"; Kerstin Aumann, Ellen Galinsky, and Kenneth Matos, *The New Male Mystique* (Families and Work Institute, 2011), http://familiesandwork.org.

112. Bruce M. Smyth et al., "Fathers in Australia: A Contemporary Snapshot," in *Fathers in Cultural Context*, ed. David W. Shwalb, Barbara J. Shwalb, and Michael E. Lamb (New York: Routledge, 2013).

113. W. Van den Berg and T. Makusha, "State of South Africa's Fathers 2018" (Cape Town: Sonke Gender Justice & Human Sciences Research Council, 2018).

114. M. E. Connor and J. L. White, "Fatherhood in Contemporary Black America: An Invisible Presence," *Black Scholar* 37, no. 2 (2007), https://doi.org/10.1080/00064246.2007.11413389.

115. Stephanie Kramer, "U.S. Has World's Highest Rate of Children Living in Single-Parent Households," Pew Research Center, December 12, 2019, https://www.pewresearch.org/short-reads/2019/12/12/u-s-children-more-likely-than-children-in-other-countries-to-live-with-just-one-parent/.

116. Li, "Fathers' Involvement in Chinese Societies"; Li and Lamb, "Fathers in Chinese Culture"; Chaundhary, "The Father's Role in the Indian Family"; Bastos et al., "Fathering in Brazil"; Madhavan et al., "'Absent Breadwinners'"; Nicholas W. Townsend, "The Complications of Fathering in Southern Africa: Separation, Uncertainty, and Multiple Responsibilities," in *Fathers in Cultural Context*, ed. David W. Shwalb, Barbara J. Shwalb, and Michael E. Lamb (New York: Routledge, 2013); Ultrata et al., "Men on the Margins"; Brainerd, "Mortality in Russia"; Gulya, "Adolescent Pregnancy in Russia"; Ahmed, "The Father's Role in the Arab World"; Smyth et al., "Fathers in Australia;" McFadden and Tamis-LeMonda, "Fathers in the U.S."

117. E. K. Holmes et al., "Do Responsible Fatherhood Programs Work? A Comprehensive Meta-AnalyticStudy," *Family Relations* 69, no. 5 (2020), https://doi.org/10.1111/fare.12435; J. B. Henry et al., "Fatherhood Matters: An Integrative Review of Fatherhood Intervention Research," *Journal of School Nursing* 36, no. 1 (2020), https://doi.org/10.1177/1059840519873380.

118. M. H. van IJzendoorn et al., "Improving Parenting, Child Attachment, and Externalizing Behaviors: Meta-Analysis of the First 25 Randomized Controlled Trials on the Effects of Video-Feedback Intervention to Promote Positive Parenting and Sensitive Discipline," *Development and Psychopathology* (2022), https://doi.org/10.1017/S0954579421001462.

119. C. Panter-Brick et al., "Practitioner Review: Engaging Fathers—Recommendations for a Game Change in Parenting Interventions Based on a Systematic Review of the Global Evidence," *Journal of Child Psychology and Psychiatry* 55, no. 11 (2014), https://doi.org/10.1111/jcpp.12280.

120. Sophie S. Havighurst et al., "Dads Tuning into Kids: A Randomized Controlled Trial of an Emotion Socialization Program for Fathers," *Social Development* 28 (2019), https://doi.org/10.1111/sode.12375.

121. A. Chacko et al., "Engaging Fathers in Effective Parenting for Preschool Children Using Shared Book Reading: A Randomized Controlled Trial," *Journal of Clinical Child and Adolescent Psychology* 47, no. 1 (2018), https://doi.org/10.1080/15374416.2016.1266648.

122. C. Hechler et al., "Prenatal Predictors of Postnatal Quality of Caregiving Behavior in Mothers and Fathers," *Parenting-Science and Practice* 19, no. 1–2 (2019), https://doi.org/10.1080/15295192.2019.1556010.

123. W. J. Doherty, M. F. Erickson, and R. LaRossa, "An Intervention to Increase Father Involvement and Skills with Infants during the Transition to Parenthood," *Journal of Family Psychology* 20, no. 3 (2006), https://doi.org/10.1037/0893-3200.20.3.438.

124. K. Alyousefi-van Dijk et al., "Development and Feasibility of the Prenatal Video-Feedback Intervention to Promote Positive Parenting for Expectant Fathers," *Journal of Reproductive and Infant Psychology* (2021), https://doi.org/10.1080/02646838.2021.188625.

125. R. S. M. Buisman et al., "Fathers' Sensitive Parenting Enhanced by Prenatal Video-Feedback: A Randomized Controlled Trial Using Ultrasound Imaging," *Pediatric Research* 93, no. 4 (2022), https://doi.org/10.1038/s41390-022-02183-9.

Chapter 7

1. K. Modig et al., "Payback Time? Influence of Having Children on Mortality in Old Age," *Journal of Epidemiology and Community Health* 71, no. 5 (2017), https://doi.org/10.1136/jech-2016-207857.

2. K. Ning et al., "Parity Is Associated with Cognitive Function and Brain Age in Both Females and Males," *Scientific Reports* 10, no. 1 (2020), https://doi.org/10.1038/s41598-020-63014-7.

3. M. Brandel, E. Melchiorri, and C. Ruini, "The Dynamics of Eudaimonic Well-Being in the Transition to Parenthood: Differences between Fathers and Mothers," *Journal of Family Issues* 39, no. 9 (2018), https://doi.org/10.1177/0192513x18758344.

4. A. P. Wingo et al., "Purpose in Life Is a Robust Protective Factor of Reported Cognitive Decline among Late Middle-Aged Adults: The Emory Healthy Aging Study," *Journal of Affective Disorders* 263 (2020), https://doi.org/10.1016/j.jad.2019.11.124.

5. J. K. Rilling and C. Hadley, "A Mixed Methods Study of the Challenges and Rewards of Fatherhood in a Diverse Sample of U.S. Fathers," *Sage Open* 13, no. 3 (2023), https://doi.org/10.1177/21582440231193939.

Bibliography

Abernathy, K., L. J. Chandler, and J. J. Woodward. "Alcohol and the Prefrontal Cortex." *International Review of Neurobiology* 91 (2010), 289–320. https://doi.org/10.1016/S0074-7742(10)91009-X.

Abraham, E., T. Hendler, I. Shapira-Lichter, Y. Kanat-Maymon, O. Zagoory-Sharon, and R. Feldman. "Father's Brain Is Sensitive to Childcare Experiences." *Proceedings of the National Academy of Sciences* 111, no. 27 (2014), 9792–9777. https://doi.org/10.1073/pnas.1402569111.

Abraham, E., G. Raz, O. Zagoory-Sharon, and R. Feldman. "Empathy Networks in the Parental Brain and Their Long-Term Effects on Children's Stress Reactivity and Behavior Adaptation." *Neuropsychologia* 16 (2018), 75–85. https://doi.org/10.1016/j.neuropsychologia.2017.04.015.

Abu-Akel, A., S. Palgi, E. Klein, J. Decety, and S. Shamay-Tsoory. "Oxytocin Increases Empathy to Pain When Adopting the Other—But Not the Self-Perspective." *Society for Neuroscience* 10, no. 1 (2015), 7–15. https://doi.org/10.1080/17470919.2014.948637.

Agrawal, J., B. Ludwig, B. Roy, and Y. Dwivedi. "Chronic Testosterone Increases Impulsivity and Influences the Transcriptional Activity of the Alpha-2a Adrenergic Receptor Signaling Pathway in Rat Brain." *Molecular Neurobiology* 56, no. 6 (2019), 4061–4071. https://doi.org/10.1007/s12035-018-1350-z.

Ahmed, Ramadan A. "The Father's Role in the Arab World: Cultural Perspectives." In *Fathers in Cultural Context*, edited by David W. Shwalb, Barbara J. Shwalb, and Michael E. Lamb, 122–147. New York: Routledge, 2013.

Ahnert, L., F. Deichmann, M. Bauer, B. Supper, and B. Piskernik. "Fathering Behavior, Attachment, and Engagement in Childcare Predict Testosterone and Cortisol." *Developmental Psychobiology* 63, no. 6 (2021). https://doi.org/10.1002/dev.22149.

Aiello, L. C., and P. Wheeler. "The Expensive-Tissue Hypothesis: The Brain and the Digestive-System in Human and Primate Evolution." *Current Anthropology* 36, no. 2 (1995), 199–221. https://doi.org/.

Ainsworth, M. D. S., S. M. Bell, and D. J. Stayton. "Infant-Mother Attachment and Social Development." In *The Introduction of the Child into a Social World*, edited by M. P. Richards, 99–135. Cambridge: Cambridge University Press, 1974.

Alexander, R. D., and K. M. Nonnan. "Concealment of Ovulation, Parental Care and Human Social Evolution." In *Evolutionary Biology and Human Social Behavior: An Anthropological Perspective*, edited by N. A. Chagnon and W. G. Irons, 436–453. North Scituate, MA: Duxbury Press, 1979.

Alger, I., P. L. Hooper, D. Cox, J. Stieglitz, and H. S. Kaplan. "Paternal Provisioning Results from Ecological Change." *Proceedings of the National Academy of Sciences* 117, no. 20 (2020), 10746–10754. https://doi.org/10.1073/pnas.1917166117.

Alonso-Alvarez, C., S. Bertrand, B. Faivre, O. Chastel, and G. Sorci. "Testosterone and Oxidative Stress: The Oxidation Handicap Hypothesis." *Philosophical Transactions of the Royal Society B: Biological Sciences* 274, no. 1611 (2007), 819–825. https://doi.org/10.1098/rspb.2006.3764.

Aluja, A., L. F. Garcia, M. Marti-Guiu, E. Blanco, O. Garcia, J. Fibla, and A. Blanch. "Interactions among Impulsiveness, Testosterone, Sex Hormone Binding Globulin and Androgen Receptor Gene Cag Repeat Length." *Physiology and Behavior* 147 (2015), 91–96. https://doi.org/10.1016/j.physbeh.2015.04.022.

Alvarado, L. C., M. N. Muller, M. E. Thompson, M. Klimek, I. Nenko, and G. Jasienska. "The Paternal Provisioning Hypothesis: Effects of Workload and Testosterone Production on Men's Musculature." *American Journal of Physical Anthropology* 158, no. 1 (2015), 19–35. https://doi.org/10.1002/ajpa.22771.

Alvergne, A., C. Faurie, and M. Raymond. "Variation in Testosterone Levels and Male Reproductive Effort: Insight from a Polygynous Human Population." *Hormones and Behavior* 56, no. 5 (2009), 491–497. https://doi.org/10.1016/j.yhbeh.2009.07.013.

Alyousefi-van Dijk, K., N. de Waal, IJzendoorn M. H. van, and M. J. Bakermans-Kranenburg. "Development and Feasibility of the Prenatal Video-Feedback Intervention to Promote Positive Parenting for Expectant Fathers." *Journal of Reproductive and Infant Psychology*(2021), 1–14. https://doi.org/10.1080/02646838.2021.1886258.

Amadei, E. A., Z. V. Johnson, Y. J. Kwon, A. C. Shpiner, V. Saravanan, W. D. Mays, S. J. Ryan, et al. "Dynamic Corticostriatal Activity Biases Social Bonding in Monogamous Female Prairie Voles." *Nature* 546, no. 7657 (2017), 297–301. https://doi.org/10.1038/nature22381.

Amato, P. R., and J. G. Gilbreth. "Nonresident Fathers and Children's Well-Being: A Meta-Analysis." *Journal of Marriage and the Family* 61, no. 3 (1999), 557–573. https://doi.org/10.2307/353560.

Anderson, K. G., H. Kaplan, and J. Lancaster. "Paternal Care by Genetic Fathers and Stepfathers I: Reports from Albuquerque Men." *Evolution and Human Behavior* 20, no. 6 (1999), 405–431. https://doi.org/10.1016/S1090-5138(99)00023-9.

Anderson, K. G., H. Kaplan, and J. B. Lancaster. "Confidence of Paternity, Divorce, and Investment in Children by Albuquerque Men." *Evolution and Human Behavior* 28, no. 1 (2007), 1–10. https://doi.org/10.1016/j.evolhumbehav.2006.06.004.

Apfelbeck, B., H. Flinks, and W. Goymann. "Variation in Circulating Testosterone during Mating Predicts Reproductive Success in a Wild Songbird." *Frontiers in Ecology and Evolution* 4 (2016), https://doi.org/10.3389/fevo.2016.00107.Apicella, C. L., and F. W. Marchlowe. "Men's Reproductive Investment Decisions: Mating, Parenting, and Self-Perceived Mate Value." *Human Nature* 18, no. 1 (2007), 22–34. https://doi.org/10.1007/BF02820844.

Apter-Levi, Y., O. Zagoory-Sharon, and R. Feldman. "Oxytocin and Vasopressin Support Distinct Configurations of Social Synchrony." *Brain Research* 1580 (2014), 124–132. https://doi.org/10.1016/j.brainres.2013.10.052.

Archer, J. "Testosterone and Human Aggression: An Evaluation of the Challenge Hypothesis." *Neuroscience and Biobehavior Reviews* 30, no. 3 (2006), 319–445. https://doi.org/10.1016/j.neubiorev.2004.12.007.

Atzil, S., T. Hendler, and R. Feldman. "Specifying the Neurobiological Basis of Human Attachment: Brain, Hormones, and Behavior in Synchronous and Intrusive Mothers." *Neuropsychopharmacology* 36, no. 13 (2011), 2603–2615. https://doi.org/10.1038/npp.2011.172

Atzil, S., A. Touroutoglou, T. Rudy, S. Salcedo, R. Feldman, J. M. Hooker, B. C. Dickerson, C. Catana, and L. F. Barrett. "Dopamine in the Medial Amygdala Network Mediates Human Bonding." *Proceedings of the National Academy of Sciences* 114, no. 9 (2017, 2361–2366.:https://doi.org/10.1073/pnas.1612233114.

Aumann, Kerstin, Ellen Galinsky, and Kenneth Matos. *The New Male Mystique*. Hillsborough, NJ: Families and Work Institute, 2011.

Bakermans-Kranenburg, M. J., and M. H. van Ijzendoorn. "Oxytocin Receptor (OXTR) and Serotonin Transporter (5-HTT) Genes Associated with Observed Parenting." *Social Cognitive and Affective Neuroscience* 3, no. 2 (2008), 128–134. https://doi.org/10.1093/scan/nsn004.

Bales, K., J. Dietz, A. Baker, K. Miller, and S. D. Tardif. "Effects of Allocare-Givers on Fitness of Infants and Parents in Callitrichid Primates." *Folia Primatologica* 71, no. 1–2 (2000), 27–38. https://doi.org/10.1159/000021728.

Ball, G. F., L. V. Riters, and J. Balthazart. "Neuroendocrinology of Song Behavior and Avian Brain Plasticity: Multiple Sites of Action of Sex Steroid Hormones." *Frontiers in Neuroendocrinology* 23, no. 2 (2002), https://doi.org/10.1006/frne.2002.0230.

Balshine, Sigal. "Patterns of Parental Care in Vertebrates." In *The Evolution of Parental Care*, edited by Nick J. Royle, Per T. Smiseth, and Mathias Kölliker. New York: Oxford University Press, 2012.

Bar-On, Y. M., R. Phillips, and R. Milo. "The Biomass Distribution on Earth." *Proceedings of the National Academy of Sciences* 115, no. 25 (2018), 6506–6511. https://doi.org/10.1073/pnas.1711842115.

Barber, N. "Single Parenthood as a Predictor of Cross-National Variation in Violent Crime." *Cross-Cultural Research* 38, no. 4 (2004), 343–358. https://doi.org/10.1177/1069397104267479.

Barr, R. G. "Preventing Abusive Head Trauma Resulting from a Failure of Normal Interaction between Infants and Their Caregivers." *Proceedings of the National Academy of Sciences* 109, suppl. 2 (2012), 17294–17301. https://doi.org/10.1073/pnas.1121267109.

Bartz, J. A., J. Zaki, N. Bolger, and K. N. Ochsner. "Social Effects of Oxytocin in Humans: Context and Person Matter." *Trends in Cognitive Sciences* 15, no. 7 (2011), 301–309. https://doi.org/10.1016/j.tics.2011.05.002.

Bastos, Ana Cecilia de Sousa, Vivian Volkmer-Pontes, Pedro Gomes Brasileiro, and Helena Marchtinelli Serra. "Fathering in Brazil: A Diverse and Unknown Reality." In *Fathers in Cultural Context*, edited by David W. Shwalb, Barbara J. Shwalb, and Michael E. Lamb, 228–249. New York: Routledge, 2013.

Baumrind, D. "Authoritarian vs. Authoritative Parental Control." *Adolescence* 3, no. 11 (1968), 255–722.

Becker, M., and V. Hesse. "Minipuberty: Why Does It Happen?" *Hormone Research in Paediatrics* 93, no. 2 (2020), 76–84. https://doi.org/10.1159/000508329.

Beckerman, Stephen, and Paul Valentine, eds. *Cultures of Multiple Fathers: The Theory and Practice of Partible Paternity in Lowland South America*. Gainesville: University of Florida Press, 2002.

Belsky, J., and M. H. van IJzendoorn. "Genetic Differential Susceptibility to the Effects of Parenting." *Current Opinion in Psychology* 15 (2017), 125–130. https://doi.org/10.1016/j.copsyc.2017.02.021.

Berg, S. J., and K. E. Wynne-Edwards. "Changes in Testosterone, Cortisol, and Estradiol Levels in Men Becoming Fathers." *Mayo Clinic Proceedings* 76, no. 6 (2001), 582–592. https://doi.org/10.4065/76.6.582.

Bernard, K., G. Nissim, S. Vaccaro, J. L. Harris, and O. Lindhiem. "Association between Maternal Depression and Maternal Sensitivity from Birth to 12 Months: A Meta-Analysis." *Attachment and Human Development* 20, no. 6 (2018), 578–599. https://doi.org/10.1080/14616734.2018.1430839.

Bishop, Joseph Bucklin. *Theodore Roosevelt and His Time Shown in His Own Letters.* 2 vols. New York: Scribner's, 1920.

Bjornebekk, A., T. Kaufmann, L. E. Hauger, S. Klonteig, I. R. Hullstein, and L. T. Westlye. "Long-Term Anabolic-Androgenic Steroid Use Is Associated with Deviant Brain Aging." *Biological Psychiatry: Cognitive Neuroscience and Neuroimaging* 6, no. 5 (2021), 579–589. https://doi.org/10.1016/j.bpsc.2021.01.001.

Blom, N., and B. Hewitt. "Becoming a Female-Breadwinner Household in Australia: Changes in Relationship Satisfaction." *Journal of Marriage and Family* 82, no. 4 (2020), 1340–1357. https://doi.org/10.1111/jomf.12653.

Blount, B. G. "Issues in Bonobo (*Pan paniscus*) Sexual-Behavior." *American Anthropologist* 92, no. 3 (1990), 702–514. https://doi.org/10.1525/aa.1990.92.3.02a00100.

Boesch, C., C. Crockford, I. Herbinger, R. Wittig, Y. Moebius, and E. Normand. "Intergroup Conflicts among Chimpanzees in Tai National Park: Lethal Violence and the Female Perspective." *American Journal of Primatology* 70, no. 6 (2008), 519–532. https://doi.org/10.1002/ajp.20524.

Booth, A., and J. M. Dabbs. "Testosterone and Men's Marriages." *Social Forces* 72, no. 2 (1993), 463–477. https://doi.org/10.2307/2579857.

Borland, J. M., J. K. Rilling, K. J. Frantz, and H. E. Albers. "Sex-Dependent Regulation of Social Reward by Oxytocin: An Inverted U Hypothesis." *Neuropsychopharmacology* 44, no. 1 (2019), 97–110. https://doi.org/10.1038/s41386-018-0129-2.

Bornstein, M. H., D. L. Putnick, P. Rigo, G. Esposito, J. E. Swain, J. T. D. Suwalsky, X. Su, et al. "Neurobiology of Culturally Common Maternal Responses to Infant Cry." *Proceedings of the National Academy of Sciences* 114, no. 45 (2017), E9465–E9473. https://doi.org/10.1073/pnas.1712022114.

Borries, C., K. Launhardt, C. Epplen, J. T. Epplen, and P. Winkler. "Males as Infant Protectors in Hanuman Langurs (*Presbytis entellus*) Living in Multimale Groups— Defence Pattern, Paternity and Sexual Behaviour." *Behavioral Ecology and Sociobiology* 46, no. 5 (1999): 350–56. https://doi.org/10.1007/s002650050629.

Bosch, O. J. "Maternal Aggression in Rodents: Brain Oxytocin and Vasopressin Mediate Pup Defence." *Philosophical Transactions of the Royal Society B: Biological Sciences* 368, no. 1631 (2013), 20130085. https://doi.org/10.1098/rstb.2013.0085.

Bosch, O. J., and L. J. Young. "Oxytocin and Social Relationships: From Attachment to Bond Disruption." *Current Topics in Behavioral Neuroscience* 35 (2018), 97–117. https://doi.org/10.1007/7854_2017_10.

Bowles, Samuel, and Herbert Gintis. *A Cooperative Species: Human Reciprocity and Its Evolution.* Princeton, NJ: Princeton University Press, 2011.

Boyd, R., P. J. Richerson, and J. Henrich. "The Cultural Niche: Why Social Learning Is Essential for Human Adaptation." 108, suppl. 2 (2011), 10918–1025. https://doi.org/10.1073/pnas.1100290108.

Boyette, A. H., S. Lew-Levy, and L. T. Gettler. "Dimensions of Fatherhood in a Congo Basin Village: A Multimethod Analysis of Intracultural Variation in Men's Parenting and Its Relevance for Child Health." *Current Anthropology* 59, no. 6 (2018), 839–847. https://doi.org/10.1086/700717.

Boyette, A. H., S. Lew-Levy, M. S. Sarma, and L. T. Gettler. "Testosterone, Fathers as Providers and Caregivers, and Child Health: Evidence from Fisher-Farmers in the Republic of the Congo." *Hormones and Behavior* 107 (2019), 35–45. https://doi.org/10.1016/j.yhbeh.2018.09.006.

Bradley, A. J., I. R. McDonald, and A. K. Lee. "Stress and Mortality in a Small Marsupial (*Antechinus stuartii*, Macleay)." *General and Comparative Endocrinology* 40, no. 2 (1980), 188–200. https://doi.org/10.1016/0016–6480(80)90122–7.

Brainerd, E. "Mortality in Russia since the Fall of the Soviet Union." *Comparative Economic Studies* 63, no. 4 (2021), 557–576. https://doi.org/10.1057/s41294-021-00169-w.

Brandel, M., E. Melchiorri, and C. Ruini. "The Dynamics of Eudaimonic Well-Being in the Transition to Parenthood: Differences between Fathers and Mothers." *Journal of Family Issues* 39, no. 9 (2018), 2572–2589. https://doi.org/10.1177/0192513x18758344.

Bribiescas, Richard G. *Men—Evolutionary and Life History.* Cambridge, MA: Harvard University Press, 2006.

Broude, G. J. "Protest Masculinity—a Further Look at the Causes and the Concept." *Ethos* 18, no. 1 (1990), 103–122. https://doi.org/10.1525/eth.1990.18.1.02a00040.

Brown, G. L., S. M. Kogan, and J. Kim. "From Fathers to Sons: The Intergenerational Transmission of Parenting Behavior among African American Young Men." *Family Process* 57, no. 1 (2018), 165–180. https://doi.org/10.1111/famp.12273.

Brown, G. L., S. C. Mangelsdorf, A. Shigeto, and M. S. Wong. "Associations between Father Involvement and Father–Child Attachment Security: Variations Based on Timing and Type of Involvement." *Journal of Family Psychology* 32, no. 8 (2018), 1015–1024. https://doi.org/10.1037/fam0000472.

Brown, G. L., S. J. Schoppe-Sullivan, S. C. Mangelsdorf, and C. Neff. "Observed and Reported Supportive Coparenting as Predictors of Infant-Mother and Infant-Father Attachment Security." *Early Child Development and Care* 180, no. 1–2 (2010), 121–137. https://doi.org/10.1080/03004430903415015.

Brown, Geoffrey L., and Hasan Alp Aytuglu. "Father–Child Attachment Relationships." In *Handbook of Fathers and Child Development*, edited by H. E. Fitzgerald, K. von Klitzing, N. Cabrera, J. Scarano de Mendonça, and Th. Skjøthaug. New York: Springer, 2020.

Brown, R. E. "Social and Hormonal Factors Influencing Infanticide and Its Suppression in Adult Male Long-Evans Rats (*Rattus*)." *Journal of Comparative Psychology* 100, no. 2 (1986), 155–161. https://doi.org/10.1037/0735-7036.100.2.155.

Brown, R. E., T. Murdoch, P. R. Murphy, and W. H. Moger. "Hormonal Responses of Male Gerbils to Stimuli from Their Mate and Pups." *Hormones and Behavior* 29, no. 4 (1995). 474–491. https://doi.org/10.1006/hbeh.1995.1275.

Brummelte, S., and L. A. Galea. "Postpartum Depression: Etiology, Treatment and Consequences for Maternal Care." *Hormones and Behavior* 77 (January 2016), 153–166. https://doi.org/10.1016/j.yhbeh.2015.08.008.

Buchan, J. C., S. C. Alberts, J. B. Silk, and J. Altmann. "True Paternal Care in a Multi-Male Primate Society." *Nature* 425, no. 6954 (2003), 179–181.

Buisman, R. S. M., K. Alyousefi-van Dijk, N. de Waal, A. R. Kesarlal, M. W. F. T. Verhees, M. H. van IJzendoorn, and M. J. Bakermans-Kranenburg. "Fathers' Sensitive Parenting Enhanced by Prenatal Video-Feedback: A Randomized Controlled Trial Using Ultrasound Imaging." *Pediatric Research* 93, no. 4 (2022), 1024–1030. https://doi.org/10.1038/s41390-022-02183-9.

Bullock, Alan. *Hitler and Stalin: Parallel Lives*. New York: Knopf, 1992.

Burgin, C. J., J. P. Colella, P. L. Kahn, and N. S. Upham. "How Many Species of Mammals Are There?" *Journal of Mammalogy* 99, no. 1 (2018), 1–14. https://doi.org/10.1093/jmammal/gyx147.

Cabrera, N. J., C. S. Tamisk-LeMonda, R. H. Bradley, S. Hofferth, and M. E. Lamb. "Fatherhood in the Twenty-First Century." *Child Development* 71, no. 1 (2000), 127–136. https://doi.org/10.1111/1467-8624.00126.

Campbell, B. C., P. B. Gray, D. T. Eisenberg, P. Ellison, and M. D. Sorenson. "Androgen Receptor CAG Repeats and Body Composition among Ariaal Men." *International Journal of Andrology* 32, no. 2 (2009), 140–148. https://doi.org/10.1111/j.1365-2605.2007.00825.x.

Cao, Y., R. Wu, F. Tai, X. Zhang, P. Yu, X. An, X. Qiao, and P. Hao. "Neonatal Paternal Deprivation Impairs Social Recognition and Alters Levels of Oxytocin and

Estrogen Receptor Alpha mRNA Expression in the MeA and NAcc, and Serum Oxytocin in Mandarin Voles." *Hormones and Behavior* 65, no. 1 (2014), 57–65. https://doi .org/10.1016/j.yhbeh.2013.11.005.

Carmichael, M. S., R. Humbert, J. Dixen, G. Palmisano, W. Greenleaf, and J. M. Davidson. "Plasma Oxytocin Increases in the Human Sexual-Response." *Journal of Clinical Endocrinology and Metabolism* 64, no. 1 (1987), 27–31. https://doi.org/10 .1210/jcem-64-1-27.

Carson, D. S., S. W. Berquist, T. H. Trujillo, J. P. Garner, S. L. Hannah, S. A. Hyde, R. D. Sumiyoshi, et al. "Cerebrospinal Fluid and Plasma Oxytocin Concentrations Are Positively Correlated and Negatively Predict Anxiety in Children." *Molecular Psychiatry* 20 (2014). https://doi.org/10.1038/mp.2014.132.

Casto, K. V., and D. A. Edwards. "Testosterone, Cortisol, and Human Competition." *Hormones and Behavior* 82 (2016), 21–37. https://doi.org/10.1016/j.yhbeh.2016.04 .004.

Chacko, A., G. A. Fabiano, G. L. Doctoroff, and B. Fortson. "Engaging Fathers in Effective Parenting for Preschool Children Using Shared Book Reading: A Randomized Controlled Trial." *Journal of Clinical Child and Adolescent Psychology* 47, no. 1 (2018), 79–93. https://doi.org/10.1080/15374416.2016.1266648.

Champagne, F., J. Diorio, S. Sharma, and M. J. Meaney. "Naturally Occurring Variations in Maternal Behavior in the Rat Are Associated with Differences in Estrogen-Inducible Central Oxytocin Receptors." *Proceedings of the National Açademy of Sciences* 98, no. 22 (2001), 12736–12741. https://doi.org/10.1073/pnas.221224598.

Chang, Jung, and Jon Halliday. *Mao: The Unknown Story*. New York: Anchor Books, 2006.

Chaundhary, Nandita. "The Father's Role in the Indian Family: A Story That Must Be Told." In *Fathers in Cultural Context*, edited by David W. Shwalb, Barbara J. Shwalb, and Michael E. Lamb, 68–94. New York: Routledge, 2013.

Chen, Z. Q., R. T. Corlett, X. G. Jiao, S. J. Liu, T. Charles-Dominique, S. C. Zhang, H. Li, et al. "Prolonged Milk Provisioning in a Jumping Spider." *Science* 362, no. 6418 (2018), 1052–1055. https://doi.org/10.1126/science.aat3692.

Cherry, K. E., and E. D. Gerstein. "Fathering and Masculine Norms: Implications for the Socialization of Children's Emotion Regulation." *Journal of Family Theory and Review* 13, no. 2 (2021), 149–163. https://doi.org/10.1111/jftr.12411.

Choi, J. K., and H. S. Pyun. "Nonresident Fathers' Financial Support, Informal Instrumental Support, Mothers' Parenting, and Child Development in Single-Mother Families with Low Income." *Journal of Family Issues* 35, no. 4 (2014), 526–546. https://doi .org/10.1177/0192513x13478403.

Christiansen, K., and E. M. Winkler. "Hormonal, Anthropometrical, and Behavioral-Correlates of Physical Aggression in Kung-San Men of Namibia." *Aggressive Behavior* 18, no. 4 (1992), 271–280. https://doi.org/10.1002/1098-2337(1992)18:4<271::AID-AB2480180403>3.0.CO;2-6.

Chudek, M., and J. Henrich. "Culture-Gene Coevolution, Norm-Psychology and the Emergence of Human Prosociality." *Trends in Cognitive Sciences* 15, no. 5 (2011), 218–226. https://doi.org/10.1016/j.tics.2011.03.003.

Chung, E. O., A. Hagaman, K. LeMasters, N. Andrabi, V. Baranov, L. M. Bates, J. A. Gallis, et al. "The Contribution of Grandmother Involvement to Child Growth and Development: An Observational Study in Rural Pakistan." *BMJ Global Health* 5, no. 8 (2020). https://gh.bmj.com/content/5/8/e002181.

Cieri, R. L., S. E. Churchill, R. G. Franciscus, J. Z. Tan, and B. Hare. "Craniofacial Feminization, Social Tolerance, and the Origins of Behavioral Modernity." *Current Anthropology* 55, no. 4 (2014), 419–443. https://doi.org/10.1086/677209.

Clark, M. M., and B. G. Galef. "A Testosterone-Mediated Trade-Off between Parental and Sexual Effort in Male Mongolian Gerbils (*Meriones unguiculatus*)." *Journal of Comparative Psychology* 113, no. 4 (1999), 388–395. https://doi.org/10.1037/0735-7036.113.4.388.

———. "Why Some Male Mongolian Gerbils May Help at the Nest: Testosterone, Asexuality and Alloparenting." *Animal Behaviour* 59 (2000), 801–806. https://doi.org/10.1006/anbe.1999.1365.

Clutton-Brock, Timothy H. *The Evolution of Parental Care.* Princeton, NJ: Princeton University Press, 1991.

Cohen-Bendahan, C. C., R. Beijers, L. J. van Doornen, and C. de Weerth. "Explicit and Implicit Caregiving Interests in Expectant Fathers: Do Endogenous and Exogenous Oxytocin and Vasopressin Matter?" *Infant Behavior and Development* 41 (2015), 26–37. https://doi.org/10.1016/j.infbeh.2015.06.007.

Cohen, D., R. E. Nisbett, B. F. Bowdle, and N. Schwarz. "Insult, Aggression, and the Southern Culture of Honor: An 'Experimental Ethnography.'" *Journal of Personal and Social Psychology* 70, no. 5 (1996), 945–959. https://doi.org/10.1037//0022-3514.70.5.945.

Coles, R. L. "Single-Father Families: A Review of the Literature." *Journal of Family Theory and Review* 7, no. 2 (2015), 144–166. https://doi.org/10.1111/jftr.12069.

Coley, R. L., and B. L. Medeiros. "Reciprocal Longitudinal Relations between Nonresident Father Involvement and Adolescent Delinquency." *Child Development* 78, no. 1 (2007), 132–147. https://doi.org/10.1111/j.1467–8624.2007.00989.x.

Comhaire, F. "Hormone Replacement Therapy and Longevity." *Andrologia* 48, no. 1 (2016), 65–68. https://doi.org/10.1111/and.12419.

Cong, X., S. M. Ludington-Hoe, N. Hussain, R. M. Cusson, S. Walsh, V. Vazquez, C. E. Briere, and D. Vittner. "Parental Oxytocin Responses during Skin-to-Skin Contact in Pre-Term Infants." *Early Human Development* 91, no. 7 (2015), 401–406. https://doi.org/10.1016/j.earlhumdev.2015.04.012.

Connor, M. E., and J. L. White. "Fatherhood in Contemporary Black America: An Invisible Presence." *Black Scholar* 37, no. 2 (2007), 2–8. https://doi.org/10.1080/00064246.2007.11413389.

Cools, S., J. H. Fiva, and L. J. Kirkeboen. "Causal Effects of Paternity Leave on Children and Parents." *Scandinavian Journal of Economics* 117, no. 3 (2015), 801–828. https://doi.org/10.1111/sjoe.12113.

Cossette, L., A. Pomerleau, G. Malcuit, and J. Kaczorowski. "Emotional Expressions of Female and Male Infants in a Social and a Nonsocial Context." *Sex Roles* 35, no. 11–12 (1996), 693–709. https://doi.org/10.1007/Bf01544087.

Costa, R. M., and S. Brody. "Women's Relationship Quality Is Associated with Specifically Penile-Vaginal Intercourse Orgasm and Frequency." *Journal of Sex and Marital Therapy* 33, no. 4 (2007), 319–327. https://doi.org/10.1080/00926230701385548.

Crittenden, A. N., and F. W. Marlowe. "Allomaternal Care among the Hadza of Tanzania." *Human Nature* 19, no. 3 (2008), 249–262. https://doi.org/10.1007/s12110-008-9043-3.

Daly, Martin, and Margo Wilson. *Homicide*. New York: Aldine de Gruyter, 1988.

Damasio, Antonio R. *Descartes' Error: Emotion, Reason, and the Human Brain*. New York: Putnam, 1994.

Davies, Nicholas B., John R. Krebs, and Stuart A. West. *An Introduction to Behavioural Ecology* 4th ed. West Sussex, UK: Wiley-Blackwell, 2012.

De Dreu, C. K., L. L. Greer, M. J. Handgraaf, S. Shalvi, G. A. Van Kleef, M. Baas, F. S. Ten Velden, E. Van Dijk, and S. W. Feith. "The Neuropeptide Oxytocin Regulates Parochial Altruism in Intergroup Conflict among Humans." *Science* 328, no. 5984 (2010), 1408–1411. https://doi.org/10.1126/science.1189047.

de Jong, T. R., M. Chauke, B. N. Harris, and W. Saltzman. "From Here to Paternity: Neural Correlates of the Onset of Paternal Behavior in California Mice (*Peromyscus californicus*)." *Hormones and Behavior* 56, no. 2 (2009), 220–231. https://doi.org/10.1016/j.yhbeh.2009.05.001.

de Waal, F. B. *Chimpanzee Politics*. London: Jonathan Cape, 1982.

———. *Bonobo: The Forgotten Ape*. Berkeley: University of California Press, 1997.

de Waal, F. B. M. "A Century of Getting to Know the Chimpanzee." *Nature* 437, no. 7055 (2005), 56–59. https://doi.org/10.1038/nature03999.

Dinh, H., A. R. Cooklin, L. S. Leach, E. M. Westrupp, J. M. Nicholson, and L. Strazdins. "Parents' Transitions into and out of Work–family Conflict and Children's Mental Health: Longitudinal Influence Via Family Functioning." *Social Science and Medicine* 194 (2017), 42–50. https://doi.org/10.1016/j.socscimed.2017.10.017.

Ditzen, B., M. Schaer, B. Gabriel, G. Bodenmann, U. Ehlert, and M. Heinrichs. "Intranasal Oxytocin Increases Positive Communication and Reduces Cortisol Levels during Couple Conflict." *Biological Psychiatry* 65, no. 9 (2009), 728–731. https://doi .org/S0006-3223(08)01240-7.

Dixson, B. J., and P. L. Vasey. "Beards Augment Perceptions of Men's Age, Social Status, and Aggressiveness, But Not Attractiveness." *Behavioral Ecology* 23, no. 3 (2012), 481–490. https://doi.org/10.1093/beheco/arr214.

Doherty, W. J., M. F. Erickson, and R. LaRossa. "An Intervention to Increase Father Involvement and Skills with Infants during the Transition to Parenthood." *Journal of Family Psychology* 20, no. 3 (2006), 438–447. https://doi.org/10.1037/0893-3200 .20.3.438.

Dollar, Cindy Brooks. "Sex Ratios, Single Motherhood, and Gendered Structural Relations: Examining Female-Headed Families across Racial-Ethnic Populations." *Sociological Focus* 50, no. 4 (2017), 375–390. https://doi.org/10.1080/00380237.2017 .1313100.

Dolotovskaya, S., C. Roos, and E. W. Heymann. "Genetic Monogamy and Mate Choice in a Pair-Living Primate." *Scientific Reports* 10, no. 1 (2020). https://doi.org /10.1038/s41598-020-77132-9.

Domes, G., M. Heinrichs, A. Michel, C. Berger, and S. C. Herpertz. "Oxytocin Improves 'Mind-Reading' in Humans." *Biological Psychiatry* 61, no. 6 (2007), 731– 733. https://doi.org/10.1016/j.biopsych.2006.07.015. h.

Domonkos, E., J. Hodosy, D. Ostatnikova, and P. Celec. "On the Role of Testosterone in Anxiety-Like Behavior across Life in Experimental Rodents." *Frontiers in Endocrinology* 9 (2018), 441. https://doi.org/10.3389/fendo.2018.00441.

Donahue, C. J., M. F. Glasser, T. M. Preuss, J. K. Rilling, and D. C. Van Essen. "Quantitative Assessment of Prefrontal Cortex in Humans Relative to Nonhuman Primates." *Proceedings of the National Academy of Sceences*115, no. 22 (2018), E5183–E5192. https://doi.org/10.1073/pnas.1721653115.

Draper, Patricia, and Henry Harpending. "Father Absence and Reproductive Strategy: An Evolutionary Perspective." *Journal of Anthropological Research* 38 (1982), 255–273. https://www.jstor.org/stable/3629848.

Dreher, J. C., S. Dunne, A. Pazderska, T. Frodl, J. J. Nolan, and J. P. O'Doherty. "Testosterone Causes Both Prosocial and Antisocial Status-Enhancing Behaviors in Human Males." *Proceedings of the National Academy of Sciences* 113, no. 41 (2016), 11633–11638. https://doi.org/10.1073/pnas.1608085113.

Dudley, D. "Paternal Behavior in the California Mouse, *Peromyscus californicus*." *Behavioral Biology* 11, no. 2 (1974), 247–252. https://doi.org/10.1016/s0091-6773(74)90433-7.

Dufty, A. M. "Testosterone and Survival—a Cost of Aggressiveness." *Hormones and Behavior* 23, no. 2 (1989), 185–193. https://doi.org/10.1016/0018-506x(89)90059-7.

Dunbar, R. I. M. "The Mating System of Callitrichid Primates: II. The Impact of Helpers." *Animal Behaviour* 50 (1995), 1071–1089. https://doi.org/10.1016/0003-3472(95)80107-3.

Ebinger, M., C. Sievers, D. Ivan, H. J. Schneider, and G. K. Stalla. "Is There a Neuroendocrinological Rationale for Testosterone as a Therapeutic Option in Depression?" *Journal of Psychopharmacology* 23, no. 7 (2009), 841–853. https://doi.org/10.1177/0269881108092337.

Edelstein, R. S., W. J. Chopik, D. E. Saxbe, B. M. Wardecker, A. C. Moors, and O. P. LaBelle. "Prospective and Dyadic Associations between Expectant Parents' Prenatal Hormone Changes and Postpartum Parenting Outcomes." *Developmental Psychobiology* 59, no. 1 (2017), 77–90. https://doi.org/10.1002/dev.21469.

Eisenegger, C., J. Haushofer, and E. Fehr. "The Role of Testosterone in Social Interaction." *Trends in Cognitive Sciences* 15, no. 6 (2011), 263–271. https://doi.org/10.1016/j.tics.2011.04.008.

Ekberg, J., R. Eriksson, and G. Friebel. "Parental Leave—a Policy Evaluation of the Swedish `Daddy-Month' Reform." *Journal of Public Economics* 97 (2013), 131–143. https://doi.org/10.1016/j.jpubeco.2012.09.001.

Ellerbe, C. Z., J. B. Jones, and M. J. Carlson. "Race/Ethnic Differences in Nonresident Fathers' Involvement after a Nonmarital Birth." *Social Science Quarterly* 99, no. 3 (2018), 1158–1182. https://doi.org/10.1111/ssqu.12482.

Elliott, K. "Caring Masculinities: Theorizing an Emerging Concept." *Men and Masculinities* 19, no. 3 (2016), 240–259. https://doi.org/10.1177/1097184x15576203.

Ellis, B. J., J. E. Bates, K. A. Dodge, D. M. Fergusson, L. J. Horwood, G. S. Pettit, and L. Woodward. "Does Father Absence Place Daughters at Special Risk for Early Sexual Activity and Teenage Pregnancy?" *Child Development* 74, no. 3 (2003), 801–821. https://doi.org/10.1111/1467-8624.00569.

Ellis, B. J., A. J. Horn, C. S. Carter, M. H. van IJzendoorn, and M. J. Bakermans-Kranenburg. "Developmental Programming of Oxytocin through Variation in

Early-Life Stress: Four Meta-Analyses and a Theoretical Reinterpretation." *Clinical Psychology Review* 86 (2021). https://doi.org/10.1016/j.cpr.2021.101985.

Ellsworth, Ryan. "Book Review: The Human That Never Evolved." *Evolutionary Psychology* 9, no. 3 (2011), 325–335.

Ember, C. R., and M. Ember. "Father Absence and Male Aggression: A Re-Examination of the Comparative Evidence." *Ethos* 29, no. 3 (2001), 296–314. https://doi.org/10.1525/eth.2001.29.3.296.

Emery Thompson, M., J. H. Jones, A. E. Pusey, S. Brewer-Marsden, J. Goodall, D. Marsden, T. Matsuzawa, et al. "Aging and Fertility Patterns in Wild Chimpanzees Provide Insights into the Evolution of Menopause." *Current Biology* 17, no. 24 (2007), 2150–2156. https://doi.org/10.1016/j.cub.2007.11.033.

Estrada, M., A. Varshney, and B. E. Ehrlich. "Elevated Testosterone Induces Apoptosis in Neuronal Cells." *Journal of Biological Chemistry* 281, no. 35 (2006), 25492–25501. https://doi.org/10.1074/jbc.M603193200.

Evans, S. L., O. Dal Monte, P. Noble, and B. B. Averbeck. "Intranasal Oxytocin Effects on Social Cognition: A Critique." *Brain Research* 1580 (2013): 69–77. https://doi.org/10.1016/j.brainres.2013.11.008.

Fagan, J., and M. Barnett. "The Relationship between Maternal Gatekeeping, Paternal Competence, Mothers' Attitudes about the Father Role, and Father Involvement." *Journal of Family Issues* 24, no. 8 (2003), 1020–1043. https://doi.org/10.1177/0192513x03256397.

Fales, M. R., K. A. Gildersleeve, and M. G. Haselton. "Exposure to Perceived Male Rivals Raises Men's Testosterone on Fertile Relative to Nonfertile Days of Their Partner's Ovulatory Cycle." *Hormones and Behavior* 65, no. 5 (2014), 454–460. https://doi.org/10.1016/j.yhbeh.2014.04.002.

Fedurek, P., K. E. Slocombe, D. K. Enigk, M. Emery Thompson, R. W. Wrangham, and M. N. Muller. "The Relationship between Testosterone and Long-Distance Calling in Wild Male Chimpanzees." *Behavioral Ecology and Sociobiology* 70, no. 5 (2016), 659–672. https://doi.org/10.1007/s00265-016-2087-1.

Fehr, E., and U. Fischbacher. "The Nature of Human Altruism." *Nature* 425, no. 6960 (2003), 785–791. http://www.ncbi.nlm.nih.gov/entrez/query.fcgi?cmd=Retrieve&db=PubMed&dopt=Citation&list_uids=14574401

Feldblum, J. T., E. E. Wroblewski, R. S. Rudicell, B. H. Hahn, T. Paiva, M. Cetinkaya-Rundel, A. E. Pusey, and I. C. Gilby. "Sexually Coercive Male Chimpanzees Sire More Offspring." *Current Biology* 24, no. 23 (2014), 2855–2860. https://doi.org/10.1016/j.cub.2014.10.039.

Feldman, R., K. Braun, and F. A. Champagne. "The Neural Mechanisms and Consequences of Paternal Caregiving." *Nature Reviews Neuroscience* 20, no. 4 (April 2019), 205–224. https://doi.org/10.1038/s41583-019-0124-6.

Feldman, R., I. Gordon, I. Schneiderman, O. Weisman, and O. Zagoory-Sharon. "Natural Variations in Maternal and Paternal Care Are Associated with Systematic Changes in Oxytocin Following Parent-Infant Contact." *Psychoneuroendocrinology* 35, no. 8 (2010), 1133–1141. https://doi.org/10.1016/j.psyneuen.2010.01.013.

Feldman, R., I. Gordon, and O. Zagoory-Sharon. "Maternal and Paternal Plasma, Salivary, and Urinary Oxytocin and Parent-Infant Synchrony: Considering Stress and Affiliation Components of Human Bonding." *Developmental Science* 14, no. 4 (2011), 752–761. https://doi.org/10.1111/j.1467-7687.2010.01021.x.

Feng, C., P. D. Hackett, A. C. DeMarco, X. Chen, S. Stair, E. Haroon, B. Ditzen, G. Pagnoni, and J. K. Rilling. "Oxytocin and Vasopressin Effects on the Neural Response to Social Cooperation Are Modulated by Sex in Humans." *Brain Imaging Behavior* 9, no. 4 (2015), 754–764. https://doi.org/10.1007/s11682-014-9333-9.

Ferguson, J. N., L. J. Young, E. F. Hearn, M. M. Matzuk, T. R. Insel, and J. T. Winslow. "Social Amnesia in Mice Lacking the Oxytocin Gene." *Nature Genetics* 25, no. 3 (2000), 284–288. https://doi.org/10.1038/77040.

Fernandez-Balsells, M. M., M. H. Murad, M. Lane, J. F. Lampropulos, F. Albuquerque, R. J. Mullan, N. Agrwal, et al. "Clinical Review 1: Adverse Effects of Testosterone Therapy in Adult Men: A Systematic Review and Meta-Analysis." *Journal of Clinical Endocrinology and Metabolism* 95, no. 6 (2010), 2560–2575. https://doi.org/10.1210/jc.2009-2575.

Fine, Cordelia. *Testosterone Rex: Myths of Sex, Science, and Society.* New York: Norton, 2017.

Finkenwirth, C., E. Martins, T. Deschner, and J. M. Burkart. "Oxytocin Is Associated with Infant-Care Behavior and Motivation in Cooperatively Breeding Marmoset Monkeys." *Hormones and Behavior* 80 (2016), 10–18. https://doi.org/10.1016/j.yhbeh.2016.01.008.

Fischer, A., K. Prufer, J. M. Good, M. Halbwax, V. Wiebe, C. Andre, R. Atencia, et al. "Bonobos Fall within the Genomic Variation of Chimpanzees." *PloS One* 6, no. 6 (2011). https://doi.org/ARTNe2160510.1371/journal.pone.0021605.

Fleming, Alison S., Carl Corter, Joy Stallings, and Meir Steiner. "Testosterone and Prolactin Are Associated with Emotional Responses to Infant Cries in New Fathers." *Hormones and Behavior* 42, no. 4 (2002). https://doi.org/10.1006/hbeh.2002.1840.

Flinn, M. V., and B. G. England. "Social Economics of Childhood Glucocorticoid Stress Response and Health." *American Journal of Physical Anthropology* 102, no. 1 (1997), 33–53.

Flinn, M. V., D. Ponzi, and M. P. Muehlenbein. "Hormonal Mechanisms for Regulation of Aggression in Human Coalitions." *Human Nature* 23, no. 1 (2012), 68–88. https://doi.org/10.1007/s12110-012-9135-y.

Flouri, E., and A. Buchanan. "Father Involvement in Childhood and Trouble with the Police in Adolescence—Findings from the 1958 British Cohort." *Journal of Interpersonal Violence* 17, no. 6 (2002), 689–701. https://doi.org/10.1177/0886260502017006006.

Folch A., D. A. Christie, F. Jutglar, and E. F. G. Garcia. "Southern Cassowary (*Casuarius casuarius*)." In *Handbook of the Birds of the World*, edited by J. A. Elliott del Hoyo, J. Sargatal, D. A. Christie, and E. de Juana. Barcelona: Lynx Edicions, 2017.

Fouts, H. N. "Father Involvement with Young Children among the Aka and Bofi Foragers." *Cross-Cultural Research* 42, no. 3 (2008), 290–312. https://doi.org/10.1177/1069397108317484.

Franssen, C. L., M. Bardi, E. A. Shea, J. E. Hampton, R. A. Franssen, C. H. Kinsley, and K. G. Lambert. "Fatherhood Alters Behavioural and Neural Responsiveness in a Spatial Task." *Journal of Neuroendocrinology* 23, no. 11 (2011), 1177–1187. https://doi.org/10.1111/j.1365-2826.2011.02225.x.

French, J. A., J. Cavanaugh, A. C. Mustoe, S. B. Carp, and S. L. Womack. "Social Monogamy in Nonhuman Primates: Phylogeny, Phenotype, and Physiology." *Journal of Sex Research* 55, no. 4–5 (2018), 410–434. https://doi.org/10.1080/00224499.2017.1339774

Galinsky, Ellen, Kerstin Aumann, and James T. Bond. *Times Are Changing*. Hillsborough, NJ: Families and Work Institute, 2011.

Garber, P. A. "One for All and Breeding for One: Cooperation and Competition as a Tamarin Reproductive Strategy." *Evolutionary Anthropology* 5, no. 6 (1997), 187–199. https://doi.org/10.1002/(SICI)1520-6505(1997)5:6<187::AID-EVAN1>3.0.CO;2-A.

Garratt, M., H. Try, and R. C. Brooks. "Access to Females and Early Life Castration Individually Extend Maximal But Not Median Lifespan in Male Mice." *Geroscience* 43, no. 3 (2021), 1437–1246. https://doi.org/10.1007/s11357-020-00308-8.

Gaudino, J. A., Jr., B. Jenkins, and R. W. Rochat. "No Fathers' Names: A Risk Factor for Infant Mortality in the State of Georgia, USA." *Social Science and Medicine* 48, no. 2 (1999), 253–265. https://doi.org/10.1016/s0277-9536(98)00342-6.

Gavrilets, S. "Human Origins and the Transition from Promiscuity to Pair-Bonding." *Proceedings of the National Academy of Sciences* 109, no. 25 (2012), 9923–9928. https://doi.org/10.1073/pnas.1200717109.

Gelstein, S., Y. Yeshurun, L. Rozenkrantz, S. Shushan, I. Frumin, Y. Roth, and N. Sobel. "Human Tears Contain a Chemosignal." *Science* 331, no. 6014 (2011), 226–230. https://doi.org/10.1126/science.1198331.

Geniole, S. N., and J. M. Carre. "Human Social Neuroendocrinology: Review of the Rapid Effects of Testosterone." *Hormones and Behavior* 104 (2018), 192–205. https://doi.org/10.1016/j.yhbeh.2018.06.001.

Gershoff, E. T., and A. Grogan-Kaylor. "Spanking and Child Outcomes: Old Controversies and New Meta-Analyses." *Journal of Family Psychology* 30, no. 4 (2016), 453–469. https://doi.org/10.1037/fam0000191.

Gettler, L. T. "Applying Socioendocrinology to Evolutionary Models: Fatherhood and Physiology." *Evolutionary Anthropology* 23, no. 4 (2014), 146–160. https://doi.org/10.1002/evan.21412.

Gettler, L. T., P. X. Kuo, M. S. Sarma, B. C. Trumble, J. E. Burke Lefever, and J. M. Braungart-Rieker. "Fathers' Oxytocin Responses to First Holding Their Newborns: Interactions with Testosterone Reactivity to Predict Later Parenting Behavior and Father-Infant Bonds." *Developmental Psychobiology* 63, no. 5 (2021), 1384–1398. https://doi.org/10.1002/dev.22121.

Gettler, L. T., T. W. McDade, A. B. Feranil, and C. W. Kuzawa. "Longitudinal Evidence That Fatherhood Decreases Testosterone in Human Males." *Proceedings of the National Academy of Sciences* 108, no. 39 (2011), 16194–16199. https://doi.org/10.1073/pnas.1105403108.

Gettler, L. T., J. J. McKenna, T. W. McDade, S. S. Agustin, and C. W. Kuzawa. "Does Cosleeping Contribute to Lower Testosterone Levels in Fathers? Evidence from the Philippines." *PloS One* 7, no. 9 (2012). https://doi.org/10.1371/journal.pone.0041559.

Gettler, L. T., C. P. Ryan, D. T. A. Eisenberg, M. Rzhetskaya, M. G. Hayes, A. B. Feranil, S. A. Bechayda, and C. W. Kuzawa. "The Role of Testosterone in Coordinating Male Life History Strategies: The Moderating Effects of the Androgen Receptor CAG Repeat Polymorphism." *Hormones and Behavior* 87 (2017)), 164–175. https://doi.org/10.1016/j.yhbeh.2016.10.012.

Gettler, L. T., M. S. Sarma, R. G. Gengo, R. C. Oka, and J. J. McKenna. "Adiposity, CVD Risk Factors and Testosterone Variation by Partnering Status and Residence with Children in US Men." *Evolution Medicine and Public Health*, no. 1 (2017), 67–80. https://doi.org/10.1093/emph/eox005.

Glasper, E. R., W. M. Kenkel, J. Bick, and J. K. Rilling. "More Than Just Mothers: The Neurobiological and Neuroendocrine Underpinnings of Allomaternal Caregiving." *Frontiers in Neuroendocrinology* 53 (2019), 100741. https://doi.org/10.1016/j.yfrne.2019.02.005.

Glocker, M. L., D. D. Langleben, K. Ruparel, J. W. Loughead, J. N. Valdez, M. D. Griffin, N. Sachser, and R. C. Gur. "Baby Schema Modulates the Brain Reward System in Nulliparous Women." *Proceedings of the National Academy of Sciences* 106, no. 22 (2009), 9115–9119. https://doi.org/10.1073/pnas.0811620106.

Golombok, S., L. Mellish, S. Jennings, P. Casey, F. Tasker, and M. E. Lamb. "Adoptive Gay Father Families: Parent–Child Relationships and Children's Psychological Adjustment." *Child Development* 85, no. 2 (2014), 456–468. https://doi.org/10.1111/cdev.12155.

Gomes, C. M., and C. Boesch. "Wild Chimpanzees Exchange Meat for Sex on a Long-Term Basis." *PloS One* 4, no. 4 (2009). https://doi.org/10.1371/journal.pone.0005116.

Goodall, Jane. *The Chimpanzees of Gombe: Patterns of Behavior*. Cambridge, MA: Harvard University Press, 1986.

Gordon, I., O. Zagoory-Sharon, J. F. Leckman, and R. Feldman. "Oxytocin and the Development of Parenting in Humans." *Biological Psychiatry* 68, no. 4 (2010), 377–382. https://doi.org/10.1016/j.biopsych.2010.02.005.

Gordon, N. S., S. Burke, H. Akil, S. J. Watson, and J. Panksepp. "Socially-Induced Brain 'Fertilization': Play Promotes Brain Derived Neurotrophic Factor Transcription in the Amygdala and Dorsolateral Frontal Cortex in Juvenile Rats." *Neuroscience Letters* 341, no. 1 (2003), 17–20. https://doi.org/10.1016/s0304-3940(03)00158-7.

Gottfried, H., L. Vigilant, R. Mundry, V. Behringer, and M. Surbeck. "Aggression by Male Bonobos against Immature Individuals Does Not Fit with Predictions of Infanticide." *Aggressive Behavior* 45, no. 3 (2019), 300–309. https://doi.org/10.1002/ab.21819.

Goymann, W., and P. F. Davila. "Acute Peaks of Testosterone Suppress Paternal Care: Evidence from Individual Hormonal Reaction Norms." *Philosophical Transactions of the Royal Society B: Biological Sciences* 284, no. 1857 (2017). https://doi.org/10.1098/rspb.2017.0632.

Grabowski, M., K. G. Hatala, W. L. Jungers, and B. G. Richmond. "Body Mass Estimates of Hominin Fossils and the Evolution of Human Body Size." *Journal of Human Evolution* 85 (2015), 75–93. https://doi.org/10.1016/j.jhevol.2015.05.005.

Gray, P. B. "Marriage, Parenting, and Testosterone Variation among Kenyan Swahili Men." *American Journal of Physical Anthropology* 122, no. 3 (2003), 279–286. https://doi.org/10.1002/ajpa.10293.

Gray, Peter B., and Kermyt G. Anderson. *Fatherhood: Evolution and Human Paternal Behavior*. Cambridge, MA: Harvard University Press, 2010.

Gray, Peter, and Alyssa Crittenden. "Father Darwin: Effects of Children on Men, Viewed from an Evolutionary Perspective." *Fathering* 12, no. 2 (2014), 121–142.

Gray, P. B., A. A. Straftis, B. M. Bird, T. S. McHale, and S. Zilioli. "Human Reproductive Behavior, Life History, and the Challenge Hypothesis: A 30-Year Review, Retrospective and Future Directions." *Hormones and Behavior* 123 (2020). https://doi.org/10.1016/j.yhbeh.2019.04.017.

Graziano, P. A., R. D. Reavis, S. P. Keane, and S. D. Calkins. "The Role of Emotion Regulation in Children's Early Academic Success." *Journal of School Psychology* 45, no. 1 (2007), 3–19. https://doi.org/10.1016/j.jsp.2006.09.002.

Gregory, R., H. Cheng, H. A. Rupp, D. R. Sengelaub, and J. R. Heiman. "Oxytocin Increases VTA Activation to Infant and Sexual Stimuli in Nulliparous and Postpartum Women." *Hormones and Behavior* 69 (2015), 82–88. https://doi.org/10.1016/j.yhbeh.2014.12.009.

Greiling, H., and D. M. Buss. "Women's Sexual Strategies: The Hidden Dimension of Extra-Pair Mating." *Personality and Individual Differences* 28, no. 5 (2000), 929–963. https://doi.org/10.1016/S0191-8869(99)00151-8.

Grewen, K. M., S. S. Girdler, J. Amico, and K. C. Light. "Effects of Partner Support on Resting Oxytocin, Cortisol, Norepinephrine, and Blood Pressure before and after Warm Partner Contact." *Psychosomatic Medicine* 67, no. 4 (2005), 531–538. https://doi.org/10.1097/01.psy.0000170341.88395.47.

Grossman, Karin, Klaus E. Grossman, Elisabeth Fremmer-Bombik, Heinz Kindler, Hermann Scheuerer-Englisch, and Peter Zimmerman. "The Uniqueness of the Child-Father Attachment Relationship: Fathers' Sensitive and Challenging Play as a Pivotal Variable in a 16-Year Longitudinal Study." *Social Development* 11, no. 3 (2002), 301–337. https://doi.org/10.1111/1467-9507.00202.

Gruber, T., and Z. Clay. "A Comparison between Bonobos and Chimpanzees: A Review and Update." *Evolutionary Anthropology* 25, no. 5 (2016), 239–252. https://doi.org/10.1002/evan.21501.

Guastella, A. J., P. B. Mitchell, and M. R. Dadds. "Oxytocin Increases Gaze to the Eye Region of Human Faces." *Biological Psychiatry* 63, no. 1 (2008), 3–5. https://doi.org/10.1016/j.biopsych.2007.06.026.

Gubernick, D. J., and T. Teferi. "Adaptive Significance of Male Parental Care in a Monogamous Mammal." *Philosophical Transactions of the Royal Society B: Biological Sciences* 267, no. 1439 (2000), 147–150. https://doi.org/10.1098/rspb.2000.0979.

Gulya, Lisa. "Adolescent Pregnancy in Russia." In *International Handbook of Adolescent Pregnancy*, edited by A. Cherry and M. Dillon. Boston: Springer, 2014.

Gurven, M. D., and R. J. Davison. "Periodic Catastrophes over Human Evolutionary History Are Necessary to Explain the Forager Population Paradox." *Proceedings of the National Academy of Sciences* 116, no. 26 (2019), 12758–12766. https://doi.org/10.1073/pnas.1902406116.

Gurven, M., and K. Hill. "Why Do Men Hunt? A Reevaluation of `Man the Hunter' and the Sexual Division of Labor." *Current Anthropology* 50, no. 1 (2009), 51–74. https://doi.org/10.1086/595620.

Gutierrez-Galve, L., A. Stein, L. Hanington, J. Heron, and P. Ramchandani. "Paternal Depression in the Postnatal Period and Child Development: Mediators and Moderators." *Pediatrics* 135, no. 2 (2015), e339–e347. https://doi.org/10.1542/peds.2014-2411.

Haas, Linda, and C. Philip Hwang. "The Impact of Taking Parental Leave on Fathers' Participation in Childcare and Relationships with Children: Lessons from Sweden." *Community, Work and Family* 11, no. 1 (2008), 85–104. https://doi.org/10.1080/13668800701785346.

Hammock, E. A., and L. J. Young. "Oxytocin, Vasopressin and Pair Bonding: Implications for Autism." *Philosophical Transactions of the Royal Society B: Biological Sciences* 361, no. 1476 (2006), 2187–2198. https://doi.org/10.1098/rstb.2006.1939.

Handelsman, D. J., A. L. Hirschberg, and S. Bermon. "Circulating Testosterone as the Hormonal Basis of Sex Differences in Athletic Performance." *Endocrine Reviews* 39, no. 5 (2018), 803–829. https://doi.org/10.1210/er.2018-00020.

Hare, B., V. Wobber, and R. Wrangham. "The Self-Domestication Hypothesis: Evolution of Bonobo Psychology Is Due to Selection against Aggression." *Animal Behaviour* 83, no. 3 (2012), 573–585. https://doi.org/10.1016/j.anbehav.2011.12.007.

Hare, Brian, and Richard Wrangham. "Equal, Similar, But Different: Convergent Bonobos and Conserved Chimpanzees." In *Chimpanzees and Human Evolution*, edited by Martin Muller, Richard Wrangham, and David Pilbean,142–173. Cambridge, MA: Harvard University Press, 2017.

Harkness, S., and C. M. Super. "The Cultural Foundations of Fathers' Roles: Evidence from Kenya and the United States." In *Father–Child Relations: Cultural and Biosocial Contexts* edited by B. S. Hewlett, 191–211, New York: Aldine de Gruyter, 1992.

Harris, K. M., and J. K. Marchmer. "Poverty, Paternal Involvement, and Adolescent Well-Being." *Journal of Family Issues* 17, no. 5 (1996), 614–640. https://doi.org/10.1177/019251396017005003.

Hartung, J. "Matrilineal Inheritance: New Theory and Analysis." *Behavioral and Brain Sciences* 8, no. 4 (1985), 661–670. https://doi.org/10.1017/S0140525x00045520.

Havighurst, Sophie S., Katherine R. Wilson, Ann E. Harley, and Christiane E. Kehoe. "Dads Tuning into Kids: A Randomized Controlled Trial of an Emotion Socialization Program for Fathers." *Social Development* 28 (2019), 979–997. https://doi.org/10.1111/sode.12375.

Hawkes, K. "Showing Off: Tests of an Hypothesis about Men's Foraging Goals." *Ethology and Sociobiology* 12, no. 1 (1991), 29–54. https://doi.org/10.1016/0162-3095(91)90011-E.

Hawkes, K., and J. E. Coxworth. "Grandmothers and the Evolution of Human Longevity: A Review of Findings and Future Directions." *Evolutionary Anthropology* 22, no. 6 (2013), 294–302. https://doi.org/10.1002/evan.21382.

Hawkes, K., J. F. O'Connell, and N. G. B. Jones. "Hadza Women's Time Allocation, Offspring Provisioning, and the Evolution of Long Postmenopausal Life Spans." *Current Anthropology* 38, no. 4 (1997), 551–577. https://doi.org/10.1086/204646.

Hawkes, K., J. F. O'Connell, N. G. Jones, H. Alvarez, and E. L. Charnov. "Grandmothering, Menopause, and the Evolution of Human Life Histories." *Proceedings of the National Academy of Sciences* 95, no. 3 (1998), 1336–1339. https://doi.org/10.1073/pnas.95.3.1336.

Hechler, C., R. Beijers, M. Riksen-Walraven, and C. de Weerth. "Prenatal Predictors of Postnatal Quality of Caregiving Behavior in Mothers and Fathers." *Parenting-Science and Practice* 19, no. 1–2 (2019), 101–119. https://doi.org/10.1080/15295192.2019.1556010.

Hegner, R. E., and J. C. Wingfield. "Effects of Experimental Manipulation of Testosterone Levels on Parental Investment and Breeding Success in Male House Sparrows." *Auk* 104, no. 3 (1987), 462–469. https://doi.org/0.2307/4087545.

Heidinger, B. J., S. P. Slowinski, A. E. Sirman, J. Kittilson, N. M. Gerlach, and E. D. Ketterson. "Experimentally Elevated Testosterone Shortens Telomeres across Years in a Free-Living Songbird." *Molecular Ecology* (2021). https://doi.org/10.1111/mec.15819.

Heim, C., L. J. Young, D. J. Newport, T. Mletzko, A. H. Miller, and C. B. Nemeroff. "Lower CSF Oxytocin Concentrations in Women with a History of Childhood Abuse." *Molecular Psychiatry* 14, no. 10 (2009), 954–958. https://doi.org/10.1038/mp.2008.112.

Heinrichs, M., T. Baumgartner, C. Kirschbaum, and U. Ehlert. "Social Support and Oxytocin Interact to Suppress Cortisol and Subjective Responses to Psychosocial Stress." *Biological Psychiatry* 54, no. 12 (2003), 1389–1398. https://doi.org/10.1016/s0006-3223(03)00465-7.

Helfrecht, C., J. W. Roulette, A. Lane, B. Sintayehu, and C. L. Meehan. "Life History and Socioecology of Infancy." *American Journal of Physical Anthropology* 173, no. 4 (2020a), 619–629. https://doi.org/10.1002/ajpa.24145.

Helmeke, C., K. Seidel, G. Poeggel, T. W. Bredy, A. Abraham, and K. Braun. "Paternal Deprivation during Infancy Results in Dendrite- and Time-Specific Changes of Dendritic Development and Spine Formation in the Orbitofrontal Cortex of the Biparental Rodent *Octoberodon degus*." *Neuroscience* 163, no. 3 (2009), 790–798. https://doi.org/10.1016/j.neuroscience.2009.07.008.

Henrich, J. "Cultural Group Selection, Coevolutionary Processes and Large-Scale Cooperation." *Journal of Economic Behavior and Organization* 53, no. 1 (2004), 3–35. https://doi.org/10.1016/S0167-2681(03)00094-5.

Henrich, J., S. J. Heine, and A. Norenzayan. "Most People Are Not Weird." *Nature* 466, no. 7302 (2010): 29. https://doi.org/10.1038/466029a.

———. "The Weirdest People in the World?" *Behavioral and Brain Science* 33, no. 2–3 (2010), 61–135. https://doi.org/10.1017/S0140525X0999152X.

Henry, J. B., W. A. Julion, D. T. Bounds, and J. Sumo. "Fatherhood Matters: An Integrative Review of Fatherhood Intervention Research." *Journal of School Nursing* 36, no. 1 (2020), 19–32. https://doi.org/10.1177/1059840519873380.

Hermans, E. J., P. Putman, and J. van Honk. "Testosterone Administration Reduces Empathetic Behavior: A Facial Mimicry Study." *Psychoneuroendocrinology* 31, no. 7 (2006), 859–866. https://doi.org/10.1016/j.psyneuen.2006.04.002.

Hewlett, B. S. "Demography and Child-Care in Preindustrial Societies." *Journal of Anthropological Research* 47, no. 1 (1991), 1–37. https://doi.org/10.1086/jar.47.1.3630579.

———"Culture, History, and Sex: Anthropological Contributions to Conceptualizing Father Involvement." *Marriage and Family Review* 29, no. 2–3 (2000), 59–73. https://doi.org/10.1300/J002v29n02_05.

Hewlett, Barry S. *Intimate Fathers: The Nature and Context of Aka Pygmy Paternal Infant Care*. Ann Arbor: University of Michigan, 1991.

Hewlett, Barry S., Jean Hudson, Adam H. Boyette, and Hillary N. Fouts. "Intimate Living: Sharing Space among Aka and Other Hunter-Gatherers." In *Towards a Broader View of Hunter-Gatherer Sharing*, edited by Noa Lavi and David E. Friesem, 39–56. Cambridge: McDonald Institute for Archaeological Research, 2019.

Hewlett, B. S., and MacFarlan, S. J. "Fathers' Roles in Hunter-Gatherers and Other Small-Scale Cultures." In *The Role of the Father in Child Development* edited by M. E. Lamb, 413–434. Hoboken, NJ: Wiley, 2010.

Hill, K. "Altruistic Cooperation during Foraging by the Ache, and the Evolved Human Predisposition to Cooperate." *Human Nature: An Interdisciplinary Biosocial Perspective* 13, no. 1 (2002), 105–128. https://doi.org.10.1007/s12110-002-1016-3.

Hill, Kim, and Magdalena Hurtado. *Ache Life History: The Ecology and Demography of a Foraging People*. London: Routledge, 1996.

Hiser, J., and M. Koenigs. "The Multifaceted Role of the Ventromedial Prefrontal Cortex in Emotion, Decision Making, Social Cognition, and Psychopathology." *Biological Psychiatry* 83, no. 8 (2018), 638–647. https://doi.org/10.1016/j.biopsych.2017.10.030.

Hoekzema, E., E. Barba-Muller, C. Pozzobon, M. Picado, F. Lucco, D. Garcia-Garcia, J. C. Soliva, et al. "Pregnancy Leads to Long-Lasting Changes in Human Brain Structure." *Nature Neuroscience* 20, no. 2 (2017), 287–2296. https://doi.org/10.1038/nn.4458.

Hoekzema, E., C. K. Tamnes, P. Berns, E. Barba-Muller, C. Pozzobon, M. Picado, F. Lucco, et al. "Becoming a Mother Entails Anatomical Changes in the Ventral Striatum of the Human Brain That Facilitate Its Responsiveness to Offspring Cues." *Psychoneuroendocrinology* 112 (2020), 104507. https://doi.org/10.1016/j.psyneuen.2019.104507.

Hofferth, S. L., J. H. Pleck, and C. K. Vesely. "The Transmission of Parenting from Fathers to Sons." *Parenting-Science and Practice* 12, no. 4 (2012), 282–305. https://doi.org/10.1080/15295192.2012.709153.

Hoffman, K. A., S. P. Mendoza, M. B. Hennessy, and W. A. Mason. "Responses of Infant Titi Monkeys, *Callicebus moloch*, to Removal of One or Both Parents—Evidence for Paternal Attachment." *Developmental Psychobiology* 28, no. 7 (1995), 399–407. https://doi.org/10.1002/dev.420280705.

Hohmann-Marchriott, B. "Coparenting and Father Involvement in Married and Unmarried Coresident Couples." *Journal of Marriage and Family* 73, no. 1 (2011), 296–309. https://doi.org/10.1111/j.1741-3737.2010.00805.x.

Hohmann, G., and B. Fruth. "Intra- and Inter-Sexual Aggression by Bonobos in the Context of Mating." *Behaviour* 140 (2003), 1389–1413. https://doi.org/10.1163/156853903771980648.

Holmes, E. K., B. M. Egginton, A. J. Hawkins, N. L. Robbins, and K. Shafer. "Do Responsible Fatherhood Programs Work? A Comprehensive Meta-Analytic Study." *Family Relations* 69, no. 5 (2020), 967–982. https://doi.org/10.1111/fare.12435.

Holt-Lunstad, J., W. C. Birmingham, and K. C. Light. "Relationship Quality and Oxytocin: Influence of Stable and Modifiable Aspects of Relationships." *Journal of Social and Personal Relationships* 32, no. 4 (2015), 472–490. https://doi.org/10.1177/0265407514536294.

Hooven, Carole. *T: The Story of Testosterone, the Hormone That Dominates and Divides Us*. New York: Holt, 2021.

Horton, B. M., I. T. Moore, and D. L. Maney. "New Insights into the Hormonal and Behavioural Correlates of Polymorphism in White-Throated Sparrows, *Zonotrichia albicollis*." *Animal Behaviour* 93 (2014), 207–219. https://doi.org/10.1016/j.anbehav.2014.04.015.

Horwitz, H., J. T. Andersen, and K. P. Dalhoff. "Health Consequences of Androgenic Anabolic Steroid Use." *Journal of Internal Medicine* 285, no. 3 (2019), 333–340. https://doi.org/10.1111/joim.12850.

Hossin, M. Z. "The Male Disadvantage in Life Expectancy: Can We Close the Gender Gap?" *International Health* 13, no. 5 (2021), 482–484. https://doi.org/10.1093/inthealth/ihaa106.

Hrdy, S. B. "Male-Male Competition and Infanticide among the Langurs (*Presbytis entellus*) of Abu, Rajasthan." *Folia Primatology* 22, no. 1 (1974), 19–58. https://doi.org/10.1159/000155616.

———. "Infanticide as a Primate Reproductive Strategy." *American Scientist* 65, no. 1 (1977), 40–49. https://www.jstor.org/stable/27847641.

Hrdy, Sarah. "Infanticide among Animals: A Review, Classification and Examination of the Implications for the Reproductive Strategies of Females." *Ethology and Sociobiology* 1 (1979), 13–40. https://doi.org/10.1016/0162-3095(79)90004-9.

———. *Mother Nature*. New York: Ballantine Books, 1999.

———. "The Optimal Number of Fathers: Evolution, Demography and History in the Shaping of Female Mate Preferences." In *Evolutionary Psychology: Alternative Approaches*, edited by Steven J. Scher and Frederick Rauscher, 111–133. Norwell, MA: Kluwer, 2003.

———. *Mothers and Others*. Cambridge, MA: Harvard University Press, 2009.

Huerta, M. C., W. Adema, J. Baxter, W. J. Han, M. Lausten, R. Lee, and J. Waldfogel. "Fathers' Leave and Fathers' Involvement: Evidence from Four OECD Countries." *European Journal of Social Security* 16, no. 4 (2014), 308–346. https://doi.org/10.1177/138826271401600403.

Hughes, V. L., and S. E. Randolph. "Testosterone Depresses Innate and Acquired Resistance to Ticks in Natural Rodent Hosts: A Force for Aggregated Distributions of Parasites." *Journal of Parasitology* 87, no. 1 (2001), 49–54. https://doi.org/10.1645/0022-3395(2001)087[0049:TDIAAR]2.0.CO;2.

Hurlemann, R., A. Patin, O. A. Onur, M. X. Cohen, T. Baumgartner, S. Metzler, I. Dziobek, et al. "Oxytocin Enhances Amygdala-Dependent, Socially Reinforced

Learning and Emotional Empathy in Humans." *Journal of Neuroscience* 30, no. 14 (2010), 4999–5007. https://doi.org/10.1523/JNEUROSCI.5538-09.2010.

Hyer, M. M., T. J. Hunter, J. Katakam, T. Wolz, and E. R. Glasper. "Neurogenesis and Anxiety-Like Behavior in Male California Mice during the Mate's Postpartum Period." *European Journal of Neuroscience* 43, no. 5 (2016), 703–709. https://doi.org /10.1111/ejn.13168.

Iersel, J. J. A. Van. "An Analysis of the Parental Behaviour of the Male Three-Spined Stickleback (*Gasterosteus aculeatus l.*)." *Behaviour* suppl. 3 (1953), 1–159.

Imamura, M., J. Tucker, P. Hannaford, M. O. da Silva, M. Astin, L. Wyness, K. W. M. Bloemenkamp, et al. "Factors Associated with Teenage Pregnancy in the European Union Countries: A Systematic Review." *European Journal of Public Health* 17, no. 6 (2007), 630–636. https://doi.org/10.1093/eurpub/ckm014.

Inhorn, Marcia C., Wendy Chavkin, and Jose-Alberto Navarro, eds. *Globalized Fatherhood*. New York: Berghahn, 2015.

Islamiah, N., S. Breinholst, M. A. Walczak, and B. H. Esbjorn. "The Role of Fathers in Children's Emotion Regulation Development: A Systematic Review." *Infant and Child Development* 32, no. 2 (2023). https://doi.org/10.1002/icd.2397.

Isler, K., and C. P. van Schaik. "Allomaternal Care, Life History and Brain Size Evolution in Mammals." *Journal of Human Evolution* 63, no. 1 (2012), 52–63. https://doi .org/10.1016/j.jhevol.2012.03.009.

James, P. J., and J. G. Nyby. "Testosterone Rapidly Affects the Expression of Copulatory Behavior in House Mice (*Mus musculus*)." *Physiology and Behavior* 75, no. 3 (2002), 287–294. https://doi.org/10.1016/S0031-9384(01)00666-7.

Jeong, J., D. C. McCoy, A. K. Yousafzai, C. Salhi, and G. Fink. "Paternal Stimulation and Early Child Development in Low- and Middle-Income Countries." *Pediatrics* 138, no. 4 (2016). https://doi.org/10.1542/peds.2016-1357.

Jia, R., F. Tai, S. An, and X. Zhang. "Neonatal Paternal Deprivation or Early Deprivation Reduces Adult Parental Behavior and Central Estrogen Receptor Alpha Expression in Mandarin Voles (*Microtus mandarinus*)." *Behavioural Brain Research* 224, no. 2 (2011), 279–289. https://doi.org/10.1016/j.bbr.2011.05.042.

Jia, R., F. D. Tai, S. C. An, and X. Zhang. "Neonatal Paternal Deprivation or Early Deprivation Reduces Adult Parental Behavior and Central Estrogen Receptor Alpha Expression in Mandarin Voles (*Microtus mandarinus*)." *Behavioural Brain Research* 224, no. 2 (2011), 279–289. https://doi.org/10.1016/j.bbr.2011.05.042.

Jia, R., F. D. Tai, S. C. An, X. Zhang, and H. Broders. "Effects of Neonatal Paternal Deprivation or Early Deprivation on Anxiety and Social Behaviors of the Adults in

Mandarin Voles." *Behavioural Processes* 82, no. 3 (2009), 271–278. https://doi.org/10.1016/j.beproc.2009.07.006.

Johns, S. E. "Perceived Environmental Risk as a Predictor of Teenage Motherhood in a British Population." *Health and Place* 17, no. 1 (2011), 122–131. https://doi.org/10.1016/j.healthplace.2010.09.006.

Johnson, W. E. "Paternal Involvement among Unwed Fathers." *Children and Youth Services Review* 23, no. 6–7 (2001), 513–536. https://doi.org/10.1016/S0190-7409(01)00146-3.

Jones, J., and W. D. Mosher. "Fathers' Involvement with Their Children: United States, 2006–2010." *National Health Statistics Report*, no. 71 (2013), 1–21. https://www.ncbi.nlm.nih.gov/pubmed/24467852.

Jones, J. S., and K. E. Wynne-Edwards. "Paternal Hamsters Mechanically Assist the Delivery, Consume Amniotic Fluid and Placenta, Remove Fetal Membranes, and Provide Parental Care during the Birth Process." *Hormones and Behavior* 37, no. 2 (2000), 116–125. https://doi.org/10.1006/hbeh.1999.1563.

Jonker, F. A., C. Jonker, P. Scheltens, and E. J. A. Scherder. "The Role of the Orbitofrontal Cortex in Cognition and Behavior." *Reviews in the Neurosciences* 26, no. 1 (2015), 1–11. https://doi.org/10.1515/revneuro-2014-0043.

Jordan-Young, Rebecca M., and Katrina Alicia Karkazis. *Testosterone: An Unauthorized Biography*. Cambridge, MA: Harvard University Press, 2019.

Kaplan, H., K. Hill, J. Lancaster, and A. M. Hurtado. "A Theory of Human Life History Evolution: Diet, Intelligence, and Longevity." *Evolutionary Anthropology* 9, no. 4 (2000): 156–185. https://doi.org/10.1002/1520-6505(2000)9:4%3C156::AID-EVAN5%3E3.0.CO;2-7.

Kaplan, H. S., P. L. Hooper, and M. Gurven. "The Evolutionary and Ecological Roots of Human Social Organization." *Philosophical Transactions of the Royal Society B: Biological Sciences* 364, no. 1533 (2009), 3289–3299. https://doi.org/10.1098/rstb.2009.0115.

Katz, M. M., and M. J. Konner. "The Role of the Father: An Anthropological Perspective." In *The Role of the Father in Child Development*, edited by M. E. Lamb, 155–186. New York: Wiley, 1981.

Kendall, Joshua. "Obama's Most Unusual Legacy? Being a Good Dad." *Washington Post*, June 19, 2016.

Kenkel, W. M., J. Paredes, J. R. Yee, H. Pournajafi-Nazarloo, K. L. Bales, and C. S. Carter. "Neuroendocrine and Behavioural Responses to Exposure to an Infant in Male Prairie Voles." *Journal of Neuroendocrinology* 24, no. 6 (2012), 874–886. https://doi.org/10.1111/j.1365-2826.2012.02301.x.

Kenkel, W. M., G. Suboc, and C. S. Carter. "Autonomic, Behavioral and Neuroendocrine Correlates of Paternal Behavior in Male Prairie Voles." *Physiological Behavior* 128 (2014), 252–259. https://doi.org/10.1016/j.physbeh.2014.02.006.

Ketterson, E. D., V. Nolan, L. Wolf, and C. Ziegenfus. "Testosterone and Avian Life Histories: Effects of Experimentally Elevated Testosterone on Behavior and Correlates of Fitness in the Dark-Eyed Junco (*Junco hyemalis*)." *American Naturalist* 140, no. 6 (1992), 980–999. https://doi.org/10.1086/285451.

Ketterson, E. D., V. Nolan Jr., L. Wolf, C. Ziegenfus, A. M. Dufty Jr., G. F. Ball, and T. S. Johnsen. "Testosterone and Avian Life Histories: The Effect of Experimentally Elevated Testosterone on Corticosterone and Body Mass in Dark-Eyed Juncos." *Hormones and Behavior* 25, no. 4 (1991), 489–503. https://doi.org/10.1016/0018-506x(91)90016-b.

Kettunen, J. A., U. M. Kujala, J. K. Aprilio, H. Backmand, M. Peltonen, J. G. Eriksson, and S. Sarna. "All-Cause and Disease-Specific Mortality among Male, Former Elite Athletes: An Average 50-Year Follow-Up." *British Journal of Sports Medicine* 49, no. 13 (2015), 893–897. https://doi.org/10.1136/bjsports-2013-093347.

Kim, P., J. F. Leckman, L. C. Mayes, R. Feldman, X. Wang, and J. E. Swain. "The Plasticity of Human Maternal Brain: Longitudinal Changes in Brain Anatomy during the Early Postpartum Period." *Behavioral Neuroscience* 124, no. 5 (2010), 695–700. https://doi.org/10.1037/a0020884.

Kim, P., P. Rigo, J. F. Leckman, L. C. Mayes, P. M. Cole, R. Feldman, and J. E. Swain. "A Prospective Longitudinal Study of Perceived Infant Outcomes at 18–24 Months: Neural and Psychological Correlates of Parental Thoughts and Actions Assessed during the First Month Postpartum." *Frontiers in Psychology* 6 (2015), 1772. https://doi.org/10.3389/fpsyg.2015.01772.

Kim, P., P. Rigo, L. C. Mayes, R. Feldman, J. F. Leckman, and J. E. Swain. "Neural Plasticity in Fathers of Human Infants." *Society for Neuroscience* 9, no. 5 (2014), 522–535. https://doi.org/10.1080/17470919.2014.933713.

Kim, S., P. Fonagy, O. Koos, K. Dorsett, and L. Strathearn. "Maternal Oxytocin Response Predicts Mother-to-Infant Gaze." *Brain Research* 1580 (2014), 133–142. https://doi.org/10.1016/j.brainres.2013.10.050.

King, L. B., H. Walum, K. Inoue, N. W. Eyrich, and L. J. Young. "Variation in the Oxytocin Receptor Gene Predicts Brain Region-Specific Expression and Social Attachment." *Biological Psychiatry* 80, no. 2 (2016), 160–169. https://doi.org/10.1016/j.biopsych.2015.12.008.

Klahr, A. M., K. Klump, and S. A. Burt. "A Constructive Replication of the Association between the Oxytocin Receptor Genotype and Parenting." *Journal of Family Psychology* 29, no. 1 (2015), 91–99. https://doi.org/10.1037/fam0000034.

Klimas, C., U. Ehlert, T. J. Lacker, P. Waldvogel, and A. Walther. "Higher Testosterone Levels Are Associated with Unfaithful Behavior in Men." *Biological Psychology* 146 (2019). https://doi.org/10.1016/j.biopsycho.2019.107730.

Knoester, C., R. J. Petts, and B. Pragg. "Paternity Leave-Taking and Father Involvement among Socioeconomically Disadvantaged U.S. Fathers." *Sex Roles* 81, no. 5–6 (2019), 257–271. https://doi.org/10.1007/s11199-018-0994-5.

Koenig, Walter, and Janis Dickinson. *Ecology and Evolution of Cooperatively Breeding Birds*. Cambridge: Cambridge University Press, 2004.

Kohl, J., A. E. Autry, and C. Dulac. "The Neurobiology of Parenting: A Neural Circuit Perspective." *Bioessays* 39, no. 1 (2017), 1–11. https://doi.org/10.1002/bies .201600159.

Kohl, J., and C. Dulac. "Neural Control of Parental Behaviors." *Current Opinion in Neurobiology* 49 (2018), 116–122. https://doi.org/10.1016/j.conb.2018.02.002.

Kong, K. A., and S. I. Kim. "Mental Health of Single Fathers Living in an Urban Community in South Korea." *Comprehensive Psychiatry* 56 (2015), 188–197. https:// doi.org/10.1016/j.comppsych.2014.09.012.

Konner, M. "Is History the Same as Evolution? No. Is It Independent of Evolution? Certainly Not." *Evolutionary Psychology* 20, no. 1 (2022). https://doi.org/10.1177 /14747049211069137.

Konner, Melvin. "Hunter-Gatherer Infancy and Childhood." In *Hunter-Gatherer Childhoods*, edited by Barry S. Hewlett and Michael E. Lamb, 46. London: Routledge, 2005.

———. *Women after All: Sex, Evolution, and the End of Male Supremacy*. New York: Norton, 2015.

Konner, Melvin J. *The Evolution of Childhood: Relationships, Emotion, Mind*. Cambridge, MA: Harvard University Press, 2010.

Koops, J. C., A. C. Liefbroer, and A. H. Gauthier. "Having a Child within a Cohabiting Union in Europe and North America: What Is the Role of Parents' Socio-Economic Status?" *Population Space and Place* 27, no. 6 (2021). https://doi.org/10 .1002/psp.2434.

Korb, S., J. Malsert, L. Strathearn, P. Vuilleumier, and P. Niedenthal. "Sniff and Mimic—Intranasal Oxytocin Increases Facial Mimicry in a Sample of Men." *Hormones and Behavior* 84 (2016), 64–74. https://doi.org/10.1016/j.yhbeh.2016.06.003.

Kosfeld, M., M. Heinrichs, P. J. Zak, U. Fischbacher, and E. Fehr. "Oxytocin Increases Trust in Humans." *Nature* 435, no. 7042 (2005), 673–676. https://doi.org/10.1038/ nature03701.

Kozorovitskiy, Y., M. Hughes, K. Lee, and E. Gould. "Fatherhood Affects Dendritic Spines and Vasopressin V1a Receptors in the Primate Prefrontal Cortex." *Nature Neuroscience* 9, no. 9 (2006), 1094–1095. https://doi.org/10.1038/nn1753.

Kramer, K. L. "Cooperative Breeding and Its Significance to the Demographic Success of Humans." *Annual Review of Anthropology* 39 (2010), 417–436. https://doi.org/10.1146/annurev.anthro.012809.105054.

Kringelbach, M. L., and E. T. Rolls. "The Functional Neuroanatomy of the Human Orbitofrontal Cortex: Evidence from Neuroimaging and Neuropsychology." *Progress in Neurobiology* 72, no. 5 (2004), 341–372. https://doi.org/10.1016/j.pneurobio.2004.03.006.

Krol, K. M., R. G. Moulder, T. S. Lillard, T. Grossmann, and J. J. Connelly. "Epigenetic Dynamics in Infancy and the Impact of Maternal Engagement." *Science Advances* 5, no. 10 (2019), eaay0680. https://doi.org/10.1126/sciadv.aay0680.

Kruger, D. J. "Male Facial Masculinity Influences Attributions of Personality and Reproductive Strategy." *Personal Relationships* 13, no. 4 (2006), 451–463. https://doi.org/10.1111/j.1475–6811.2006.00129.x.

Kuo, P. X., J. Carp, K. C. Light, and K. M. Grewen. "Neural Responses to Infants Linked with Behavioral Interactions and Testosterone in Fathers." *Biological Psychiatry* 91, no. 2 (2012), 302–306. https://doi.org/10.1016/j.biopsycho.2012.08.002.

Kuo, P. X., E. K. Saini, E. Thomason, O. C. Schultheiss, R. Gonzalez, and B. L. Volling. "Individual Variation in Fathers' Testosterone Reactivity to Infant Distress Predicts Parenting Behaviors with Their 1-Year-Old Infants." *Developmental Psychobiology* 58, no. 3 (2016), 303–314. https://doi.org/10.1002/dev.21370.

Kurokawa, H., Y. Kinari, H. Okudaira, K. Tsubouchi, Y. Sai, M. Kikuchi, H. Higashida, and F. Ohtake. "Oxytocin-Trust Link in Oxytocin-Sensitive Participants and Those without Autistic Traits." *Frontiers in Neuroscience* 15 (2021), 659737. https://doi.org/10.3389/fnins.2021.659737.

LaFrance, M., M. A. Hecht, and E. L. Paluck. "The Contingent Smile: A Meta-Analysis of Sex Differences in Smiling." *Psychological Bulletin* 129, no. 2 (2003), 305–334. https://doi.org/10.1037/0033-2909.129.2.305.

LaMantia, A. S., and P. Rakic. "Axon Overproduction and Elimination in the Cerebral Cortex of the Developing Rhesus Monkey." *Journal of Neuroscience* 10 (1990), 2156–2175.

Lamb, M. E. "Father-Infant and Mother-Infant Interaction in 1st Year of Life." *Child Development* 48, no. 1 (1977), 167–181. https://doi.org/10.2307/1128896.

———, ed. *The Role of the Father in Child Development*. Hoboken, NJ: Wiley, 2010.

Lambert, C. T., A. C. Sabol, and N. G. Solomon. "Genetic Monogamy in Socially Monogamous Mammals Is Primarily Predicted by Multiple Life History Factors: A Meta-Analysis." *Frontiers in Ecology and Evolution* 6 (2018). https://doi.org/10.3389/fevo.2018.00139.

Lamborn, S. D., N. S. Mounts, L. Steinberg, and S. M. Dornbusch. "Patterns of Competence and Adjustment among Adolescents from Authoritative, Authoritarian, Indulgent, and Neglectful Families." *Child Development* 62, no. 5 (1991), 1049–1065. https://doi.org/10.1111/j.1467–8624.1991.tb01588.x.

Lancet Public Health. "Single Fathers: Neglected, Growing, and Important." *Lancet Public Health* 3, no. 3 (2018), e100. https://doi.org/10.1016/S2468-2667(18)30032-X.

Langos, D., L. Kulik, A. Ruiz-Lambides, and A. Widdig. "Does Male Care, Provided to Immature Individuals, Influence Immature Fitness in Rhesus Macaques?" *PloS One* 10, no. 9 (2015). https://doi.org/10.1371/journal.pone.0137841.

Lappan, S. "Male Care of Infants in a Siamang (*Symphalangus syndactylus*) Population Including Socially Monogamous and Polyandrous Groups." *Behavioral Ecology and Sociobiology* 62, no. 8 (2008), 1307–1317. https://doi.org/10.1007/s00265-008-0559-7.

Laurent, H. K., and J. C. Ablow. "A Cry in the Dark: Depressed Mothers Show Reduced Neural Activation to Their Own Infant's Cry." *Social Cognitive and Affective Neuroscience* 7, no. 2 (2012), 125–134. https://doi.org/10.1093/scan/nsq091.

———. "A Face a Mother Could Love: Depression-Related Maternal Neural Responses to Infant Emotion Faces." *Social Neuroscience* 8, no. 3 (2013), 228–339. https://doi.org/10.1080/17470919.2012.762039.

Laurent, H. K., A. Stevens, and J. C. Ablow. "Neural Correlates of Hypothalamic-Pituitary-Adrenal Regulation of Mothers with Their Infants." *Biological Psychiatry* 70, no. 9 (2011), 826–832. https://doi.org/10.1016/j.biopsych.2011.06.011.

Lee, A. W., and R. E. Brown. "Medial Preoptic Lesions Disrupt Parental Behavior in Both Male and Female California Mice (*Peromyscus californicus*)." *Behavioral Neuroscience* 116, no. 6 (2002), 968–975. https://doi.org/10.1037//0735-7044.116.6.968.

Lee, H. J., A. H. Macbeth, J. H. Pagani, and W. S. Young. "Oxytocin: The Great Facilitator of Life." *Progress in Neurobiology* 88, no. 2 (2009), 127–151. https://doi.org/10.1016/j.pneurobio.2009.04.001.

Lee, M. R., T. A. Shnitko, S. W. Blue, A. V. Kaucher, A. J. Winchell, D. W. Erikson, K. A. Grant, and L. Leggio. "Labeled Oxytocin Administered via the Intranasal Route Reaches the Brain in Rhesus Macaques." *Nature Communications* 11, no. 1 (2020), 2783. https://doi.org/10.1038/s41467-020-15942-1.

Lee, Richard B., and Richard Daly, eds. *The Cambridge Encyclopedia of Hunters and Gatherers*. Cambridge: Cambridge University Press, 2004.

Lee, Shawna J., Joyce Y. Lee, and Olivia D. Chang. "The Characteristics and Lived Experience of Modern Stay-at-Home Fathers." In *Handbook of Fathers and Child Development*, edited by Hiram Fitzgerald, Natasha Cabrera, Thomas Skjøthaug, Kai von Klitzing, and Júlia Scarano de Mendonça, 537–549. Berlin: Springer, 2020.

Lehmann, J., G. Fickenscher, and C. Boesch. "Kin Biased Investment in Wild Chimpanzees." *Behaviour* 143 (2006), 931–955. https://doi.org/10.1163/156853906776823635.

Lemaitre, J. F., V. Ronget, M. Tidiere, D. Allaine, V. Berger, A. Cohas, F. Colchero, et al. "Sex Differences in Adult Lifespan and Aging Rates of Mortality across Wild Mammals." *Proceedings of the National Academy of Sciences* 117, no. 15 (2020), 8546–8553. https://doi.org/10.1073/pnas.1911999117.

Leng, G., and M. Ludwig. "Intranasal Oxytocin: Myths and Delusions." *Biological Psychiatry* 79, no. 3 (2016), 243–250. https://doi.org/10.1016/j.biopsych.2015.05.003.

Lenzi, D., C. Trentini, P. Pantano, E. Macaluso, M. Iacoboni, G. L. Lenzi, and M. Ammaniti. "Neural Basis of Maternal Communication and Emotional Expression Processing during Infant Preverbal Stage." *Cerebral Cortex* 19, no. 5 (2009), 1124–1133. https://doi.org/10.1093/cercor/bhn153.

Li, T., X. Chen, J. Mascaro, E. Haroon, and J. K. Rilling. "Intranasal Oxytocin, But Not Vasopressin, Augments Neural Responses to Toddlers in Human Fathers." *Hormones and Behavior* 93 (2017), 193–202. https://doi.org/10.1016/j.yhbeh.2017.01.006.

Li, T., M. Horta, J. S. Mascaro, K. Bijanki, L. H. Arnal, M. Adams, R. G. Barr, and J. K. Rilling. "Explaining Individual Variation in Paternal Brain Responses to Infant Cries." *Physiology and Behavior* 193, pt. A (2018), 43–54. https://doi.org/10.1016/j.physbeh.2017.12.033.

Li, X. "Fathers' Involvement in Chinese Societies: Increasing Presence, Uneven Progress." *Child Development Perspectives* 14, no. 3 (2020), 150–156. https://doi.org/10.1111/cdep.12375.

Li, Xuan, and Michael Lamb. "Fathers in Chinese Culture." In *Fathers in Cultural Context*, edited by David W. Shwalb, Barbara J. Shwalb, and Michael E. Lamb, 15–41. New York: Routledge, 2013.

Ligon, J. D., R. Thornhill, M. Zuk, and K. Johnson. "Male-Male Competition, Ornamentation and the Role of Testosterone in Sexual Selection in Red Jungle Fowl." *Animal Behaviour* 40 (1990), 367–373. https://doi.org/10.1016/S0003-3472(05)80932-7.

Limerick, S. "Courtship Behavior and Oviposition of the Poison-Arrow Frog *Dendrobates pumilio*." *Herpetologica* 36, no. 1 (1980), 69–71.

Lincoln, G. A., F. Guinness, and R. V. Short. "Way in Which Testosterone Controls the Social and Sexual Behavior of the Red Deer Stag (*Cervus elaphus*)." *Hormones and Behavior* 3, no. 4 (1972), 375–396. https://doi.org/=10.1016/0018-506x(72)90027-X.

"Rhinoderma Darwinii." Animal Diveristy Web, 2000, accessed January 30, 2024. https://animaldiversity.org/accounts/Rhinoderma_darwinii/.

Logothetis, N. K. "The Neural Basis of the Blood-Oxygen-Level-Dependent Functional Magnetic Resonance Imaging Signal." *Philosophical Transactions of the Royal Society B: Biological Sciences* 357, no. 1424 (2002), 1003–1037.

Lorberbaum, J. P., J. D. Newman, A. R. Horwitz, J. R. Dubno, R. B. Lydiard, M. B. Hamner, D. E. Bohning, and M. S. George. "A Potential Role for Thalamocingulate Circuitry in Human Maternal Behavior." *Biological Psychiatry* 51, no. 6 (2002), 431–445. https://doi.org/10.1016/S0006-3223(01)01284-7.

Lovejoy, C. O. "The Origin of Man." *Science* 211, no. 4480 (1981), 341–350. https://doi.org/10.1126/science.211.4480.341.

Lovejoy, M. C., P. A. Graczyk, E. O'Hare, and G. Neuman. "Maternal Depression and Parenting Behavior: A Meta-Analytic Review." *Clinical Psychology Review* 20, no. 5 (2000), 561–592. https://doi.org/10.1016/S0272-7358(98)00100-7.

Lowe, A. E., C. Hobaiter, C. Asiimwe, K. Zuberbuhler, and N. E. Newton-Fisher. "Intra-Community Infanticide in Wild, Eastern Chimpanzees: A 24-Year Review." *Primates* 61, no. 1 (2020), 69–82. https://doi.org/10.1007/s10329-019-00730-3.

Lowe, A. E., C. Hobaiter, and N. E. Newton-Fisher. "Countering Infanticide: Chimpanzee Mothers Are Sensitive to the Relative Risks Posed by Males on Differing Rank Trajectories." *American Journal of Physical Anthropology* 168, no. 1 (2019), 3–9. https://doi.org/10.1002/ajpa.23723.

Lucassen, N., A. Tharner, M. H. Van IJzendoorn, M. J. Bakermans-Kranenburg, B. L. Volling, F. C. Verhulst, M. P. Lambregtse-Van den Berg, and H. Tiemeier. "The Association between Paternal Sensitivity and Infant-Father Attachment Security: A Meta-Analysis of Three Decades of Research." *Journal of Family Psychology* 25, no. 6 (2011), 986–992. https://doi.org/10.1037/a0025855.

Lukas, D., and T. Clutton-Brock. "Evolution of Social Monogamy in Primates Is Not Consistently Associated with Male Infanticide." *Proceedings of the National Academy of Sciences* 111, no. 17 (2014), E1674. https://doi.org/10.1073/pnas.1401012111.

Lunn, S. F., and A. S. McNeilly. "Failure of Lactation to Have a Consistent Effect on Interbirth Interval in the Common Marmoset, *Callithrix jacchus jacchus*." *Folia Primatology* 37 (1982), 99–105.Luyckx, K., E. A. Tildesley, B. Soenens, J. A. Andrews, S. E. Hampson, M. Peterson, and B. Duriez. "Parenting and Trajectories of Children's Maladaptive Behaviors: A 12-Year Prospective Community Study." *Journal of Clinical*

Child and Adolescent Psychology 40, no. 3 (2011), 468–478. https://doi.org/10.1080 /15374416.2011.563470.

Lynch, K. A., and J. B. Greenhouse. "Risk-Factors for Infant-Mortality in 19th-Century Sweden." *Population Studies* 48, no. 1 (1994), 117–133. https://doi.org/10 .1080/0032472031000147506.

Macdonald, D. W., L. A. D. Campbell, J. F. Kamler, J. Marchino, G. Werhahn, and C. Sillero-Zubiri. "Monogamy: Cause, Consequence, or Corollary of Success in Wild Canids?" *Frontiers in Ecology and Evolution* 7 (2019). https://doi.org/10.3389/fevo .2019.00341.

Macdonald, K., and D. Feifel. "Oxytocins Role in Anxiety: A Critical Appraisal." *Brain Research* (2014). https://doi.org/10.1016/j.brainres.2014.01.025.

Macdonald, K., and R. D. Parke. "Bridging the Gap—Parent-Child Play Interaction and Peer Interactive Competence." *Child Development* 55, no. 4 (1984), 1265–1277. https://doi.org/10.2307/1129996.

MacLean, Paul D. *The Triune Brain in Evolution: Role in Paleocerebral Functions.* New York: Plenum Press, 1990.

Macrides, F., A. Bartke, and S. Dalterio. "Strange Females Increase Plasma Testosterone Levels in Male Mice." *Science* 189, no. 4208 (1975), 1104–1106. https://doi.org /10.1126/science.1162363.

Madden, J. R., and T. H. Clutton-Brock. "Experimental Peripheral Administration of Oxytocin Elevates a Suite of Cooperative Behaviours in a Wild Social Mammal." *Philosophical Transactions of the Royal Society B: Biological Sciences* 278, no. 1709 (2011), 1189–1194. https://doi.org/10.1098/rspb.2010.1675.

Madhavan, S., N. W. Townsend, and A. I. Garey. "'Absent Breadwinners' Father–Child Connections and Paternal Support in Rural South Africa." *Journal of Southern African Studies* 34 (2008), 647–663. https://doi.org/10.1080/03057070802259902.

Mah, B. L., M. J. Bakermans-Kranenburg, M. H. Van Ijzendoorn, and R. Smith. "Oxytocin Promotes Protective Behavior in Depressed Mothers: A Pilot Study with the Enthusiastic Stranger Paradigm." *Depression and Anxiety* 32, no. 2 (2014), 76–81. https://doi.org/10.1002/da.22245.

Majdandzic, M., E. L. Moller, W. de Vente, S. M. Bogels, and D. C. van den Boom. "Fathers' Challenging Parenting Behavior Prevents Social Anxiety Development in Their 4-Year-Old Children: A Longitudinal Observational Study." *Journal of Abnormal Child Psychology* 42, no. 2 (2014), 301–310. https://doi.org/10.1007/s10802-013 -9774-4.

Mak, G. K., and S. Weiss. "Paternal Recognition of Adult Offspring Mediated by Newly Generated CNS Neurons." *Nature Neuroscience* 13, no. 6 (2010), 753-U134. https://doi.org/10.1038/nn.2550.

Malo, A. F., E. R. S. Roldan, J. J. Garde, A. J. Soler, J. Vicente, C. Gortazar, and M. Gomendio. "What Does Testosterone Do for Red Deer Males?" *Philosophical Transactions of the Royal Society B: Biological Sciences* 276, no. 1658 (2009), 971–980. https://doi.org/10.1098/rspb.2008.1367.

Manson, J. H., and R. W. Wrangham. "Intergroup Aggression in Chimpanzees and Humans." *Current Anthropology* 32, no. 4 (1991), 369–390. https://doi.org/10.1086/203974.

Marler, P., S. Peters, G. F. Ball, A. M. Dufty, and J. C. Wingfield. "The Role of Sex Steroids in the Acquisition and Production of Birdsong." *Nature* 336, no. 6201 (1988), 770–772. https://doi.org/10.1038/336770a0.

Marlowe, F. W. "Paternal Investment and the Human Mating System." *Behavioural Processes* 51, no. 1–3 (2000), 45–61. https://doi.org/10.1016/S0376–6357(00)00118–2

———. "A Critical Period for Provisioning by Hadza Men: Implications for Pair Bonding." *Evolution and Human Behavior* 24, no. 3 (2003), 217–229. https://doi.org/10.1016/S1090–5138(03)00014-X.

Marshall, John. "The Hunters." Documentary Educational Resources, 1958.

Marsiglio, W. "Paternal Engagement Activities with Minor Children." *Journal of Marriage and the Family* 53, no. 4 (1991), 973–986. https://doi.org/10.2307/353001.

Marsiglio, William, and Ramon Hinojosa. "Stepfathers' Lives." In *The Role of the Father in Child Development.* Hoboken, NJ: Wiley, 2010.

Martin N. Muller, Melissa Emery Thompson, Sonya M. Kahlenberg sand Richard W. Wrangham "Sexual Coercion by Male Chimps Shows That Female Choice May Be More Apparent Than Real." *Behavioral Ecology and Sociobiology* 65 (2011, 921–933. https://doi.org/10.1007/s00265-010-1093-y.

Martinez-Garcia, M., M. Paternina-Die, M. Desco, O. Vilarroya, and S. Carmona. "Characterizing the Brain Structural Adaptations across the Motherhood Transition." *Frontiers in Global Women's Health* 2 (2021), 742775. https://doi.org/10.3389/fgwh.2021.742775.

Mascaro, J. S., P. D. Hackett, H. Gouzoules, A. Lori, and J. K. Rilling. "Behavioral and Genetic Correlates of the Neural Response to Infant Crying among Human Fathers." *Social and Cognitive Affective Neuroscience* 9, no. 11 (2014), 1704–1712. https://doi.org/10.1093/scan/nst166.

Mascaro, J. S., P. D. Hackett, and J. K. Rilling. "Testicular Volume Is Inversely Correlated with Nurturing-Related Brain Activity in Human Fathers." *Proceedings of the*

National Academy of Sciences 110, no. 39 (2013), 15746–15751. https://doi.org/10 .1073/pnas.1305579110.

———. "Differential Neural Responses to Child and Sexual Stimuli in Human Fathers and Non-Fathers and Their Hormonal Correlates." *Psychoneuroendocrinology* 46 (2014), 153–163. https://doi.org/10.1016/j.psyneuen.2014.04.014.

Mascaro, J. S., K. E. Rentscher, P. D. Hackett, M. R. Mehl, and J. K. Rilling. "Child Gender Influences Paternal Behavior, Language, and Brain Function." *Behavioral Neuroscience* 131, no. 3 (2017), 262–273. https://doi.org/10.1037/bne0000199.

Mathur, M. B., E. Epel, S. Kind, M. Desai, C. G. Parks, D. P. Sandler, and N. Khazeni. "Perceived Stress and Telomere Length: A Systematic Review, Meta-Analysis, and Methodologic Considerations for Advancing the Field." *Brain, Behavior, and Immunity* 54 (2016), 158–169. https://doi.org/10.1016/j.bbi.2016.02.002.

Mazur, Allan. *Biosociology of Dominance and Deference.* New York: Rowman and Littlefield, 2005.

McFadden, Karen E., and Catherine S. Tamis-LeMonda. "Fathers in the U.S." In *Fathers in Cultural Context*, edited by David W. Shwalb, Barbara J. Shwalb, and Michael E. Lamb, 250–276. New York: Routledge, 2013.

Mchenry, H. M. "Behavioral Ecological Implications of Early Hominid Body-Size." *Journal of Human Evolution* 27, no. 1–3 (994), 77–87. https://doi.org/10.1006/jhev .1994.1036.

McLanahan, S., L. Tach, and D. Schneider. "The Causal Effects of Father Absence." *Annual Review of Sociology* 39 (2013), 399–427. https://doi.org/10.1146/annurev-soc -071312-145704.

McNeilly, A. S., I. C. Robinson, M. J. Houston, and P. W. Howie. "Release of Oxytocin and Prolactin in Response to Suckling." *British Medical Journal (Clinical Research Edition)* 286, no. 6361 (1983), 257–259. https://doi.org/10.1136/bmj.286.6361.257.

McNulty, J. K., C. A. Wenner, and T. D. Fisher. "Longitudinal Associations among Relationship Satisfaction, Sexual Satisfaction, and Frequency of Sex in Early Marriage." *Archives of Sexual Behavior* 45, no. 1 (2016), 85–97. https://doi.org/10.1007 /s10508-014-0444-6.

Meehan, C. L. "The Effects of Residential Locality on Parental and Alloparental Investment among the Aka Foragers of the Central African Republic." *Human Nature* 16, no. 1 (2005), 58–80. https://doi.org/10.1007/s12110-005-1007-2.

Mehta, P. H., and R. A. Josephs. "Testosterone Change after Losing Predicts the Decision to Compete Again." *Hormones and Behavior* 50, no. 5 (2006), 684–692. https:// doi.org/10.1016/j.yhbeh.2006.07.001.

————. "Testosterone and Cortisol Jointly Regulate Dominance: Evidence for a Dual-Hormone Hypothesis." *Hormones and Behavior* 58, no. 5 (2010), 898–906. https://doi.org/10.1016/j.yhbeh.2010.08.020.

Meisenberg, G., and W. H. Simmons. "Minireview. Peptides and the Blood-Brain Barrier." *Life Sciences* 32, no. 23 (1983), 2611–2623. https://doi.org/10.1016/0024-3205(83)90352-1.

Menard, N., F. von Segesser, W. Scheffrahn, J. Pastorini, D. Vallet, B. Gaci, R. D. Martin, and A. Gautier-Hion. "Is Male-Infant Caretaking Related to Paternity and/or Mating Activities in Wild Barbary Macaques (*Macaca sylvanus*)?" *Comptes Rendus de l'Académie des Sciences, series III,* 324, no. 7 (2001), 601–610. https://doi.org/10.1016/s0764-4469(01)01339-7.

Meyer-Lindenberg, A., G. Domes, P. Kirsch, and M. Heinrichs. "Oxytocin and Vasopressin in the Human Brain: Social Neuropeptides for Translational Medicine." *Nature Reviews Neuroscience* 12, no. 9 (2011), 524–238. https://doi.org/10.1038/nrn3044.

Michalska, K. J., J. Decety, C. Liu, Q. Chen, M. E. Martz, S. Jacob, A. E. Hipwell, et al. "Genetic Imaging of the Association of Oxytocin Receptor Gene (OXTR) Polymorphisms with Positive Maternal Parenting." *Frontiers in Behavioral Neuroscience* 8 (2014), 21. https://doi.org/10.3389/fnbeh.2014.00021.

Miller, Benjamin Graham, Stephanie Kors, and Jenny Macfie. "No Differences? Meta-Analytic Comparisons of Psychological Adjustment in Children of Gay Fathers and Heterosexual Parents." *Psychology of Sexual Orientation and Gender Diversity* 4, no. 1 (2017), 14–22. https://doi.org/10.1037/sgd0000203.

Miller, S. L., and J. K. Maner. "Scent of a Woman: Men's Testosterone Responses to Olfactory Ovulation Cues." *Psychological Science* 21, no. 2 (2010), 276–283. https://doi.org/10.1177/0956797609357733.

Miller, S. L., J. K. Maner, and J. K. McNulty. "Adaptive Attunement to the Sex of Individuals at a Competition: The Ratio of Opposite- to Same-Sex Individuals Correlates with Changes in Competitors' Testosterone Levels." *Evolution and Human Behavior* 33, no. 1 (2012), 57–63. https://doi.org/10.1016/j.evolhumbehav.2011.05.006.

Min, K. J., C. K. Lee, and H. N. Park. "The Lifespan of Korean Eunuchs." *Current Biology* 22, no. 18 (2012), R792-R793. https://doi.org/0.1016/j.cub.2012.06.036.

Miner, H. "Body Ritual among the Nacirema." *American Anthropologist* 58, no. 3 (1956), 503–507. https://doi.org/10.1525/aa.1956.58.3.02a00080.

Minge, C., A. Berghanel, O. Schulke, and J. Ostner. "Patterns and Consequences of Male-Infant Relationships in Wild Assamese Macaques (*Macaca assamensis*)." *International Journal of Primatology* 37, no. 3 (2016), 350–370. https://doi.org/10.1007/s10764-016-9904-2.

Minkov, M., and K. Beaver. "A Test of Life History Strategy Theory as a Predictor of Criminal Violence across 51 Nations." *Personality and Individual Differences* 97 (2016), 186–192. https://doi.org/10.1016/j.paid.2016.03.063.

Mishra, S. K., B. Ram, A. Singh, and A. Yadav. "Birth Order, Stage of Infancy and Infant Mortality in India." *Journal of Biosocial Science* 50, no. 5 (2018), 604–625. https://doi.org/10.1017/S0021932017000487.

Mitani, J. C., and D. Watts. "The Evolution of Non-maternal Caretaking among Anthropoid Primates: Do Helpers Help?" *Behavioral Ecology and Sociobiology* 40, no. 4 (1997): 213–220. https://doi.org/10.1007/s002650050335.

Mitani, J. C., D. P. Watts, and S. J. Amsler. "Lethal Intergroup Aggression Leads to Territorial Expansion in Wild Chimpanzees." *Current Biology* 20, no. 12 (2010), R507-R508. https://doi.org/10.1016/j.cub.2010.04.021.

Modig, K., M. Talback, J. Torssander, and A. Ahlbom. "Payback Time? Influence of Having Children on Mortality in Old Age." *Journal of Epidemiology and Community Health* 71, no. 5 (2017), 424–430. https://doi.org/10.1136/jech-2016-207857.

Moll, J., R. Zahn, R. de Oliveira-Souza, F. Krueger, and J. Grafman. "Opinion: The Neural Basis of Human Moral Cognition." *Nature Reviews Neuroscience* 6, no. 10 (2005), 799–809. https://doi.org/10.1038/nrn1768.

Moller, A. P. "The Evolution of Monogamy: Mating Relationships, Parental Care, and Sexual Selection." In *Monogamy: Mating Strategies and Partnerships in Birds, Humans, and Other Mammals*, edited by U. H. Reichard and C. Boesch, 29–41. Cambridge: Cambridge University Press, 2003.

Moller, A. P., and J. J. Cuervo. "The Evolution of Paternity and Paternal Care in Birds." *Behavioral Ecology* 11, no. 5 (2000), 472–485. https://doi.org/10.1093/beheco/11.5.472.

Moller, E. L., M. Nikolic, M. Majdandzic, and S. M. Bogels. "Associations between Maternal and Paternal Parenting Behaviors, Anxiety and Its Precursors in Early Childhood: A Meta-Analysis." *Clinical Psychology Review* 45 (2016), 17–33. https://doi.org/10.1016/j.cpr.2016.03.002.

Moore, L. A. "Population Ecology of the Southern Cassowary *Casuarius casuarius johnsonii*, Mission Beach North Queensland." *Journal of Ornithology* 148, no. 3 (2007), 357–366. https://doi.org/10.1007/s10336-007-0137-1.

Morris, A. R., A. Turner, C. H. Gilbertson, G. Corner, A. J. Mendez, and D. E. Saxbe. "Physical Touch during Father-Infant Interactions Is Associated with Paternal Oxytocin Levels." *Infant Behavior and Development* 64 (2021), 101613. https://doi.org/10.1016/j.infbeh.2021.101613.

Morsey, Leila, and Richard Rothstein. *Mass Incarceration and Children's Outcomes: Criminal Justice Policy Is Education Policy*. Washington, DC: Economic Policy Institute 2016.

Mougeot, F., J. R. Irvine, L. Seivwright, S. M. Redpath, and S. Piertney. "Testosterone, Immunocompetence, and Honest Sexual Signaling in Male Red Grouse." *Behavioral Ecology* 15, no. 6 (2004), 930–937. https://doi.org/10.1093/beheco/arh087.

Moynihan, Daniel Patrick. *The Negro Family: The Case for National Action*. Washington, DC: US Department of Labor, 1965.

Muehlenbein, M. P., and D. P. Watts. "The Costs of Dominance: Testosterone, Cortisol and Intestinal Parasites in Wild Male Chimpanzees." *BioPsychoSocial Medicine* 4 (2010), 21. https://doi.org/10.1186/1751-0759-4-21.

Muehlenbein, M. P., D. P. Watts, and P. L. Whitten. "Dominance Rank and Fecal Testosterone Levels in Adult Male Chimpanzees (*Pan troglodytes schweinfurthii*) at Ngogo, Kibale National Park, Uganda." *American Journal of Primatology* 64, no. 1 (2004), 71–82. https://doi.org/10.1002/ajp.20062.

Muldal, A. M., J. D. Moffatt, and R. J. Robertson. "Parental Care of Nestlings by Male Red-Winged Blackbirds." *Behavioral Ecology and Sociobiology* 19, no. 2 (1986), 105–114. https://doi.org/10.1007/Bf00299945.

Muller, M. N. "Testosterone and Reproductive Effort in Male Primates." *Hormones and Behavior* 91 (2017), 36–51. https://doi.org/10.1016/j.yhbeh.2016.09.001.

Muller, M. N., F. W. Marlowe, R. Bugumba, and P. T. Ellison. "Testosterone and Paternal Care in East African Foragers and Pastoralists." *Philosophical Transactions of the Royal Society B: Biological Sciences* 276, no. 1655 (2009), 347–354. https://doi.org/10.1098/rspb.2008.1028.

Muller, M. N., and R. W. Wrangham. "Dominance, Aggression and Testosterone in Wild Chimpanzees: A Test of the 'Challenge Hypothesis.'" *Animal Behavior* 67 (2004), 113–123. https://doi.org/10.1016/j.anbehav.2003.03.013.

———. "Mortality Rates among Kanyawara Chimpanzees." *Journal of Human Evolution* 66 (2014), 107–114. https://doi.org/10.1016/j.jhevol.2013.10.004.

Muller, Martin N., Richard W. Wrangham, and David R. Pilbeam, eds. *Chimpanzees and Human Evolution*. Cambridge, MA: Belknap Press of Harvard University Press, 2017a.

———. *Chimpanzees and Human Evolution*. Cambridge, MA: Belknap Press of Harvard University Press, 2017b.

Mumme, R. L. "Do Helpers Increase Reproductive Success: An Experimental-Analysis in the Florida Scrub Jay." *Behavioral Ecology and Sociobiology* 31, no. 5 (1992), 319–328. https://doi.org/10.1007/BF00177772.

Murphy, M. R., P. D. MacLean, and S. C. Hamilton. "Species-Typical Behavior of Hamsters Deprived from Birth of the Neocortex." *Science* 213, no. 4506 (1981), 459–461. https://doi.org/10.1126/science.7244642.

Murray, C. M., M. A. Stanton, E. V. Lonsdorf, E. E. Wroblewski, and A. E. Pusey. "Chimpanzee Fathers Bias Their Behaviour towards Their Offspring." *Royal Society Open Science* 3, no. 11 (2016), 160441. https://doi.org/10.1098/rsos.160441.

Murray, Henry. *Analysis of The Personality of Adolph Hitler.* Washington, DC: U.S. Office of Strategic Services, 1943.

Naber, F., M. H. van Ijzendoorn, P. Deschamps, H. van Engeland, and M. J. Bakermans-Kranenburg. "Intranasal Oxytocin Increases Fathers' Observed Responsiveness during Play with Their Children: A Double-Blind within-Subject Experiment." *Psychoneuroendocrinology* 35, no. 10 (2010), 1583–1586. https://doi.org/10.1016/j.psyneuen.2010.04.007.

Nagasawa, M., S. Mitsui, S. En, N. Ohtani, M. Ohta, Y. Sakuma, T. Onaka, K. Mogi, and T. Kikusui. "Social Evolution. Oxytocin-Gaze Positive Loop and the Coevolution of Human-Dog Bonds." *Science* 348, no. 6232 (2015), 333–336. https://doi.org/10.1126/science.1261022.

Nave, G., C. Camerer, and M. McCullough. "Does Oxytocin Increase Trust in Humans? A Critical Review of Research." *Perspectives on Psychological Science* 10, no. 6 (2015), 772–789. https://doi.org/10.1177/1745691615600138.

Naylor, R., S. J. Richardson, and B. M. McAllan. "Boom and Bust: A Review of the Physiology of the Marsupial Genus Antechinus." *Journal of Comparative Physiology B: Biochemical Systems and Environmental Physiology* 178, no. 5 (2008), 545–562. https://doi.org/10.1007/s00360-007-0250-8.

Nettle, D. "Dying Young and Living Fast: Variation in Life History across English Neighborhoods." *Behavioral Ecology* 21, no. 2 (2010), 387–395. https://doi.org/10.1093/beheco/arp202.

Nielsen, R., C. Bustamante, A. G. Clark, S. Glanowski, T. B. Sackton, M. J. Hubisz, A. Fledel-Alon, et al. "A Scan for Positively Selected Genes in the Genomes of Humans and Chimpanzees." *PLoS Biology* 3, no. 6 (2005), e170. https://doi.org/10.1371/journal.pbio.0030170.

Ning, K., L. Zhao, M. Franklin, W. Matloff, I. Batta, N. Arzouni, F. Sun, and A. W. Toga. "Parity Is Associated with Cognitive Function and Brain Age in Both Females and Males." *Scientific Reports* 10, no. 1 (2020), 6100. https://doi.org/10.1038/s41598-020-63014-7.

Nishiyama, K., and J. V. Johnson. "Karoshi—Death from Overwork: Occupational Health Consequences of Japanese Production Management." *International Journal of Health Services* 27, no. 4 (1997), 625–641. https://doi.org/0.2190/1jpc-679v-Dynt -Hj6g.

Novak, M., and H. Harlow. "Social Recovery of Monkeys Isolated for the First Year of Life." *Developmental Psychology* 11 (1975), 453–465. https://doi.org/10.1037 /h0078077.

Numan, M. "Motivational Systems and the Neural Circuitry of Maternal Behavior in the Rat." *Developmental Psychobiology* 49, no. 1 (2007), 12–21. https://doi.org/10 .1002/dev.20198.

Numan, Michael. *The Parental Brain: Mechanisms, Development and Evolution.* New York: Oxford University Press, 2020.

Nunes, S., J. E. Fite, and J. A. French. "Variation in Steroid Hormones Associated with Infant Care Behaviour and Experience in Male Marmosets (*Callithrix kuhlii*)." *Animal Behaviour* 60 (2000), 857–865. https://doi.org/10.1006/anbe.2000.1524.

Nunes, S., J. E. Fite, K. J. Patera, and J. A. French. "Interactions among Paternal Behavior, Steroid Hormones, and Parental Experience in Male Marmosets (*Callithrix kuhlii*)." *Hormones and Behavior* 39, no. 1 (Febuary 2001), 70–82. https://doi.org/10 .1006/hbeh.2000.1631.

O'Connor, J. J. M., K. Pisanski, C. C. Tigue, P. J. Fraccaro, and D. R. Feinberg. "Perceptions of Infidelity Risk Predict Women's Preferences for Low Male Voice Pitch in Short-Term over Long-Term Relationship Contexts." *Personality and Individual Differences* 56 (2014), 73–77. https://doi.org/10.1016/j.paid.2013.08.029.

Ohba, S. Y., N. Okuda, and S. Kudo. "Sexual Selection of Male Parental Care in Giant Water Bugs." *Royal Society Open Science* 3, no. 5 (May 2016), 150720. https://doi.org /10.1098/rsos.150720.

Olazabal, D. E. "Role of Oxytocin in Parental Behaviour." *Journal of Neuroendocrinology* 30, no. 7 (2018). https://doi.org/10.1111/jne.12594.

Olweus, D., A. Mattsson, D. Schalling, and H. Low. "Testosterone, Aggression, Physical, and Personality Dimensions in Normal Adolescent Males." *Psychosomatic Medicine* 42, no. 2 (March 1980), 253–69. https://doi.org/10.1097/00006842-198003000 -00003.

Opie, C., Q. D. Atkinson, R. I. Dunbar, and S. Shultz. "Male Infanticide Leads to Social Monogamy in Primates." *Proceedings of the National Academy of Sciences* 110, no. 33 (13 2013), 13328–13332. https://doi.org/10.1073/pnas.1307903110.

Ottenbacher, K. J., M. E. Ottenbacher, A. J. Ottenbacher, A. A. Acha, and G. V. Ostir. "Androgen Treatment and Muscle Strength in Elderly Men: A Meta-Analysis." *Journal of the American Geriatrics Society* 54, no. 11 (2006), 1666–1673. https://doi.org/10 .1111/j.1532-5415.2006.00938.x.

Owens, I. P. F. "Male-Only Care and Classical Polyandry in Birds: Phylogeny, Ecology and Sex Differences in Remating Opportunities." *Philosophical Transactions of the Royal Society B: Biological Sciences* 357, no. 1419 (2002a), 283–293. https://doi.org/10.1098/rstb.2001.0929.

———. "Sex Differences in Mortality Rate." *Science* 297, no. 5589 (2002b), 2008–2009. https://doi.org/0.1126/science.1076813.

Pace, G. T., S. J. Lee, and A. Grogan-Kaylor. "Spanking and Young Children's Socioemotional Development in Low- and Middle-Income Countries." *Child Abuse and Neglect* 88 (2019), 84–95. https://doi.org/10.1016/j.chiabu.2018.11.003.

Pancsofar, N., and L. Vernon-Feagans. "Mother and Father Language Input to Young Children: Contributions to Later Language Development." *Journal of Applied Developmental Psychology* 27, no. 6 (2006), 571–587. https://doi.org/10.1016/j.appdev.2006.08.003.

Panksepp, Jaak. "Play, ADHD, and the Construction of the Social Brain: Should the First Class Each Day Be Recess?" *American Journal of Play* 1, no. 1 (2008), 55–79. https://files.eric.ed.gov/fulltext/EJ1069040.pdf.

Panter-Brick, C., A. Burgess, M. Eggerman, F. McAllister, K. Pruett, and J. F. Leckman. "Practitioner Review: Engaging Fathers—Recommendations for a Game Change in Parenting Interventions Based on a Systematic Review of the Global Evidence." *Journal of Child Psychology and Psychiatry* 55, no. 11 (2014), 1187–1212. https://doi.org/10.1111/jcpp.12280.

Paquette, Daniel. "Theorizing the Father–Child Relationship: Mechanisms and Developmental Outcomes." *Human Development* 47 (2004), 193–219. https://doi.org/10.1159/000078723.

Paquette, Daniel, Carole Gagnon, and Julio Macario de Medeiros. "Fathers and the Activation Relationship." In *Handbook of Fathers and Child Development*, edited by H. E. Fitzgerald, K. von Klitzing, N. Cabrera, J. Scarano de Mendonça, and Th. Skjøthaug, Berlin: Springer, 2020.

Parr, L. A., M. Heintz, E. Lonsdorf, and E. Wroblewski. "Visual Kin Recognition in Nonhuman Primates (*Pan troglodytes* and *Macaca mulatta*) Inbreeding Avoidance or Male Distinctiveness?" *Journal of Comparative Psychology* 124, no. 4 (2010), 343–350. https://doi.org/10.1037/a0020545.

Parssinen, M., U. Kujala, E. Vartiainen, S. Sarna, and T. Seppala. "Increased Premature Mortality of Competitive Powerlifters Suspected to Have Used Anabolic Agents." *International Journal of Sports Medicine* 21, no. 3 (2003), 225–227. https://doi.org/10.1055/s-2000-304.

Passingham, Richard E. *What Is Special about the Human Brain.* Oxford: Oxford University Press, 2008.

Paternina-Die, M., M. Marchtinez-Garcia, C. Pretus, E. Hoekzema, E. Barba-Muller, D. Marchtin de Blas, C. Pozzobon, et al. "The Paternal Transition Entails Neuroanatomic Adaptations That Are Associated with the Father's Brain Response to His Infant Cues." *Cerebral Cortex Communications* 1, no. 1 (2020), tgaa082. https://doi.org/10.1093/texcom/tgaa082.

Patterson, Charlotte. "Parents' Sexual Orientation and Children's Development." *Child Development Perspectives* 11, no. 1 (2016), 45–49. https://doi.org/10.1111/cdep.12207.

Paulk, Amber, and Ryan Zayac. "Attachment Style as a Predictor of Risky Sexual Behavior in Adolescents." *Journal of Social Sciences* 9, no. 2 (2013), 42–47. https://doi.org/10.3844/jssp.2013.42.47.

Paulson, J. F., and S. D. Bazemore. "Prenatal and Postpartum Depression in Fathers and Its Association with Maternal Depression: A Meta-Analysis." *JAMA* 303, no. 19 (2010), 1961–1969. https://doi.org/10.1001/jama.2010.605.

Paulson, James F., Kelsey T. Ellis, and Regina Alexander. "Paternal Prenatal Depression and Postpartum Depression." In *Handbook of Fathers and Child Development*, edited by H. E. Fitzgerald, K. von Klitzing, N. Cabrera, J. Scarano de Mendonça, and Th. Skjøthaug. Berlin: Springer, 2020.

Paus, Tomas, Alex Zijdenbos, H. Keith Worsely, D. Louis Collins, Jonathan Blumenthal, Jay N. Giedd, Judith L. Rapoport, and Alan C. Evans. "Structural Maturation of Neural Pathways in Children and Adolescents: In Vivo Study." *Science* 283 (1999), 1908–1911. https://doi.org/10.1126/science.283.5409.1908.

Pawluski, J. L., E. Hoekzema, B. Leuner, and J. S. Lonstein. "Less Can Be More: Fine Tuning the Maternal Brain." *Neuroscience and Biobehavioral Reviews* 133 (2022), 104475. https://doi.org/10.1016/j.neubiorev.2021.11.045.

Pawluski, J. L., J. S. Lonstein, and A. S. Fleming. "The Neurobiology of Postpartum Anxiety and Depression." *Trends in Neurosciences* 40, no. 2 (2017), 106–120. https://doi.org/10.1016/j.tins.2016.11.009.

Pedersen, C. A., J. D. Caldwell, C. Walker, G. Ayers, and G. A. Mason. "Oxytocin Activates the Postpartum Onset of Rat Maternal Behavior in the Ventral Tegmental and Medial Preoptic Areas." *Behavioral and Brain Sciences* 108, no. 6 (1994), 1163–1171. https://doi.org/10.1037//0735-7044.108.6.1163.

Pellis, Sergio M., Vivien C. Pellis, and Heather C. Bell. "The Function of Play in the Development of the Social Brain." *American Journal of Play* 2, no. 3 (2010), 278–296.

Pepper, G. V., and D. Nettle. "The Behavioural Constellation of Deprivation: Causes and Consequences." *Behavioral and Brain Sciences* 40 (2017). https://doi.org/10.1017/S0140525X1600234X.

Perkeybile, A. M., C. S. Carter, K. L. Wroblewski, M. H. Puglia, W. M. Kenkel, T. S. Lillard, T. Karaoli, et al. "Early Nurture Epigenetically Tunes the Oxytocin Receptor." *Psychoneuroendocrinology* 99 (2019), 128–136. https://doi.org/10.1016/j.psyneuen .2018.08.037.

Perrigo, G., W. C. Bryant, and F. S. Vomsaal. "Fetal, Hormonal and Experiential Factors Influencing the Mating-Induced Regulation of Infanticide in Male House Mice." *Physiology and Behavior* 46, no. 2 (1989), 121–128. https://doi.org/10.1016 /0031–9384(89)90244–8.

Peters, M., L. W. Simmons, and G. Rhodes. "Testosterone Is Associated with Mating Success But Not Attractiveness or Masculinity in Human Males." *Animal Behaviour* 76 (2008), 297–303. https://doi.org/10.1016/j.anbehav.2008.02.008.

Petts, R. J., and C. Knoester. "Are Parental Relationships Improved If Fathers Take Time Off of Work after the Birth of a Child?" *Social Forces* 98, no. 3 (2020), 1223– 1256. https://doi.org/10.1093/sf/soz014.

Phalane, K. G., C. Tribe, H. C. Steel, M. C. Cholo, and V. Coetzee. "Facial Appearance Reveals Immunity in African Men." *Scientific Reports* 7 (2017). https://doi.org /744310.1038/s41598-017-08015-9.

Pickrell, John. "How the Earliest Mammals Thrived Alongside Dinosaurs." *Nature* 574 (2019), 468–472. https://www.nature.com/articles/d41586-019-03170-7.

Pitts, Leonard. *Becoming Dad: Black Men and the Journey to Fatherhood*. Marietta, GA: Longstreet, 2006.

Plavcan, J. Michael. "Sexual Dimorphism in Hominin Ancestors." In *The International Encyclopedia of Anthropology*, edited by Hilary Callan. Hoboken, NJ: Wiley, 2018.

Pleck, J. H. "Fatherhood and Masculinity." In *The Role of the Father in Child Development*, edited by Michael Lamb, 27–57. Hoboken, NJ: Wiley, 2010.

Pollet, T. V., L. van der Meij, K. D. Cobey, and A. P. Buunk. "Testosterone Levels and Their Associations with Lifetime Number of Opposite Sex Partners and Remarriage in a Large Sample of American Elderly Men and Women." *Hormones and Behavior* 60, no. 1 (2011), 72–77. https://doi.org/10.1016/j.yhbeh.2011.03.005.

Pope, S. K., L. Whiteside, J. Brooksgunn, K. J. Kelleher, V. I. Rickert, R. H. Bradley, and P. H. Casey. "Low-Birth-Weight Infants Born to Adolescent Mothers: Effects of Coresidency with Grandmother on Child-Development." *JAMA* 269, no. 11 (1993), 1396–1400. https://doi.org/10.1001/jama.269.11.1396.

Post, C., and B. Leuner. "The Maternal Reward System in Postpartum Depression." *Archives of Women's Mental Health* 22, no. 3 (2019), 417–429. https://doi.org/10.1007 /s00737-018-0926-y.

Pougnet, E., L. A. Serbin, D. M. Stack, J. E. Ledingham, and A. E. Schwartzman. "The Intergenerational Continuity of Fathers' Absence in a Socioeconomically Disadvantaged Sample." *Journal of Marriage and Family* 74, no. 3 (2012), 540–555. https://doi.org/10.1111/j.1741-3737.2012.00962.x.

Pozzi, L., J. A. Hodgson, A. S. Burrell, K. N. Sterner, R. L. Raaum, and T. R. Disotell. "Primate Phylogenetic Relationships and Divergence Dates Inferred from Complete Mitochondrial Genomes." *Molecular Phylogenetics and Evolution* 75 (2014), 165–183. https://doi.org/10.1016/j.ympev.2014.02.023.

Prall, S. P., and B. A. Scelza. "Why Men Invest in Non-Biological Offspring: Paternal Care and Paternity Confidence among Himba Pastoralists." *Philosophical Transactions of the Royal Society B: Biological Sciences* 287, no. 1922 (2020). https://doi.org/10.1098/rspb.2019.2890.

Prudom, S. L., C. A. Broz, N. Schultz-Darken, C. T. Ferris, C. Snowdon, and T. E. Ziegler. "Exposure to Infant Scent Lowers Serum Testosterone in Father Common Marmosets (*Callithrix jacchus*)." *Biology Letters* 4, no. 6 (2008), 603–605. https://doi.org/10.1098/rsbl.2008.0358.

Purvis, K., and N. B. Haynes. "Effect of Odor of Female Rat Urine on Plasma Testosterone Concentrations in Male Rats." *Journal of Reproduction and Fertility* 53, no. 1 (1978), 63–65. https://doi.org/10.1530/jrf.0.0530063.

Pusey, Ann. "Magnitude and Sources of Variation in Female Reproductive Performance." In *The Evolution of Primate Societies*, edited by John C. Mitani, Josep Call, Peter M. Kappeler, Ryne A. Palombit, and Joan B. Silk, 343–366. Chicago: University of Chicago Press, 2012.

Ranson, Kenna E., and Liana J. Urichuk. "The Effect of Parent–Child Attachment Relationships on Child Biopsychosocial Outcomes: A Review." *Early Child Development and Care* 178, no. 2 (2008), 129–152. https://doi.org/10.1080/03004430600685282.

Rassovsky, Y., A. Harwood, O. Zagoory-Sharon, and R. Feldman. "Martial Arts Increase Oxytocin Production." *Scientific Reports* 9, no. 1 (2019), 12980. https://doi.org/10.1038/s41598-019-49620-0.

Reed, W. L., M. E. Clark, P. G. Parker, S. A. Raouf, N. Arguedas, D. S. Monk, E. Snajdr, V. Nolan, and E. D. Ketterson. "Physiological Effects on Demography: A Long-Term Experimental Study of Testosterone's Effects on Fitness." *American Naturalist* 167, no. 5 (2006), 667–683. https://doi.org/10.1086/503054.

Reichard, U. "Extra-Pair Copulations in Monogamous Wild White-Handed Gibbons (*Hylobates lar*)." *Zeitschrift Fur Saugetierkunde/International Journal of Mammalian Biology* 60, no. 3 (1995), 186–188.

Reynolds, J. D., N. B. Goodwin, and R. P. Freckleton. "Evolutionary Transitions in Parental Care and Live Bearing in Vertebrates." *Philosophical Transactions of the Royal*

Society B–Biological Sciences 357, no. 1419 (2002), 269–281. https://doi.org/10.1098/rstb.2001.0930.

Rhodes, G., J. Chan, L. A. Zebrowitz, and L. W. Simmons. "Does Sexual Dimorphism in Human Faces Signal Health?" *Philosophical Transactions of the Royal Society B: Biological Sciences* 270 (2003), S93-S95. https://doi.org/10.1098/rsbl.2003.0023.

Riem, M. M., M. J. Bakermans-Kranenburg, S. Pieper, M. Tops, M. A. Boksem, R. R. Vermeiren, M. H. van Ijzendoorn, and S. A. Rombouts. "Oxytocin Modulates Amygdala, Insula, and Inferior Frontal Gyrus Responses to Infant Crying: A Randomized Controlled Trial." *Biological Psychiatry* 70, no. 3 (2011), 291–297. https://doi.org/10.1016/j.biopsych.2011.02.006.

Riem, M. M. E., A. M. Lotz, L. I. Horstman, M. Cima, M. W. F. T. Verhees, K. Alyousefi-van Dijk, M. H. van IJzendoorn, and M. J. Bakermans-Kranenburg. "A Soft Baby Carrier Intervention Enhances Amygdala Responses to Infant Crying in Fathers: A Randomized Controlled Trial." *Psychoneuroendocrinology* 132 (2021). https://doi.org/10.1016/j.psyneuen.2021.105380.

Rigo, P., P. Kim, G. Esposito, D. L. Putnick, P. Venuti, and M. H. Bornstein. "Specific Maternal Brain Responses to Their Own Child's Face: An FMRI Meta-Analysis." *Developmental Review* 51 (2019), 58–69. https://doi.org/10.1016/j.dr.2018.12.001.

Rilling, J. K. "A Potential Role for Oxytocin in the Intergenerational Transmission of Secure Attachment." *Neuropsychopharmacology* 34, no. 13 (2009), 2621–2622. https://doi.org/10.1038/npp.2009.136.

Rilling, J. K., X. Chen, X. Chen, and E. Haroon. "Intranasal Oxytocin Modulates Neural Functional Connectivity during Human Social Interaction." *American Journal of Primatology* 80, no. 10 (2018), e22740. https://doi.org/10.1002/ajp.22740.

Rilling, J. K., and T. R. Insel. "The Primate Neocortex in Comparative Perspective Using Magnetic Resonance Imaging." *Journal of Human Evolution* 37, no. 2 (1999), 191–223. https://doi.org/10.1006/jhev.1999.0313.

Rilling, J. K., and J. S. Mascaro. "The Neurobiology of Fatherhood." *Current Opinion in Psychology* 15 (2017), 26–32. https://doi.org/10.1016/j.copsyc.2017.02.013.

Rilling, J. K., L. Richey, E. Andari, and S. Hamann. "The Neural Correlates of Paternal Consoling Behavior and Frustration in Response to Infant Crying." *Developmental Psychobiology* 63, no. 5 (2021), 1370–1383. https://doi.org/10.1002/dev.22092.

Rilling, J. K., and L. J. Young. "The Biology of Mammalian Parenting and Its Effect on Offspring Social Development." *Science* 345, no. 6198 (2014), 771–776. https://doi.org/10.1126/science.1252723.

Rizk, P. J., T. P. Kohn, A. W. Pastuszak, and M. Khera. "Testosterone Therapy Improves Erectile Function and Libido in Hypogonadal Men." *Currrent Opinion in Urolology* 27, no. 6 (2017), 511–515. https://doi.org/10.1097/MOU.0000000000000442.

Rizzolatti, G., and M. Fabbri-Destro. "Mirror Neurons: From Discovery to Autism." *Experimental Brain Research* 200, no. 3–4 (2010), 223–237. https://doi.org/10.1007/s00221-009-2002-3.

Robbins, M. M., and A. M. Robbins. "Variation in the Social Organization of Gorillas: Life History and Socioecological Perspectives." *Evolutionary Anthropology* 27, no. 5 (2018), 218–233. https://doi.org/10.1002/evan.21721.

Robbins, T. W., and B. J. Everitt. "Neurobehavioural Mechanisms of Reward and Motivation." *Current Opinion in Neurobiology* 6, no. 2 (1996), 228–236. https://doi.org/10.1016/s0959-4388(96)80077-8.

Robinson, B. E., and L. Kelley. "Adult Children of Workaholics: Self-Concept, Anxiety, Depression, and Locus of Control." *American Journal of Family Therapy* 26, no. 3 (1998), 223–238. https://doi.org/10.1080/01926189808251102.

Roellke, E., M. Raiss, S. King, J. Lytel-Sternberg, and D. M. Zeifman. "Infant Crying Levels Elicit Divergent Testosterone Response in Men." *Parenting-Science and Practice* 19, no. 1–2 (2019), 39–55. https://doi.org/10.1080/15295192.2019.1555425.

Romero-Fernandez, W., D. O. Borroto-Escuela, L. F. Agnati, and K. Fuxe. "Evidence for the Existence of Dopamine D2-Oxytocin Receptor Heteromers in the Ventral and Dorsal Striatum with Facilitatory Receptor-Receptor Interactions." *Molecular Psychiatry* 18, no. 8 (t2013), 849–950. https://doi.org/10.1038/mp.2012.103.

Ronay, R., and W. von Hippel. "The Presence of an Attractive Woman Elevates Testosterone and Physical Risk Taking in Young Men." *Social Psychological and Personality Science* 1, no. 1 (2010), 57–64. https://doi.org/10.1177/1948550609352807.

Roney, J. R., K. N. Hanson, K. M. Durante, and D. Maestripieri. "Reading Men's Faces: Women's Mate Attractiveness Judgments Track Men's Testosterone and Interest in Infants." *Philosophical Transactions of the Royal Society B: Biological Sciences* 273, no. 1598 (2006), 2169–2175. https://doi.org/10.1098/rspb.2006.3569.

Roney, J. R., A. W. Lukaszewski, and Z. L. Simmons. "Rapid Endocrine Responses of Young Men to Social Interactions with Young Women." *Hormones and Behavior* 52, no. 3 (2007), 326–333. https://doi.org/10.1016/j.yhbeh.2007.05.008.

Roney, J. R., S. V. Mahler, and D. Maestripieri. "Behavioral and Hormonal Responses of Men to Brief Interactions with Women." *Evolution and Human Behavior* 24, no. 6 (2003), 365–375. https://doi.org/10.1016/S1090-5138(03)00053-9.

Roney, James R., and Lee T. Gettler. "The Role of Testosterone in Human Romantic Relationships." *Current Opinion in Psychology* 1 (2015), 81–86. https://doi.org/10.1016/j.copsyc.2014.11.003.

Rosenbaum, S., J. B. Silk, and T. S. Stoinski. "Male-Immature Relationships in Multi-Male Groups of Mountain Gorillas (*Gorilla beringei beringei*)." *American Journal of Primatology* 73, no. 4 (2011), 356–365. https://doi.org/10.1002/ajp.20905.

Rosenbaum, S., L. Vigilant, C. W. Kuzawa, and T. S. Stoinski. "Caring for Infants Is Associated with Increased Reproductive Success for Male Mountain Gorillas." *Scientific Reports* 8 (2018). https://doi.org/10.1038/s41598-018-33380-4.

Rosenblatt, J. S., and K. Ceus. "Estrogen Implants in the Medial Preoptic Area Stimulate Maternal Behavior in Male Rats." *Hormones and Behavior* 33, no. 1 (1998), 23–30. https://doi.org/10.1006/hbeh.1997.1430.

Rosenblatt, J. S., S. Hazelwood, and J. Poole. "Maternal Behavior in Male Rats: Effects of Medial Preoptic Area Lesions and Presence of Maternal Aggression." *Hormones and Behavior* 30, no. 3 (1996), 201–215. https://doi.org/10.1006/hbeh.1996.0025.

Ross, C. N., J. A. French, and K. J. Patera. "Intensity of Aggressive Interactions Modulates Testosterone in Male Marmosets." *Physiology and Behavior* 83, no. 3 (2004), 437–445. https://doi.org/10.1016/j.physbeh.2004.08.036.

Ross, H. E., and L. J. Young. "Oxytocin and the Neural Mechanisms Regulating Social Cognition and Affiliative Behavior." *Frontiers in Neuroendocrinology* 30, no. 4 (2009), 534–547. https://doi.org/10.1016/j.yfrne.2009.05.004.

Rostad, W. L., P. Silverman, and M. K. McDonald. "Daddy's Little Girl Goes to College: An Investigation of Females' Perceived Closeness with Fathers and Later Risky Behaviors." *Journal of American College Health* 62, no. 4 (2014), 213–220. https://doi.org/10.1080/07448481.2014.887570.

Roubinov, D. S., and W. T. Boyce. "Parenting and SES: Relative Values or Enduring Principles?" *Current Opinion in Psychology* 15 (2017), 162–167. https://doi.org/10.1016/j.copsyc.2017.03.001.

Rowe, M. L., D. Coker, and B. A. Pan. "A Comparison of Fathers' and Mothers' Talk to Toddlers in Low-Income Families." *Social Development* 13, no. 2 (2004), 278–291. https://doi.org/10.1111/j.1467–9507.2004.000267.x.

Rowe, M. L., K. A. Leech, and N. Cabrera. "Going Beyond Input Quantity: Wh-Questions Matter for Toddlers' Language and Cognitive Development." *Cognitive Science* 41 (2017), 162–179. https://doi.org/10.1111/cogs.12349.

Saad, F., G. Doros, K. S. Haider, and A. Haider. "Differential Effects of 11 Years of Long-Term Injectable Testosterone Undecanoate Therapy on Anthropometric and Metabolic Parameters in Hypogonadal Men with Normal Weight, Overweight and

Obesity in Comparison with Untreated Controls: Real-World Data from a Controlled Registry Study." *International Journal of Obesity* 44, no. 6 (2020), 1264–1278. https://doi.org/10.1038/s41366-019-0517-7.

Saito, A., and K. Nakamura. "Oxytocin Changes Primate Paternal Tolerance to Off-spring in Food Transfer." *Journal of Comparative Physiology: A Neuroethology, Sensory, Neural, and Behavioral Physiology* 197, no. 4 (April 2011), 329–337. https://doi.org/10.1007/s00359-010-0617-2.

Salomon, M., J. Schneider, and Y. Lubin. "Maternal Investment in a Spider with Suicidal Maternal Care, *Stegodyphus lineatus* (Araneae, Eresidae)." *Oikos* 109, no. 3 (2005), 614–622. https://doi.org/10.1111/j.0030–1299.2005.13004.x.

Saltzman, W., L. J. Digby, and D. H. Abbott. "Reproductive Skew in Female Common Marmosets: What Can Proximate Mechanisms Tell Us about Ultimate Causes?" *Philosophical Transactions of the Royal Society B: Biological Sciences* 276, no. 1656 (2009), 389–399. https://doi.org/10.1098/rspb.2008.1374.

Samuni, L., A. Preis, R. Mundry, T. Deschner, C. Crockford, and R. M. Wittig. "Oxytocin Reactivity during Intergroup Conflict in Wild Chimpanzees." *Proceedings of the National Academy of Sciences* 114, no. 2 (2016). 268–273. https://doi.org/10.1073/pnas.1616812114.

Sapolsky, R. M. "Endocrine Aspects of Social Instability in the Olive Baboon (*Papio anubis*)." *American Journal of Primatology* 5, no. 4 (1983), 365–379. https://doi.org/10.1002/ajp.1350050406.

———. "Stress-Induced Elevation of Testosterone Concentration in High Ranking Baboons: Role of Catecholamines." *Endocrinology* 118, no. 4 (1986), 1630–1635. https://doi.org/10.1210/endo-118-4-1630.

———. "Testicular Function, Social Rank and Personality among Wild Baboons." *Psychoneuroendocrinology* 16, no. 4 (1991), 281–293. https://doi.org/10.1016/0306-4530(91)90015-L.

Sarkadi, A., R. Kristiansson, F. Oberklaid, and S. Bremberg. "Fathers' Involvement and Children's Developmental Outcomes: A Systematic Review of Longitudinal Studies." *Acta Paediatrica* 97, no. 2 (2008), 153–158. https://doi.org/10.1111/j.1651-2227.2007.00572.x.

Sato, H. "Male Participation in Nest Building in the Dung Beetle *Scarabaeus catenatus* (Coleoptera: Scarabaeidae), Mating Effort Versus Paternal Effort." *Journal of Insect Behavior* 11, no. 6 (1998), 833–843. https://doi.org/10.1023/A:1020860010165.

Saxbe, D. E., R. S. Edelstein, H. M. Lyden, B. M. Wardecker, W. J. Chopik, and A. C. Moors. "Fathers' Decline in Testosterone and Synchrony with Partner Testosterone during Pregnancy Predicts Greater Postpartum Relationship Investment." *Hormones and Behavior* 90 (2017), 39–47. https://doi.org/10.1016/j.yhbeh.2016.07.005.

Saxbe, D. E., C. D. Schetter, C. D. Simon, E. K. Adam, and M. U. Shalowitz. "High Paternal Testosterone May Protect against Postpartum Depressive Symptoms in Fathers, But Confer Risk to Mothers and Children." *Hormones and Behavior* 95 (2017), 103–112. https://doi.org/10.1016/j.yhbeh.2017.07.014.

Scaramella, T. J., and W. A. Brown. "Serum Testosterone and Aggressiveness in Hockey Players." *Psychosomatic Medicine* 40, no. 3 (1978), 262–265. https://doi.org /10.1097/00006842-197805000-00007.

Scelza, B. A. "Fathers' Presence Speeds the Social and Reproductive Careers of Sons." *Current Anthropology* 51, no. 2 (2010), 295–303. https://doi.org/10.1086/651051.

Schacht, R., and K. L. Kramer. "Are We Monogamous? A Review of the Evolution of Pair-Bonding in Humans and Its Contemporary Variation Cross-Culturally." *Frontiers in Ecology and Evolution* 7 (2019). https://doi.org/10.3389/fevo.2019.00230.

Schacht, R., and M. B. Mulder. "Sex Ratio Effects on Reproductive Strategies in Humans." *Royal Society Open Science* 2, no. 1 (2015). https://doi.org/10.1098/rsos .140402.

Scheele, D., N. Striepens, O. Gunturkun, S. Deutschlander, W. Maier, K. M. Kendrick, and R. Hurlemann. "Oxytocin Modulates Social Distance between Males and Females." *Journal of Neuroscience* 32, no. 46 (2012), 16074–16079. https://doi.org/10 .1523/JNEUROSCI.2755-12.2012.

Scheele, D., A. Wille, K. M. Kendrick, B. Stoffel-Wagner, B. Becker, O. Gunturkun, W. Maier, and R. Hurlemann. "Oxytocin Enhances Brain Reward System Responses in Men Viewing the Face of Their Female Partner." *Proceedings of the National Academy of Sciences* 110, no. 50 (2013), 20308–20313. https://doi.org/10.1073/pnas .1314190110.

Schmeer, K. "Father Absence Due to Migration and Child Illness in Rural Mexico." *Social Science and Medicine* 69, no. 8 (2009), 1281–1286. https://doi.org/10.1016/j .socscimed.2009.07.030.

Schneider, J. S., C. Burgess, T. H. Horton, and J. E. Levine. "Effects of Progesterone on Male-Mediated Infant-Directed Aggression." *Behavioural Brain Research* 199, no. 2 (2009), 340–344. https://doi.org/10.1016/j.bbr.2008.12.019.

Schneider, J. S., M. K. Stone, K. E. Wynne-Edwards, T. H. Horton, J. Lydon, B. O'Malley, and J. E. Levine. "Progesterone Receptors Mediate Male Aggression toward Infants." *Proceedings of the National Academy of Sciences* 100, no. 5 (2003), 2951–2956. https://doi.org/10.1073/pnas.0130100100.

Schneiderman, I., O. Zagoory-Sharon, J. F. Leckman, and R. Feldman. "Oxytocin during the Initial Stages of Romantic Attachment: Relations to Couples' Interactive Reciprocity." *Psychoneuroendocrinology* 37, no. 8 (2012), 1277–1285. https://doi.org /10.1016/j.psyneuen.2011.12.021.

Schoppe-Sullivan, S. J., L. E. Altenburger, M. A. Lee, D. J. Bower, and C. M. Kamp Dush. "Who Are the Gatekeepers? Predictors of Maternal Gatekeeping." *Parenting: Science and Practice* 15, no. 3 (2015), 166–186. https://doi.org/10.1080/15295192 .2015.1053321.

Schoppe-Sullivan, S. J., G. L. Brown, E. A. Cannon, S. C. Mangelsdorf, and M. S. Sokolowski. "Maternal Gatekeeping, Coparenting Quality, and Fathering Behavior in Families with Infants." *Journal of Family Psychology* 22, no. 3 (2008), 389–398. https://doi.org/10.1037/0893-3200.22.3.389.

Schrijner, S., and J. Smits. "Grandmothers and Children's Schooling in Sub-Saharan Africa." *Human Nature* 29, no. 1 (2018), 65–89. https://doi.org/10.1007/s12110-017 -9306-y.

Schroeder, E. T., M. Villanueva, D. W. D. West, and S. M. Phillips. "Are Acute Post-Resistance Exercise Increases in Testosterone, Growth Hormone, and Igf-1 Necessary to Stimulate Skeletal Muscle Anabolism and Hypertrophy?" *Medicine and Science in Sports and Exercise* 45, no. 11 (2013), 2044–2451. https://doi.org/10.1249/Mss .0000000000000147.

Sear, R. "Do Human 'Life History Strategies' Exist?" *Evolution and Human Behavior* 41, no. 6 (2020), 513–526. https://doi.org/10.1016/j.evolhumbehav.2020.09.004.

Sear, R., and R. Mace. "Who Keeps Children Alive? A Review of the Effects of Kin on Child Survival." *Evolution and Human Behavior* 29, no. 1 (2008), 1–18. https://doi.org /10.1016/j.evolhumbehav.2007.10.001.

Secrest, Alexa L. "Daddy Issues: Why Do Swedish Fathers Claim Paternity Leave at Higher Rates Than French Fathers?" *Student Publications* (2020): 784. https://cupola .gettysburg.edu/student_scholarship/784.

Servedio, M. R., T. D. Price, and R. Lande. "Evolution of Displays within the Pair Bond." *Philosophical Transactions of the Royal Society B: Biological Sciences* 280, no. 1757 (2013). https://doi.org/10.1098/rspb.2012.3020.

Shamay-Tsoory, S. G., and A. Abu-Akel. "The Social Salience Hypothesis of Oxy-tocin." *Biological Psychiatry* 79, no. 3 (1 2016): 194–202. https://doi.org/10.1016/j .biopsych.2015.07.020.

Shattuck, E. C., and M. P. Muehlenbein. "Human Sickness Behavior: Ultimate and Proximate Explanations." *American Journal of Physical Anthropology* 157, no. 1 (2015a), 1–18. https://doi.org/10.1002/ajpa.22698.

———. "Mood, Behavior, Testosterone, Cortisol, and Interleukin-6 in Adults during Immune Activation: A Pilot Study to Assess Sickness Behaviors in Humans." *American Journal of Human Biology* 27, no. 1 (2015b), 133–135. https://doi.org/10.1002/ ajhb.22608.

Sherman, G. D., J. S. Lerner, R. A. Josephs, J. Renshon, and J. J. Gross. "The Interaction of Testosterone and Cortisol Is Associated with Attained Status in Male Executives." *Journal of Personality and Social Psychology* 110, no. 6 (2016), 921–929. https://doi.org/10.1037/pspp0000063.

Shine, Richard. "Parental Care in Reptiles." In *Biology of Reptilia*, edited by Carl Gans and Raymond B. Huey. New York: Alan R. Liss, 1987.

Shore, Bradd. *Culture in Mind*. New York: Oxford University Press, 1996.

Short, Philip. *Mao: A Life*. New York: Owl Books, 2001.

Shostak, Marjorie. *Nisa, the Life and Words of a !Kung Woman*. New York: Vintage Books, 1983.

Shwalb, David W., Barbara J. Shwalb, and Michael E. Lamb, eds. *Fathers in Cultural Context*. New York: Routledge, 2013.

Sigle-Rushton, Wendy, and Sara McLanahan. "Father Absence and Child Wellbeing: A Critical Review." In *The Future of the Family*, edited by D. P. Moynihan, L. Rainwater, and T. Smeeding, 116–155. New York: Russell Sage Foundation, 2004.

Silverin, B. "Effects of Long-Acting Testosterone Treatment on Free-Living Pied Flycatchers, *Ficedula hypoleuca*, during the Breeding Period." *Animal Behaviour* 28, no. (1980), 906–912. https://doi.org/10.1016/S0003-3472(80)80152-7.

Simmons, Z. L., and J. R. Roney. "Androgens and Energy Allocation: Quasi-Experimental Evidence for Effects of Influenza Vaccination on Men's Testosterone." *American Journal of Human Biology* 21, no. 1 (2009), 133–135. https://doi.org/10.1002/ajhb.20837.

———. "Variation in CAG Repeat Length of the Androgen Receptor Gene Predicts Variables Associated with Intrasexual Competitiveness in Human Males." *Hormones and Behavior* 60, no. 3 (2011), 306–312. https://doi.org/10.1016/j.yhbeh.2011.06.006.

Singleton, I., and C. P. van Schaik. "The Social Organisation of a Population of Sumatran Orang-Utans." *Folia Primatology l*73, no. 1 (2002), 1–20. https://doi.org/10.1159/000060415.

Smith, E. A. "Why Do Good Hunters Have Higher Reproductive Success?" *Human Nature* 15, no. 4 (2004), 343–364. https://doi.org/10.1007/s12110-004-1013-9.

Smith, R. L. "Brooding Behavior of a Male Water Bug *Belostoma flumineum* (Hemiptera: Belostomatidae)." *Journal of the Kansas Entomological Society* 49 (1976), 333–343.

Smith, T. M., C. Austin, K. Hinde, E. R. Vogel, and M. Arora. "Cyclical Nursing Patterns in Wild Orangutans." *Science Advances* 3, no. 5 (2017), e1601517. https://doi.org/10.1126/sciadv.1601517.

Smuts, Barbara B. *Sex and Friendship in Baboons: With a New Preface.* Cambridge, MA: Harvard University Press, 1999.

Smyth, Bruce M., Jennifer A. Baxter, Richard J. Fletcher, and Lawrence J. Moloney. "Fathers in Australia: A Contemporary Snapshot." In *Fathers in Cultural Context*, edited by David W. Shwalb, Barbara J. Shwalb, and Michael E. Lamb, 361–382. New York: Routledge, 2013.

Sobolewski, M. E., J. L. Brown, and J. C. Mitani. "Territoriality, Tolerance and Testosterone in Wild Chimpanzees." *Animal Behaviour* 84, no. 6 (2012), 1469–1474. https://doi.org/10.1016/j.anbehav.2012.09.018.

———. "Female Parity, Male Aggression, and the Challenge Hypothesis in Wild Chimpanzees." *Primates* 54, no. 1 (2013), 81–88. https://doi.org/10.1007/s10329 -012-0332-4.

Spence-Aizenberg, A., A. Di Fiore, and E. Fernandez-Duque. "Social Monogamy, Male-Female Relationships, and Biparental Care in Wild Titi Monkeys (*Callicebus discolor*)." *Primates* 57, no. 1 (2016), 103–112. https://doi.org/10.1007/s10329-015 -0489-8.

Steinberg, L., I. Blatt-Eisengart, and E. Cauffman. "Patterns of Competence and Adjustment among Adolescents from Authoritative, Authoritarian, Indulgent, and Neglectful Homes: A Replication in a Sample of Serious Juvenile Offenders." *Journal of Research on Adolescence* 16, no. 1 (2006): 47–58. https://doi.org/10.1111/j .1532–7795.2006.00119.x.

Strassman, Beverly. "Sexual Selection, Paternal Care, and Concealed Ovulation in Humans." *Ethology and Sociobiology* 2 (1981): 31–40. https://doi.org/10.1016/0162 -3095(81)90020-0.

Strathearn, L., P. Fonagy, J. Amico, and P. R. Montague. "Adult Attachment Predicts Maternal Brain and Oxytocin Response to Infant Cues." *Neuropsychopharmacology* 34, no. 13 (2009), 2655–2666. https://doi.org/10.1038/npp.2009.103.

Strathearn, L., A. A. Mamun, J. M. Najman, and M. J. O'Callaghan. "Does Breastfeeding Protect against Substantiated Child Abuse and Neglect? A 15-Year Cohort Study." *Pediatrics* 123, no. 2 (2009), 483–493. https://doi.org/10.1542/peds.2007–3546.

Striepens, N., K. M. Kendrick, V. Hanking, R. Landgraf, U. Wullner, W. Maier, and R. Hurlemann. "Elevated Cerebrospinal Fluid and Blood Concentrations of Oxytocin Following Its Intranasal Administration in Humans." *Scientific Reports* 3 (2013), 3440. https://doi.org/10.1038/srep03440.

Stuebe, A. M., K. Grewen, and S. Meltzer-Brody. "Association between Maternal Mood and Oxytocin Response to Breastfeeding." *Journal of Women's Health* 22, no. 4 (2013), 352–361. https://doi.org/10.1089/jwh.2012.3768.

Sturgis, J. D., and R. S. Bridges. "N-Methyl-DL-Aspartic Acid Lesions of the Medial Preoptic Area Disrupt Ongoing Parental Behavior in Male Rats." *Physiology and Behavior* 62, no. 2 (1997), 305–310. https://doi.org/10.1016/s0031-9384(97)88985-8.

Surbeck, M., K. E. Langergraber, B. Fruth, L. Vigilant, and G. Hohmann. "Male Reproductive Skew Is Higher in Bonobos Than Chimpanzees." *Current Biology* 27, no. 13 (2017), R640–R641. https://doi.org/10.1016/j.cub.2017.05.039.

Sussman, R. W., and P. A. Garber. "A New Interpretation of the Social-Organization and Mating System of the Callitrichidae." *International Journal of Primatology* 8, no. 1 (1987), 73–92. https://doi.org/10.1007/Bf02737114.

Swedell, L., and T. Tesfaye. "Infant Mortality after Takeovers in Wild Ethiopian Hamadryas Baboons." *American Journal of Primatology* 60, no. 3 (2003), 113–118. https://doi.org/10.1002/ajp.10096.

Tanner, J. M. *Fetus into Man*. Cambridge, MA: Harvard University Press, 1990.

Taylor, J. H., and J. A. French. "Oxytocin and Vasopressin Enhance Responsiveness to Infant Stimuli in Adult Marmosets." *Hormones and Behavior* 75 (2015), 154–159. https://doi.org/10.1016/j.yhbeh.2015.10.002.

"Pouched Frog." Australian Museum. https://australian.museum/learn/animals/frogs /pouched-frog/.

Teicher, M. H., J. A. Samson, C. M. Anderson, and K. Ohashi. "The Effects of Child-hood Maltreatment on Brain Structure, Function and Connectivity." *Nature Reviews Neuroscience* 17, no. 10 (2016), 652–666 https://www.ncbi.nlm.nih.gov/pubmed /27640984.

Tennie, C., J. Call, and M. Tomasello. "Ratcheting up the Ratchet: On the Evolution of Cumulative Culture." *Philosophical Transactions of the Royal Society B: Biological Sciences* 364, no. 1528 (2009), 2405–2415. https://doi.org/10.1098/rstb.2009.0052.

Terborgh, J., and A. W. Goldizen. "On the Mating System of the Cooperatively Breeding Saddle-Backed Tamarin (*Saguinus fuscicollis*)." *Behavioral Ecology and Sociobi-ology* 16, no. 4 (1985), 293–399. https://doi.org/10.1007/Bf00295541.

"This Is Dad of the Year." *Daily Mail*, March 10, 2014. https://www.dailymail.co.uk /news/article-2577520/The-devoted-Chinese-father-carries-son-18-MILES-school -day-Sichuan-province.html.

Thomson, K. C., H. Romaniuk, C. J. Greenwood, P. Letcher, E. Spry, J. A. Macdonald, H. M. McAnally, et al. "Adolescent Antecedents of Maternal and Paternal Perinatal Depression: A 36-Year Prospective Cohort." *Psychological Medicine* 51, no. 12 (2021), 2126–2133. https://doi.org/10.1017/S0033291720000902.

Thornhill, R., and S. W. Gangestad. "Facial Sexual Dimorphism, Developmental Stability, and Susceptibility to Disease in Men and Women." *Evolution and Human*

Behavior 27, no. 2 (2006), 131–144. https://doi.org/10.1016/j.evolhumbehav.2005 .06.001.

Thornton, J., J. L. Zehr, and M. D. Loose. "Effects of Prenatal Androgens on Rhesus Monkeys: A Model System to Explore the Organizational Hypothesis in Primates." *Hormones and Behavior* 55, no. 5 (2009), 633–644. https://doi.org/10.1016/j.yhbeh .2009.03.015.

Thul, T. A., E. J. Corwin, N. S. Carlson, P. A. Brennan, and L. J. Young. "Oxytocin and Postpartum Depression: A Systematic Review." *Psychoneuroendocrinology* 120 (2020), 104793. https://doi.org/10.1016/j.psyneuen.2020.104793.

Townsend, Nicholas W. "The Complications of Fathering in Southern Africa: Separation, Uncertainty, and Multiple Responsibilities." In *Fathers in Cultural Context*, edited by David W. Shwalb, Barbara J. Shwalb, and Michael E. Lamb, 173–200. New York: Routledge, 2013.

Trainor, B. C., I. M. Bird, N. A. Alday, B. A. Schlinger, and C. A. Marchler. "Variation in Aromatase Activity in the Medial Preoptic Area and Plasma Progesterone Is Associated with the Onset of Paternal Behavior." *Neuroendocrinology* 78, no. 1 (2003), 36–44. https://doi.org/10.1159/000071704.

———. "Testosterone Promotes Paternal Behaviour in a Monogamous Mammal via Conversion to Oestrogen." *Philosophical Transactions of the Royal Society B: Biological Sciences* 269, no. 1493 (2002), 823–829. https://doi.org/10.1098/rspb.2001.1954.

Trevathan, Wenda, and James J. McKenna. "Evolutionary Environments of Human Birth and Infancy: Insights to Apply to Contemporary Life." *Children's Environments* 11, no. 2 (1994), 88–104. https://www.jstor.org/stable/41514918.

Trumble, B. C., D. K. Cummings, K. A. O'Connor, D. J. Holman, E. A. Smith, H. S. Kaplan, and M. D. Gurven. "Age-Independent Increases in Male Salivary Testosterone during Horticultural Activity among Tsimane Forager-Farmers." *Evolution and Human Behavior* 34, no. 5 (2013), 350–357. https://doi.org/10.1016/j.evolhumbehav .2013.06.002.

Trumble, B. C., E. A. Smith, K. A. O'Connor, H. S. Kaplan, and M. D. Gurven. "Successful Hunting Increases Testosterone and Cortisol in a Subsistence Population." *Philosophical Transactions of the Royal Society B: Biological Sciences* 281, no. 1776 (2014), 20132876. https://doi.org/10.1098/rspb.2013.2876.

Tuttle, E. M. "Alternative Reproductive Strategies in the White-Throated Sparrow: Behavioral and Genetic Evidence." *Behavioral Ecology* 14, no. 3 (2003), 425–432. https://doi.org/10.1093/beheco/14.3.425.

Ultrata, Jennifer, Jean M. Ispa, and Simone Ispa-Landa. "Men on the Margins of Family Life: Fathers in Russia." In *Fathers in Cultural Context*, edited by David W.

Shwalb, Barbara J. Shwalb, and Michael E. Lamb, 279–302. New York: Routledge, 2013.

Utami, S. S., B. Goossens, M. W. Bruford, J. R. de Ruiter, and J. A. R. A. M. van Hooff. "Male Bimaturism and Reproductive Success in Sumatran Orang-Utans." *Behavioral Ecology* 13, no. 5 (2002), 643–652. https://doi.org/10.1093/beheco/13.5.643.

Vadenkiernan, N., N. S. Ialongo, J. Pearson, and S. Kellam. "Household Family-Structure and Children's Aggressive-Behavior: A Longitudinal-Study of Urban Elementary School Children." *Journal of Abnormal Child Psychology* 23, no. 5 (1995), 553–568. https://doi.org/10.1007/Bf01447661.

Vakrat, A., Y. Apter-Levy, and R. Feldman. "Sensitive Fathering Buffers the Effects of Chronic Maternal Depression on Child Psychopathology." *Child Psychiatry and Human Development* 49, no. 5 (2018), 779–785. https://doi.org/10.1007/s10578-018-0795-7.

van Anders, S. M., R. M. Tolman, and B. L. Volling. "Baby Cries and Nurturance Affect Testosterone in Men." *Hormones and Behavior* 61, no. 1 (2012), 31–6. https://doi.org/10.1016/j.yhbeh.2011.09.012.

van IJzendoorn, M. H., C. Schuengel, Q. Wang, and M. J. Bakermans-Kranenburg. "Improving Parenting, Child Attachment, and Externalizing Behaviors: Meta-Analysis of the First 25 Randomized Controlled Trials on the Effects of Video-Feedback Intervention to Promote Positive Parenting and Sensitive Discipline." *Development and Psychopathology* 35, no. 1 (2022), 241–256. https://doi.org/10.1017/S0954579421001462.

Viktorin, A., S. Meltzer-Brody, R. Kuja-Halkola, P. F. Sullivan, M. Landen, P. Lichtenstein, and P. K. E. Magnusson. "Heritability of Perinatal Depression and Genetic Overlap with Nonperinatal Depression." *American Journal of Psychiatry* 173, no. 2 (2016), 158–165. https://doi.org/10.1176/appi.ajp.2015.15010085.

Vittner, D., J. McGrath, J. Robinson, G. Lawhon, R. Cusson, L. Eisenfeld, S. Walsh, E. Young, and X. Cong. "Increase in Oxytocin from Skin-to-Skin Contact Enhances Development of Parent-Infant Relationship." *Biological Research for Nursing* 20, no. 1 (2018), 54–62. https://doi.org/10.1177/1099800417735633.

Volkow, N. D., G. F. Koob, and A. T. McLellan. "Neurobiologic Advances from the Brain Disease Model of Addiction." *New England Journal of Medicine* 374, no. 4 (2016), 363–371. https://doi.org/10.1056/NEJMra1511480.

Volkow, N. D., G. J. Wang, and R. D. Baler. "Reward, Dopamine and the Control of Food Intake: Implications for Obesity." *Trends in Cognitive Sciences* 15, no. 1 (2011), 37–46. https://doi.org/10.1016/j.tics.2010.11.001.

Wade, T. J., S. Veldhuizen, and J. Cairney. "Prevalence of Psychiatric Disorder in Lone Fathers and Mothers: Examining the Intersection of Gender and Family

Structure on Mental Health." *Canadian Journal of Psychiatry/Revue Canadienne de Psychiatrie* 56, no. 9 (2011), 567–573. https://doi.org/10.1177/070674371105600908.

Wagner, U., K. N'Diaye, T. Ethofer, and P. Vuilleumier. "Guilt-Specific Processing in the Prefrontal Cortex." *Cerebral Cortex* 21, no. 11 (2011), 2461–2470. https://doi.org /10.1093/cercor/bhr016. https://www.ncbi.nlm.nih.gov/pubmed/21427167.

Walum, H., I. D. Waldman, and L. J. Young. "Statistical and Methodological Considerations for the Interpretation of Intranasal Oxytocin Studies." *Biological Psychiatry* 79, no. 3 (2016), 251–257. https://doi.org/10.1016/j.biopsych.2015.06.016.

Wang, B., Y. Li, R. Wu, S. Zhang, and F. Tai. "Behavioral Responses to Pups in Males with Different Reproductive Experiences Are Associated with Changes in Central OT, TH and OTR, D1R, D2R mRNA Expression in Mandarin Voles." *Hormones and Behavior* 67 (2015), 73–82. https://doi.org/10.1016/j.yhbeh.2014.11.013.

Wang, Z., C. F. Ferris, and G. J. De Vries. "Role of Septal Vasopressin Innervation in Paternal Behavior in Prairie Voles (*Microtus ochrogaster*)." *Proceedings of the National Academy of Sciences* 91, no. 1 (1994), 400–4004. https://doi.org/10.1073/pnas.91.1 .400.

Watts, D. P. "Infanticide in Mountain Gorillas—New Cases and a Reconsideration of the Evidence." *Ethology* 81, no. 1 (1989), 1–18. https://doi.org/10.1111/j.1439-0310 .1989.tb00754.x.

Weisman, O., E. Delaherche, M. Rondeau, M. Chetouani, D. Cohen, and R. Feldman. "Oxytocin Shapes Parental Motion during Father-Infant Interaction." *Biology Letters* 9, no. 6 (2013), 20130828. https://doi.org/10.1098/rsbl.2013.0828.

Weisman, O., O. Zagoory-Sharon, and R. Feldman. "Oxytocin Administration to Parent Enhances Infant Physiological and Behavioral Readiness for Social Engagement." *Biological Psychiatry* 72, no. 12 (2012), 982–989. https://doi.org/10.1016/j .biopsych.2012.06.011.

———. "Oxytocin Administration, Salivary Testosterone, and Father–Infant Social Behavior." *Progress in Neuro-Psychopharmacolog and Biological Psychiatry* 49 (2014), 47–52. https://doi.org/10.1016/j.pnpbp.2013.11.006.

Weygoldt, P. "Evolution of Parental Care in Dart Poison Frogs (Amphibia, Anura, Dendrobatidae)." *Zeitschrift Fur Zoologische Systematik Und Evolutionsforschung* 25, no. 1 (1987), 51–67. https://doi.org/10.1111/j.1439-0469.1987.tb00913.x.

Whiting, B. B. "Sex Identity Conflict and Physical Violence: A Comparative Study." *American Anthropologist* 67, no. 6 (1965), 123–140. https://www.jstor.org/stable /668842.

Whiting, John W. M., and Beatrice Whiting. "Aloofness and Intimacy of Husbands and Wives: A Cross-Cultural Study." *Ethos* 3, no. 2 (1975), 183–207. https://doi.org /10.1525/eth.1975.3.2.02a00080.

Wich, Serge A, Hans de Vries, Marc Ancrenaz, Lori Perkins,; Robert W. Shumaker,Akira Suzuki, and Carel P.van Schaik. "Orangutan Life History Variation." In *Orangutans: Geographic Variation in Behavioral Ecology and Conservation*, edited by Serge A. Wich, S. Suci Utami Atmoko, Tatang Mitra Setia, and Carel P. van Schaik, 67–68. New York: Oxford University Press, 2009.

Williams, J. M., E. V. Lonsdorf, M. L. Wilson, J. Schumacher-Stankey, J. Goodall, and A. E. Pusey. "Causes of Death in the Kasekela Chimpanzees of Gombe National Park, Tanzania." *American Journal of Primatology* 70, no. 8 (2008), 766–777. https://doi.org /10.1002/ajp.20573.

Williams, J. R., T. R. Insel, C. R. Harbaugh, and C. S. Carter. "Oxytocin Administered Centrally Facilitates Formation of a Partner Preference in Female Prairie Votes (*Microtus ochrogaster*)." *Journal of Neuroendocrinology* 6, no. 3 (1994), 247–250. https:// doi.org/10.1111/j.1365–2826.1994.tb00579.x.

Wilson, A. B., I. Ahnesjo, A. C. Vincent, and A. Meyer. "The Dynamics of Male Brooding, Mating Patterns, and Sex Roles in Pipefishes and Seahorses (Family Syngnathidae)." *Evolution* 57, no. 6 (2003), 1374–1386. https://doi.org/10.1111/j.0014 -3820.2003.tb00345.x.

Wingfield, J. C. "Androgens and Mating Systems—Testosterone-Induced Polygyny in Normally Monogamous Birds." *Auk* 101, no. 4 (1984), 665–671. https://doi.org /10.2307/4086893.

Wingfield, John C., Robert E. Hegner Jr., Alfred M. Dufty, and Gregory F. Ball. "The `Challenge Hypothesis': Theoretical Implications for Patterns of Testosterone Secretion, Mating Systems and Breeding Strategies." *American Naturalist* 136, no. 6 (1990), 829–846. https://www.jstor.org/stable/2462170.

Wingo, A. P., T. S. Wingo, W. Fan, S. Bergquist, A. Alonso, M. Marchcus, A. I. Levey, and J. J. Lah. "Purpose in Life Is a Robust Protective Factor of Reported Cognitive Decline among Late Middle-Aged Adults: The Emory Healthy Aging Study." *Journal of Affective Disorders* 263 (2020), 310–317. https://doi.org/10.1016/j.jad.2019.11.124.

Winking, J., M. Gurven, H. Kaplan, and J. Stieglitz. "The Goals of Direct Paternal Care among a South Amerindian Population." *American Journal of Physical Anthropology* 139, no. 3 (2009), 295–304. https://doi.org/10.1002/ajpa.20981.

Winking, J., and J. Koster. "The Fitness Effects of Men's Family Investments: A Test of Three Pathways in a Single Population." *Human Nature* 26, no. 3 (2015), 292–312. https://doi.org/10.1007/s12110-015-9237-4.

Witte, A. M., M. M. E. Riem, N. van der Knaap, M. H. M. de Moor, IJzendoorn M. H. van, and M. J. Bakermans-Kranenburg. "The Effects of Oxytocin and Vasopressin Administration on Fathers' Neural Responses to Infant Crying: A Randomized Controlled within-Subject Study." *Psychoneuroendocrinology* 140 (2022), 105731. https://doi.org/10.1016/j.psyneuen.2022.105731.

Woller, M. J., M. E. Sosa, Y. Chiang, S. L. Prudom, P. Keelty, J. E. Moore, and T. E. Ziegler. "Differential Hypothalamic Secretion of Neurocrines in Male Common Marchmosets: Parental Experience Effects?" *Journal of Neuroendocrinology* 24, no. 3 (2012), 413–421. https://doi.org/10.1111/j.1365-2826.2011.02252.x.

Wood, T. K. "Aggregation Behavior of *Umbonia crassicornis* (Homoptera: Membracidae)." *Canadian Entomologist* 106, no. 2 (1974), 169–173. https://doi.org/10.4039/Ent106169-2.

Wood, T.K. "Biology and Presocial Behavior of *Platycotis vittata* (Homoptera: Membracidae)." *Annals of the Entomological Society of America* 69, no. 5 (1976), 807–811. https://doi.org/10.1093/aesa/69.5.807.

Woodroffe, R., and A. Vincent. "Mother's Little Helpers: Patterns of Male Care in Mammals." *Trends in Ecology and Evolution* 9, no. 8 (1994), 294–297. https://doi.org/10.1016/0169-5347(94)90033-7.

Woolfenden, Glen Everett, and John W. Fitzpatrick. *The Florida Scrub Jay: Demography of a Cooperatively Breeding Bird*. Princeton, NJ: Princeton University Press, 1985.

Worthman, C. M., and M. J. Konner. "Testosterone Levels Change with Subsistence Hunting Effort in Kung San Men." *Psychoneuroendocrinology* 12, no. 6 (1987), 449–458. https://doi.org/10.1016/0306-4530(87)90079-5.

Wrangham, R. W. "Hypotheses for the Evolution of Reduced Reactive Aggression in the Context of Human Self-Domestication." *Frontiers in Psychology* 10 (2019). https://doi.org/10.3389/fpsyg.2019.01914.

Wrangham, R. W., and L. Glowacki. "Intergroup Aggression in Chimpanzees and War in Nomadic Hunter-Gatherers Evaluating the Chimpanzee Model." *Human Nature* 23, no. 1 (2012), 5–29. https://doi.org/10.1007/s12110-012-9132-1.

Wright, H. W. Y. "Paternal Den Attendance Is the Best Predictor of Offspring Survival in the Socially Monogamous Bat-Eared Fox." *Animal Behaviour* 71 (2006), 503–510. https://doi.org/10.1016/j.anbehav.2005.03.043.

Wright, P. C. "Biparental Care in *Aotus trivirgatus* and *Callicebus moloch*." In *Female Primates: Studies by Women Primatologists*, edited by M. F. Small, 59–75, New York: Alan R. Liss, 1984.

Wroblewski, E. E., C. M. Murray, B. F. Keele, J. C. Schumacher-Stankey, B. H. Hahn, and A. E. Pusey. "Male Dominance Rank and Reproductive Success in Chimpanzees,

Pan troglodytes schweinfurthii." *Animal Behavior* 77, no. 4 (2009), 873–885. https://doi .org/10.1016/j.anbehav.2008.12.014.

Wu, Z., A. E. Autry, J. F. Bergan, M. Watabe-Uchida, and C. G. Dulac. "Galanin Neurons in the Medial Preoptic Area Govern Parental Behaviour." *Nature* 509, no. 7500 (2014), 325–330. https://doi.org/10.1038/nature13307.

Yu, P., S. C. An, F. D. Tai, J. L. Wang, R. Y. Wu, and B. Wang. "Early Social Deprivation Impairs Pair Bonding and Alters Serum Corticosterone and the NACC Dopamine System in Mandarin Voles." *Psychoneuroendocrinology* 38, no. 12 (2013), 3128–3138. https://doi.org/10.1016/j.psyneuen.2013.09.012.

Yuan, W., Z. He, W. Hou, L. Wang, L. Li, J. Zhang, Y. Yang, et al. "Role of Oxytocin in the Medial Preoptic Area (MPOA) in the Modulation of Paternal Behavior in Mandarin Voles." *Hormones and Behavior* 110 (2019), 46–55. https://doi.org/10.1016 /j.yhbeh.2019.02.014.

Zheng, X., X. Xu, L. Xu, J. Kou, L. Luo, X. Ma, and K. M. Kendrick. "Intranasal Oxytocin May Help Maintain Romantic Bonds by Decreasing Jealousy Evoked by Either Imagined or Real Partner Infidelity." *Journal of Psychopharmacology* 35, no. 6 (2021), 668–680. https://doi.org/10.1177/0269881121991576.

Zhu, R., C. Feng, S. Zhang, X. Mai, and C. Liu. "Differentiating Guilt and Shame in an Interpersonal Context with Univariate Activation and Multivariate Pattern Analyses." *NeuroImage* 186 (2019), 476–486. https://doi.org/10.1016/j.neuroimage .2018.11.012.

Ziegler, T. E., L. J. Peterson, M. E. Sosa, and A. M. Barnard. "Differential Endocrine Responses to Infant Odors in Common Marmoset (*Callithrix jacchus*) Fathers." *Hormones and Behavior* 59, no. 2 (2011), 265–270. https://doi.org/10.1016/j.yhbeh.2010 .12.001.

Ziegler, T. E., S. L. Prudom, N. J. Schultz-Darken, A. V. Kurian, and C. T. Snowdon. "Pregnancy Weight Gain: Marmoset and Tamarin Dads Show It Too." *Biology Letters* 2, no. 2 (2 2006), 181–183. https://doi.org/10.1098/rsbl.2005.0426.

Ziegler, T. E., S. L. Prudom, and S. R. Zahed. "Variations in Male Parenting Behavior and Physiology in the Common Marmoset." *American Journal of Human Biology* 21, no. 6 (2009), 739–744. https://doi.org/10.1002/ajhb.20920.

Ziegler, T. E., N. J. Schultz-Darken, J. J. Scott, C. T. Snowdon, and C. F. Ferris. "Neuroendocrine Response to Female Ovulatory Odors Depends upon Social Condition in Male Common Marmosets, *Callithrix jacchus.*" *Hormones and Behavior* 47, no. 1 (2005), 56–64. https://doi.org/10.1016/j.yhbeh.2004.08.009.

Zielinski, W. J., and J. G. Vandenbergh. "Testosterone and Competitive Ability in Male House Mice, *Mus musculus*—Laboratory and Field Studies." *Animal Behaviour* 45, no. 5 (1993), 873–891. https://doi.org/10.1006/anbe.1993.1108.

Zilioli, S., D. Ponzi, A. Henry, K. Kubicki, N. Nickels, M. C. Wilson, and D. Maestripieri. "Interest in Babies Negatively Predicts Testosterone Responses to Sexual Visual Stimuli among Heterosexual Young Men." *Psychological Science* 27, no. 1 (2016), 114–118. https://doi.org/10.1177/0956797615615868.

Zilles, K., E. Armstrong, A. Schleicher, and H. J. Kretschmann. "The Human Pattern of Gyrification in the Cerebral Cortex." *Anatomy and Embryology* 179, no. 2 (1988), 173–179. https://doi.org/10.1007/BF00304699.

Index